The Informed Student Guide to MANAGEMENT SCIENCE

Edited by Hans Daellenbach and Robert Flood

THOMSON

Australia • Canada • Mexico • Singapore • Spain • United Kingdom • United States

THOMSON

The Informed Student Guide to Management Science

For more information, contact Thomson, High Holborn House, 50/51 Bedford Row, London, WC1R 4LR or visit us on the World Wide Web at:
http://www.thomsonlearning.co.uk

British Library Cataloguing-in-Publication Data
A catalogue record for this book is available from the British Library

ISBN 1-86152-542-7

Typeset by J&L Composition Ltd, Filey, North Yorkshire

Printed in Great Britain by TJ International, Padstow, Cornwall

Prelude

The Informed Student Guide to Management Science is an encyclopaedic collection of short essays and explanations on topics and concepts in the wide and diverse discipline of management science and operational research or MS/OR. The entries form the core of what any serious student and practitioner of MS/OR should know about these topics at least to the level presented here. Obviously, for the practice of MS/OR, in-depth knowledge, familiarity and application skills for an appropriately chosen subset of methods, tools and techniques is a must. The guide should, therefore, not be seen as a substitute for an in-depth text in the field. This is not only precluded by the limited length of the entries, but also by the breadth of topics covered. It encompasses the entire scope of MS/OR, from the 'hard' quantitative techniques, often using advanced mathematical concepts, to the 'soft' problem structuring methods which are based on philosophical, sociological and psychological concepts and principles. This dichotomy between the 'hard' and the 'soft' side of MS/OR means that most texts in the field, with a few notable exceptions, concentrate on one or the other, rarely ever referring to both. To some extent, the two sides also call for different skills from the practitioner – a solid basis in mathematics and statistics for translating the physical, technical and logical relationships into a mathematical model, and good interpersonal and facilitation skills for problem structuring interventions, where the people involved bring different perceptions of 'what the problem is' and have different personal values. However, both need a good grounding of systems concepts and their implications for problem solving, which forms their common basis and interface. The guide recognizes this aspect and gives similar coverage to both sides, as well as systems in general.

This diversity of people's perceptions of 'what the problem is' finds an echo in how people, practitioners and teachers, but particularly the latter, interpret MS/OR – its core, its philosophical basis, its underlying paradigms, and its role. While some still firmly believe that the mission of MS/OR is to find optimal solutions to problems, others see its purpose as providing decision aids for better insight into decision making, hopefully leading to better decisions, and still others interpret it as a means towards emancipation of all stakeholders in a problem situation. This is clearly reflected in the contributions made by the many contributors. You will find that the various authors emphasize different aspects and may even partially disagree in their views. However, this should not be seen as a weakness or as inconsistencies, but rather as the strength of a maturing discipline that is capable of accommodating differences in viewpoints and paradigms, and recognizes their contribution towards more informed, more insightful and more equitable and just decision making. Let experience show which paradigms will survive the test of time.

While the text strives to give a comprehensive and authoritative coverage of the subject, as the editor responsible for the final compilation of the text, I clearly kept in mind the needs and background of the target audience. The text is aimed at students with little or no background in MS/OR and who may just have begun their studies in the field, or students from other fields, as well as managers, who come across an MS/OR concept and would like to know more about it. They wish to do that without the need for extensive preliminary reading simply to be able to understand the terminology used and follow the explanations. I am also not convinced that is it helpful to overwhelm them at that stage with technical details, intricacies, and refinements that may do more to confuse than to enlighten. After careful study of an entry, the reader should come away with a reasonable understanding of the major concepts, arguments and lines of thought of the topic to be able to take part in an intelligent discussion. Hopefully, it will also kindle the desire and determination to know more about the subject. My aim was for the entries to:

- be concise, only stressing the major points, concepts, and arguments, generally limited to between 300 and 400 words;
- be easy to read, keeping the terminology to a minimum and/or providing short definitions for technical terms;
- be self-contained, in the sense that they can be read without the need to also study other entries first; and
- provide a few suitable references for further study.

It is, however, in the nature of the subject that some topics are 'easy', more self-contained and limited in extent, while others are 'hard', dealing with complex issues or techniques, and highly interconnected with other topics. This means that a few entries are a bit longer, their language even more concise, and each sentence packed with information. They may require more than one reading to get the full meaning.

In many entries you will find within the text one or a few words in **bold letters**. They signal that there is an entry in the text that discusses this topic in more detail. Although you should be able to understand the discussion without referring to these other entries, their subsequent study will without doubt deepen and enhance your understanding of the original topic, or open up new vistas that due to the word limit could not be covered there. Many of these cross-references form part of the special terminology that each discipline develops for itself. It is a mistake to assume that their technical meaning is the same as their everyday use.

For those interested in the historical development of the discipline, there are two entries under **history**, listing the year when various tools and techniques were developed or published for the first time and the person(s) mainly responsible for it.

The author of each contribution, except for a few definitional ones, is shown in *italics* at the end of each entry.

Hans G. Daellenbach
April 2002

List of contributors

Russell L. Ackoff is Anheuser-Busch Professor Emeritus of Management Science at the Wharton School, University of Pennsylvania. He received a PhD in Philosophy of Science from the University of Pennsylvania. At the Wharton School he was Chairman of the Social Systems Sciences Department and the Busch Center, which specialized in systems planning, research and design. He was President of the Operations Research Society of America, Vice President of The Institute of Management Sciences and President of the Society for General Systems Research. Dr Ackoff is the author and co-author of 22 books, his most recent being *The Democratic Corporation, Ackoff's Best* and *Re-Creating the Corporation*. He has also published over 200 articles. He has received six honorary degrees from US, British and South American universities and has been elected a member of the Academy of Natural Sciences of the Russian Federation. In September 2000 he was honoured by the establishment of the *Ackoff Center for Advancement of Systems Practices* at the University of Pennsylvania. His work has involved more than 350 corporations and 75 government agencies in the USA and abroad.

Gary C. Alexander is Associate Professor of Educational Policy and Administration at the University of Idaho, Boise. He holds a PhD from the University of Minnesota and is a Certified Public Mediator. His areas of research interest include organizational change, leadership, systems thinking and design, conflict resolution and mediation, distance learning, and multi-cultural, cross-cultural and international education. He was the recipient of the College of Education Faculty Award for Teaching (1995) and the Faculty Award for Research (2000). He was an invited scholar to Dalian University, Dalian, China (1999) and The University of Amsterdam (2001).

J. Scott Armstrong (PhD, MIT) is Professor of Marketing at the Wharton School, University of Pennsylvania. He is a founder of the *Journal of Forecasting* the *International Journal of Forecasting* and the *International Symposium on Forecasting*. In 1996, he was selected as one of the first six 'Honorary Fellows' by the International Institute of Forecasters. Along with Philip Kotler and Gerald Zaltman, he was the SAM/JAI Press Distinguished Marketing Scholar of 2000. He is the creator of the Forecasting Principles web site (forecastingprinciples.com) and editor of *Principles of Forecasting: A Handbook for Researchers and Practitioners* (Norvell, MA: Kluwer, 2001).

John Barton graduated in Mathematics from the University of Melbourne and later undertook graduate studies in economics and econometrics at La Trobe University, and executive development programmes at the Sloan School of Management, MIT. Currently he is a Director of Marshall Place Pty Ltd, a Melbourne-based strategy consultancy. Previous positions included Director of a Master of Management programme in systems thinking at Monash University, Head of the Department of Mathematics and Operations Research at the Footscray Institute of Technology, and Manager of Corporate Planning in a large water utility. He is a member of the Policy

Council of the System Dynamics Society and President of the Society's Australian–New Zealand Chapter, and actively publishes on systems thinking in management.

Stephen Batstone graduated from Management Science Honours programme at the University of Canterbury, for which he worked extensively on international network expansion problems for a large telecommunications company. He is in the final stage of submitting his PhD thesis in Management Science at the University of Canterbury. His research interests include risk management in electricity markets, with particular focus on the interactions of hydro storage, market power and long-term contracts. He currently lectures on a variety of topics, including applied management science and simulation, at both advanced undergraduate and graduate level.

Valerie Belton (PhD, Cambridge) is Professor of Management Science at the University of Strathclyde. She was co-director of the MENTOR project on computer-assisted teaching, Vice-President of the European Federation of OR Societies for 1996–2000, and is currently the president of the International Society of MCDM. She is well known for her work and numerous publications on multiple criteria analysis, and is the editor of its flagship, the *Journal of Multi-Criteria Decision Analysis*. Her research led to the development of the visual interactive sensitivity analysis software V.I.S.A., which is used worldwide for research, teaching and consultancy work. Current research focuses on group decision support systems and on the integration of problem structuring methods (in particular, cognitive mapping) and other OR/MS approaches, for example, data envelopment analysis, conflict analysis and system dynamics with multiple criteria analysis.

Peter Bennett has been Principal OR Analyst in the Economics and Operational Research Division of the Department of Health (UK) since 1996. Following a degree in physics and a PhD in Philosophy of Science, he joined the Operational Research Group at Sussex University, researching on decision making in conflict situations. This started a long-term interest in decisions involving risk, uncertainty, multiple objectives and conflict. In 1987 he moved to Strathclyde University as senior lecturer and then Reader in Management Science, where he was also involved in consultancy and applied research work for organizations ranging from multinational companies to local community groups. His current work covers various policy areas, but with a particular focus on analysis of risks to public health, including issues of risk communication and governance. Meanwhile he has continued to collaborate with academic colleagues, particularly in the development of hypergame analysis and drama theory.

C. Piet Beukman (BEng (Hons), Stellenbosch; MEng, Pretoria) is a registered engineer, holding a teaching appointment with the University of South Australia. Previous to that he was Director of the Master of Engineering Management and Senior Lecturer at the University of Canterbury. While in New Zealand he had an active interest and involvement in local economic development, particularly the management of innovation and technology, as well as the measurement of engineering organization effectiveness. He also established the Recovered Materials Foundation for Christchurch, where technology, business principles and waste recycling are combined to form a viable industry and employment creation initiative. Prior to that he worked in South Africa in marine, consulting engineering and aerospace industries.

Ken C. Bowen is Visiting Professor of Operational Research at Royal Holloway, University of London, and a member of the board of the Euro Working Group on Methodology for Complex Societal Problems. In 2001, he was the recipient of the

Operational Research Society's award of Companionship of Operational Research. His recent publications have been in game and drama theory.

John Brocklesby is Professor of Management at Victoria University. He holds a PhD from the Warwick Business School. Prior to entering academic life he worked in various management positions in the steel industry. He has published extensively in many leading academic and practitioner journals and is on the editorial board of *Systemic Practice and Action Research* and *The Journal of the Australian and New Zealand Academy of Management*. He is a past editor of *The New Zealand Journal of Business*. He is an active member of the Academy of Management. His main research interests are in the fields of organization theory, management and systems thinking. He is particularly interested in managerial and organizational applications of theories that originate in the life sciences and in the new sciences of complexity.

Ian Brooks (BA in Law, MBA (Hons), PhD, Canterbury) is a Senior Lecturer in the Department of Management at the University of Canterbury, where he teaches organizational behaviour. He also has a qualification from the NTL Institute for Applied Behavioural Science. His special areas of interest are group dynamics and applied management skills.

John T. Buchanan is Associate Professor of Management Systems in the Waikato Management School at the University of Waikato, New Zealand. He holds a PhD in Operations Research from the University of Canterbury. He has published in a number of international journals, particularly in the area of decision making with multiple criteria. Previously, he was a Senior Consultant with Price Waterhouse.

Diane Campbell-Hunt is a self-employed consultant working in a wide range of areas, including futures research, strategic thinking, ecology and science policy for a wide range of New Zealand and Australian private companies and government agencies. In 1999 she wrote, on behalf of the NZ Futures Trust, a 'Cultural Futures Pack' for use in secondary schools, a project funded by the National Commission for UNESCO. She has degrees in zoology and resource management.

Bob Cavana (PhD) is currently a senior lecturer in decision sciences with the School of Business and Public Management at Victoria University of Wellington. He co-authored (with K.E. Maani) *Systems Thinking and Modelling: Understanding Change and Complexity* (Auckland: Prentice Hall, 2000) and (with B.L. Delahaye and U. Sekaran) *Applied Business Research: Qualitative and Quantitative Methods* (Brisbane: Wiley, 2001). He is an Associate Editor of the *System Dynamics Review*, and was the President of the Operational Research Society of New Zealand. Prior to joining Victoria University in 1989 he held the position of Corporate Economist with New Zealand Railways Corporation.

Chris Chatfield is Reader in Statistics in the Department of Mathematical Sciences at the University of Bath. He has a PhD in Statistics from Imperial College, London. He is the author of five books, including *The Analysis of Time Series,* 5th edn (Chapman and Hall/CRC Press, 1996) and *Time-Series Forecasting* (Chapman and Hall/CRC Press, 2001). He has written over 50 research papers, with emphasis on time series forecasting, initial data analysis and problem solving. He is a Fellow of the Royal Statistical Society, an elected member of the International Statistical Institute, and a member of the International Institute of Forecasters.

Deb Chattopadhyay (PhD) is a Principal in the New Zealand office of Charles River Associates, a leading international business consulting firm. His area of consulting and research includes a wide range of activities in the electricity indus-try, including market design, market modelling, pricing, environmental issues and power system engineering aspects. He has done research, worked, and published extensively in international journals on the development of optimization models to address power system planning and operations problems. He taught Operations Research at the Department of Management, University of Canterbury between 1997 and 2000.

Bo Chen is lecturer in the Warwick Business School at the University of Warwick. Prior to that he was a Management Research Fellow of the Economic & Social Research Council (UK). He holds a PhD from Erasmus University. In the areas of oper-ational research, applied mathematics and computer science he has contributed to a variety of books and published numerous articles in international journals. He is Associate Editor of the *Journal of Combinatorial Optimization*.

Hans G. Daellenbach is professor emeritus at the University of Canterbury. He holds a PhD in MS/OR from the University of California, where he studied under C. West Churchman whose teaching has had a lasting influence on him. In 1970 he moved to the University of Canterbury where he introduced operations research as a disci-pline and launched its MBA programme in 1983. He was the editor of *New Zealand Operational Research*, the *Asia-Pacific Journal of OR* and is serving on a number of jour-nal advisory boards. Being a generalist, he has published extensively on a broad range of applied OR topics. He is the author and co-author of several textbooks in operations research and systems thinking.

D.P. Dash has a PhD in management studies/systems thinking from the University of Lincolnshire and Humberside (UK). He is currently Assistant Professor of General Management at Xavier Institute of Management, Bhubaneswar, India, where he teaches strategic management and corporate planning, both from a systems theoretic point of view. He is a keen student of the developments in systemic and cybernetic thinking and their reflection in the philosophies and methods of action-oriented research. His research interests include the study of interactive and collective phe-nomena in various domains, a research topic that aims to inform the design of co-ordination structures and conversational processes. He is on the Editorial Board of the journal *Systems Research and Behavioral Science*.

Kathryn Dowsland has an MSc in Pure Mathematics and a PhD in Operational Research from the University of Wales. After an early career in operational research in the steel industry, she joined the University of Wales Swansea where until recently she held a Readership in OR. She has published widely in international and national refereed journals and these have included over 30 papers in the areas or cutting, packing, scheduling and modern heuristic techniques.

William Dowsland has a PhD in Information Systems from the University of Wales, and until recently was a Senior Lecturer in OR/IT at the University of Wales Swansea. He is a Chartered Engineer, a member of the British Computer Society and the Oper-ational Research Society. He is a also director of Gower Optimal Algorithms Ltd (GOAL) (http://www.goweralg.co.uk/), a company providing solutions to cutting, packing and scheduling problems to high-profile companies worldwide.

Shane Dye is a lecturer in MS/OR with the Department of Management at the Uni-versity of Canterbury in New Zealand. He received a BSc (Hons) in Mathematics from the University of Canterbury in 1992 and a PhD in Operations Research from Massey University in 1995. His research interests are currently in stochastic programming and stochastic integer programming applications. He held a three-year post-doctoral research position, funded by Telenor (a Norwegian telecommunications company), at the Norwegian University of Science and Technology in Trondheim, where he investigated stochastic programming models for service delivery in intelligent networks.

Arun Abraham Elias is a PhD student at the School of Business and Public Management of the Victoria University of Wellington and a Lecturer at the School of Management of the Open Polytechnic of New Zealand. In his PhD research, he is analysing the dynamics of stakeholders in environmental conflict using a system dynamics approach.

Merrelyn Emery obtained her PhD from the University of New South Wales. Since 1970 she has been developing the theory and practice of Open Systems Theory (OST). She is a founding Director of the Fred Emery Institute and a Visitor at the Centre for Continuing Education at the Australian National University. She teaches intensive workshops around the world on the history and 'state of the art' of OST.

Robert Louis Flood is currently Visiting Professor at Monash University and at the Maastricht School of Management, and an independent action researcher. He has a DSc and PhD and is a Chartered Engineer and Fellow of the Institute of Measurement and Control. He is the author of several books, including *Rethinking the Fifth Discipline*, and is the Editor of the international journal *Systemic Practice and Action Research*. Previously, he was Professor of Management Sciences and Head of Department at Hull University.

Jeff Foote (Btech (Hons), Mtech (Hons)) is a systems scientist with the Institute of Environmental Science and Research Limited (New Zealand). Previously, he was involved in research that examined the factors that influenced the uptake of worldclass manufacturing practices in small manufacturing enterprises. His current research interests include systems thinking, hospital administration and environmental management.

Les Foulds is Professor of Operations Management and Director of the Centre for Supply Chain Management at the Waikato Management School, University of Waikato, New Zealand. He has a doctorate in industrial engineering and operations research and has published four books on graph theory and optimization. He has also written numerous articles on scheduling, traffic planning and facilities planning in international journals. He is a Founding Fellow of the Institute of Combinatorics and Applications and the current President of the New Zealand Operational Research Society.

John Friend, a graduate in mathematics from Cambridge University, worked for ten years as a statistician and OR scientist in the steel, civil aviation and chemical industries, before joining the Tavistock Institute in 1964. Working with social scientists on complex public planning issues kindled his interest in alternative OR approaches, which 20 years later were labelled 'problem structuring methods'. It resulted in his first co-authored book, *Local Government and Strategic Choice* (London: Tavistock, 1969; 2nd edn, Oxford: Pergamon Press, 1977), a pioneering work in soft OR, consolidated subsequently as the 'Strategic Choice Approach', the topic of *Planning under Pressure* (Oxford: Pergamon Press, 1987), co-authored with Allen Hickling. Since leaving Tavistock in 1986, he has been involved in further development of the strategic choice software, in community action and international development projects, as well as being Visiting Professor at the University of Lincoln.

Nicholas C. Georgantzas is the Coordinator of Fordham University's System Dynamic Consultancy and a Senior Consultant of Strategic Scenarios, Inc., a management systems consulting firm specializing in system dynamics simulation modelling for strategic support and business process (re)design. He holds a PhD, an MPhil and an MBA, all from CUNY. In 1987, he won the Oscar Lasdon Award for best dissertation. He is the author of *Scenario-Driven Planning: Learning to Manage Strategic Uncertainty* (Westport, CN: Greenwood Press, 1995) and has published over 70 articles in scholarly journals, edited books, and conference proceedings on systems thinking, organizational learning, and strategy design in a dynamic context. He has extensive practical industrial experience in a variety of sectors.

Ion Georgiou is Visiting Professor at the Universidade Estadual do Sudoeste da Bahia, Brazil, and Senior Lecturer at Kingston University (UK). He has also taught and undertaken research at the London School of Economics. His main interests are Bertalanffy's general system theory, phenomenology (in particular the philosophies of Husserl and Sartre), management methods and problem structuring methods. Fluent in five languages, he has consulted on commercial and academically linked public projects across Europe and Brazil and has also taught at universities in Russia and Spain.

John Giffin is a senior lecturer in MS/OR at the University of Canterbury and former subject leader in the Institute of Information Sciences and Technology at Massey University, New Zealand. He received his PhD from the University of Canterbury in 1984, and subsequently worked at the University of Arizona as an Assistant Professor of Industrial Engineering. His research and teaching interests are in the design and analysis of heuristics and metaheuristics, with applications to vehicle routing, vehicle scheduling, arc routing, facility layout and location, areas in which he has published in a number of international journals.

Davydd James Greenwood is the Goldwin Smith Professor of Anthropology and Director of the Institute for European Studies at Cornell University. He was also elected a Corresponding Member of the Spanish Royal Academy of Moral and Political Sciences. His work centres on action research, political economy, and the Spanish Basque Country. He has published and/or edited nine books in these fields.

Abeyratna Gunasekarage is a Lecturer in Finance at the Department of Accountancy, Finance and Information Systems of the University of Canterbury. He holds an MAcc and a PhD from Dundee and is an associate member of the Institute of Chartered Accountants of Sri Lanka. He teaches Financial Management and Corporate Finance.

John Holt is a Senior Consultant with HVR Consulting Services Ltd, working primarily in Defence OR Consultancy. Much of his work has been concerned with pioneering the use of soft OR applications in Defence. He has an MSc in Management Science from Imperial College, London, and PhD in Management Science from Southampton University. He was Research Fellow and Lecturer in OR at Southampton University. Prior to joining HVR-CSL, he spent approximately ten years as a consultant with Centre for OR and Defence Analysis (CORDA), now part of BAE SYSTEMS.

Don Houston is Senior Lecturer, Quality Management and Program Coordinator for the Graduate Diploma in Quality Assurance in the Institute of Technology and Engineering, Massey University, New Zealand. He has worked in education for over 20 years in research, curriculum design and evaluation, teaching and management roles. His major teaching and research interests are quality management in services and education. He is a fellow of the Quality Society of Australasia.

Rob Hyndman (BSc(Hons), PhD, A.Stat.) is Associate Professor of Statistics and Director of the Business and Economics Forecasting Unit at Monash University. He is the Editor of the *Australian and New Zealand Journal of Statistics*, and has more than 17 years statistical consulting experience working with hundreds of clients. He is best known as co-author of the international best-selling textbook *Forecasting: methods and applications* (Makridakis, Wheelwright and Hyndman, 3rd edn, 1998, New York: Wiley). He also publishes regularly in scholarly refereed journals and is an award-winning lecturer at Monash University.

Ross James is a Senior Lecturer in the Department of Management, University of Canterbury, where he teaches in the areas of operations management and management science. His main research interests are in the areas of heuristics, search heuristics and scheduling applications, where he has published in a number of international journals.

Paul Keys (BSc, PhD) is a Senior Lecturer in the Hull University Business School. His research into the process and methodology of operational research and systems thinking has been widely published in a variety of academic journals, books and conference proceedings.

Bruce Lloyd is Professor of Strategic Management of South Bank University in London. Before joining the academic world, he had 20 years of experience in industry and the City. He has published over 100 articles and papers in various journals on strategic subjects. He obtained a PhD for his work on 'The future of offices and office work'.

Hernán López-Garay is Professor and Principal Researcher of the Center for Interpretive Systemology (CSI) of the University of Los Andes, Venezuela. He has a PhD from Wharton, an MA in Systems from Lancaster University, and an MSc in Systems from Case Western Reserve University. His main research area is in interpretive systemology and its application to public institutions, particularly in Latin America. He has published in numerous American, European and Latin American journals and is co-author of the first comprehensive book on systems thinking published in Colombia and Venezuela. Currently he is leading an hermeneutic project on the Justice System in Latin America at the CSI.

Victoria Mabin (BSc(Hons), PhD (Lancaster)), is a Senior Lecturer in Management at Victoria University of Wellington, teaching a variety of soft and hard OR/MS methods and applying them across a range of contexts, including public, private sector and not-for-profit organizations. She has published in journals and books worldwide, especially on Goldratt's theory of constraints, and multicriteria decision analysis. Prior to taking up her academic position at VUW in 1991, she was an OR consultant with the leading scientific and industrial research organization in NZ.

Donald C. McNickle is a Senior Lecturer in Management Science and Head of the Management Department, University of Canterbury. He holds a PhD in Statistics from the University of Auckland. His teaching and research interests include queues, simulation, business forecasting and applied statistics. He is co-author of two text books and has published extensively in international journals.

Gerald Midgley is Director of the Centre for Systems Studies, a research institute based in the Business School at the University of Hull (UK). He has published over

150 papers in international journals, edited books and practitioner magazines, and has been involved in a wide variety of Community OR projects. He is also the author of *Systemic Intervention: Philosophy, Methodology, and Practice* (Dordrecht: Kluwer/ Plenum, 2000), and the co-author of *Operational Research and Environmental Management: A New Agenda* (Operational Research Society, 2001).

John Mingers is Professor of Operational Research and Systems at the Warwick Business School, University of Warwick. He is a past Chair of the UK Systems Society and has been a member of the Council of the OR Society. His research interests include the use of systems methodologies in problem situations, particularly the mixing of different methodologies within an intervention (multimethodology), the development of the critical systems approach, autopoiesis and its applications, and the nature of information and meaning. He has published over 80 papers in these areas, and four books, including *Self-Producing Systems: Implications and Applications of Autopoiesis*.

V. Nilakant is a Senior Lecturer in the Department of Management at the University of Canterbury. He received a PhD in Organizational Behaviour from the Case Western Reserve University. His research and teaching interests include the management of change and organization theory. He is the lead author of *Managing Organisational Change* (New Delhi: Response Books, 1998).

John F. Raffensperger has an MBA and a PhD from the University of Chicago. He has extensive experience as an industrial engineer and consultant. Currently, he is a Lecturer in Management Science in the Department of Management, University of Canterbury, where he teaches operations management and management science. His main research interests are in logistics and transportation, mathematical programming, and portfolio optimization.

E. Grant Read leads the Energy Modelling Research Group at the University of Canterbury, where he is Associate Professor in Management Science. He is actively involved as an international consultant on electricity modelling and market design issues in Australia, New Zealand, Canada, Singapore, Malaysia, and the Philippines. His interests include reservoir management, market-clearing models for real-time dispatch and pricing of energy, reserve, and transmission in electricity systems, and game theoretic models of price manipulation in such markets, areas in which he has published extensively. He holds a BSc (Hons) in mathematics and a PhD in Operations Research from the University of Canterbury. He is a past president of the Operations Research Society of New Zealand.

Norma R.A. Romm is a Senior Researcher in the Centre for Systems Studies, The University of Hull Business School. She has previously held senior positions in universities in South Africa and Swaziland. She is the author of *The Methodologies of Positivism and Marxism* (1991) and *Accountability in Social Research* (2001), co-author of *People's Education in Theoretical Perspective* (with V.I. McKay, 1992) and *Diversity Management* (with R.L. Flood, 1996), and co-editor of *Social Theory* (with M. Sarakinsky, 1994) and *Critical Systems Thinking* (with R.L. Flood, 1996). She has written over 60 articles in journals and edited books, primarily on social theory and social research (and their relationship).

Jonathan Rosenhead is Professor of Operational Research at the London School of Economics. He previously worked in the steel industry and in management consultancy, and held visiting positions at the University of Pennsylvania. A former president of the Operational Research Society, he holds its Beale, Goodeve and President's

Medals. He has a particular interest in the application of analysis to public domain problems, with an emphasis on health services and on work with community groups. His edited book *Rational Analysis for a Problematic World* (Chichester: Wiley, 1989), and its revised edition, *Rational Analysis for a Problematic World Revisited* (Chichester: Wiley, 2001), are the standard introduction to problem structuring methods.

David Ryan (PhD) is Professor of Operations Research at the University of Auckland. His main interests are in the development of combinatorial optimization models and computational solution methods for large-scale scheduling problems arising in business and industry, where he has published extensively in international journals. The application of these techniques in crew scheduling systems at Air New Zealand has been awarded the 1999 IPENZ Engineering Excellence Award in Information Technology and it has been a finalist in the prestigious international Franz Edelman Award sponsored by INFORMS in 2000. In 2001, he was awarded the inaugural Daellenbach Prize by the Operations Research Society of New Zealand. He presented the Opening Plenary Address at the Beijing IFORS meeting in 1999 and was the IFORS Distinguished Lecturer at the INFORMS meeting in San Jose in November, 2002.

Jay Sankaran is Associate Professor at the Department of Management Science and Information Systems at the University of Auckland. He holds a PhD in management science from the University of Chicago. His research interests span both deductive modelling and inductive, empirical research. He has published widely in the areas of management science, logistics and supply chain management (latterly with a strong New Zealand orientation), and R&D management. Journals that feature his work include *Mathematical Programming, International Journal of Physical Distribution & Logistics Management* and *Leadership Quarterly*.

David Smith is senior lecturer in operational research in the School of Mathematical Sciences at the University of Exeter. He teaches courses in mathematical modelling, mathematical programming and network optimization. After a mathematics degree at the University of Cambridge, he studied OR at the University of Lancaster before his appointment in Exeter. He is the author of two textbooks and has published many papers on operational research practice and methodology. He is also a regular speaker at conferences. OR applications in the not-for-profit sector and methods of sequential optimization are amongst his current research areas. He edits the *International Abstracts in Operations Research*.

Ralph D. Snyder is an Associate Professor at Monash University. His specialization is business modelling and operations research. The focus of his research is quantitative and statistical methods for sales forecasting and inventory management. His publications appear in leading international academic journals. He also consults with Australian business and industry.

Ramaswamy Sridharan (PhD, Carnegie-Mellon) is a Senior Lecturer in the Institute of Information Sciences and Technology at Massey University, New Zealand. His areas of interest in teaching and research are supply chain optimization, mathematical programming, Lagrangian relaxation, heuristics and plant location. The last three areas are also where he has a number of publications in international journals. He has held teaching appointments in India, Australia and the UK and has been involved in project work for international development agencies.

Mark Stewart holds a BSc (Hons) in Management Science from the University of Canterbury, New Zealand, which included a study of risk analysis for one of the major emergency services. He is currently completing a PhD in Management Science. His research interests include the general use of mathematical modelling in telecommunications and Internet settings.

Stephanie Stray is a Senior Lecturer in the Operational Research and Systems Group of Warwick Business School at the University of Warwick. She has a degree in Economics and Statistics from the University of York and a PhD from the University of Essex and is a Chartered Statistician. She has numerous publications in international journals and has given presentations of many papers at both national and international conferences.

Johan Strümpfer is director of a consulting firm, Systems Practice, since 1992. He has an MSc and a doctorate in Operations Research from the University of Cape Town. He has been involved in technical and organizational problem solving for 20 years, as a private consultant and senior research fellow at the University of Cape Town, the SA Institute for Futures Research and the SA Institute for Maritime Technology, as well as at the Wharton Business School. He was responsible for the search planning for the SAA 'Helderberg'. His fields of interest are systems thinking and the systems approach, strategic planning and organizational renewal. He has taught in the Business Schools at the University of Cape Town and at the Stellenbosch University.

Emmanuel Thanassoulis has been Professor in Management Sciences at Aston Business School since 1999, and prior to that was Lecturer and Senior Lecturer at the Warwick Business School. He holds an MSc and a PhD from Warwick University. His research field is in comparative performance management, particularly data envelopment analysis, where he has published extensively. He is the author of *Introduction to the Theory and Application of Data Envelopment Analysis* (Boston: Kluwer). He runs specialist training courses on DEA for practitioners. In the late 1980s he developed in collaboration with colleagues the Warwick DEA Software. He has had extensive involvement within the UK on DEA applications for regulated utilities, regulators of such utilities, the Home Office, the Treasury, the Higher Education Funding Council for England and Wales and the Department for Education and Employment.

Anastasios Tsoularis has been Lecturer in Operational Research and Statistics in the Institute of Information and Mathematical Sciences, Massey University, Auckland Campus since 1999. He holds an MSc from the University of London and a PhD from the University of Reading. He worked as a mathematical modeller/scientific programmer for the electricity industry in the UK for five years.

Werner Ulrich (Dr.rer.pol, Fribourg; PhD, Berkeley) studied economics and social sciences at the Universities of Fribourg and Zurich and philosophy of social systems design at the University of California at Berkeley with C.W. Churchman. His work on critical systems heuristics pioneered critical systems thinking (CTS). His current research programme 'CST for Professionals and Citizens' explores the ways critical systems thinking can contribute to responsible professional practice and to preparing citizens for their role in a living civil society. He has extensive experience as a researcher in the public sector and as professor of social planning, program evaluation, poverty research, and critical systems thinking at the University of Fribourg. He has held visiting professorships for CST at the University of Hull, the University of Lincolnshire & Humberside and the University of Canterbury. He is currently director of the Lugano Summer School of Systems Design at the Università della Svizzera Italiana, Lugano.

John Vargo is Senior Lecturer in Information Systems at the University of Canterbury. He has a particular interest in Strategic Planning and Telecommunication Systems and is co-author of the textbook *Telecommunications in Business: Strategy and Application* (Irwin). His current research interests include effective e-commerce systems, e-learning systems using Internet technologies, and the application of telecommunication systems to strategic business needs.

P. Venkateswarlu works as a process improvement leader for Macpac Wilderness Equipment Ltd., an outdoor equipment manufacturer based in Christchurch, New Zealand. Prior to that he was a Lecturer in the Department of Management at the University of Canterbury, teaching operations management and quality control at both the undergraduate and MBA levels. He holds a PhD from the Indian Institute of Technology in Bombay.

Alan Washburn is Professor of Operations Research at the Naval Postgraduate School in Monterey, California, where he has served as chairman of the OR Department. His publications include papers in the areas of search theory, optimization, statistics, probability analysis, and game theory. His military interest areas include mine warfare, antisubmarine warfare, munitions planning, and information warfare.

Richard Watson (MSc, PhD, MASOR) worked for many years as a Senior Research Scientist in the Australian Defence Science and Technology Organisation in Canberra (DSTO) and Melbourne where he developed expertise in OR, logistics, systems studies and advanced distributed simulation. He has also been a Lecturer in Information Technology at Swinburne University of Technology in Melbourne and since retirement from the DSTO in 2000 has been a Business Analyst with Open Telecommunications Limited in Melbourne, a firm which develops software for the telecommunications industry worldwide.

D. Clay Whybark is the Macon Patton Distinguished Professor at the Kenan-Flagler Business School, University of North Carolina. He is founding director of the Global Manufacturing Research Group and has published widely on international manufacturing, production planning and control, logistics and other operations management topics. He is the author or co-author of several books and is on a number of editorial boards. He holds a BS in Aeronautical Engineering from the University of Washington, an MBA from Cornell University and a PhD from Stanford University.

Jennifer Wilby is a Research Fellow in the NHS Centre for Research and Dissemination (CRD) at the University of York. A graduate of the University of California (BA, Politics), she holds masters degrees in Cybernetic Systems (California State University) and Public Health (Leeds). Her research interests include general systems theory, hierarchy, public health and systematic reviews in the field of healthcare. She is affiliated with the Centre for Systems Research at the University of Hull and is a member of the board of the International Society for the Systems Sciences (ISSS), editing publications for the society.

Dave Worthington is Senior Lecturer in Operational Research, and Director of the Masters programme in Operational Research in the Department of Management Science at Lancaster University. Over the last 25 years, he has been employed in, consulted on, taught and researched the application of management science in the public sector and especially to healthcare. He also researches into queue management – which more than occasionally overlaps with his healthcare interests.

A

ABC classification, Pareto analysis

The Italian economist, Vilfredo Pareto (1848–1923), observed that about 80 per cent of the national wealth belonged to fewer than 20 per cent of the population. Many years later, the quality control statistician, J.M. Juran, observing this same '80–20 principle' or the principle of the 'vital few and the trivial many' in the field of quality control, named it the *Pareto principle*. In particular, he found that a high percentage of defective parts can be attributed to a few causes.

The principle is encountered in most areas of human activity – a small proportion of something accounting for a very large proportion of something else. For example, a small number of causes account for the largest number of customer complaints, 80 per cent of sales volume is generated by 20 per cent of the customers, more than 80 per cent of the health costs in most countries are caused by less than 10 per cent of the population, and so on.

In **inventory control** the Pareto principle has been extended to the so-called ABC classification. Ranking all products in terms of their annual dollar-volume of sales, the following approximate patterns are commonly observed:

		Percentage of items	
Class	Percentage of dollar-volume	Consumer goods	Industrial goods
A	50	10	5
B	40	30	20
C	10	60	75

A items tends to be given individual, proactive control with follow-up if anything departs from plan. Stock levels are monitored on a constant basis. Replenishments are based on accurate forecasts of actual planned usage or expected demand. C items may often be controlled via an automatic *two-bin system*, where the second bin corresponds to the *safety stock* to cover demand during the **lead time**. Goods are withdrawn first from the first bin and only when it is empty from the second bin, at which time a replenishment is automatically initiated. The new replenishment, based on the **economic order quantity (EOQ) formula**, first restores the safety stock in bin 2, and only the remainder goes into bin 1. No individual stock records may be kept. For B items, control by exception replaces the individual attention given to the A items. Stock records are updated by recording each transaction, and replenishments are based on automatic forecasts.

Abstract system

An abstract **system** is a system that does not model a physical reality. It has no physical counterpart in the form of components or activities in the real world, such as people and machines that engage in activities. It consists only of abstract concepts and entities, relationships and interactions between entities, and rules for these relationships and interactions. Its inputs and outputs are abstract concepts or data. Typical examples are classification systems that classify items according to specific rules and entity **attributes**, the number system that allows us to manipulate numbers and perform operations with numbers, the legal system that governs the legal relationships between individuals and so on. Many MS/OR models, and all **optimization** techniques used in MS/OR, such as the **simplex method** of **linear programming**, are abstract systems.

Acceptance sampling

Acceptance sampling or lot acceptance sampling (LAS) is a statistical method used by buyers or consumers of goods to weed out lots shipped by a producer/supplier that contain an unacceptably high fraction of defective or bad items. It is used when one or more of the following conditions apply: (1) testing destroys the item, (2) the cost of 100 per cent inspection is very high, (3) 100 per cent inspection takes too long, and (4) the supplier has a good track record on quality. LAS is therefore a trade-off between no inspection and 100 per cent inspection. The US military pioneered the approach to test ammunition during World War II in order to avoid supplying bullets to its soldiers at the front that were prone to malfunction, with potentially disastrous results.

Assume that a firm receives a lot of size N of unknown quality (i.e. the true number of defectives). The lot is acceptable if it contains fewer than, say, 2 per cent defectives. The decision of whether or not to accept a lot is made on the following sampling plan: Take a sample of size n; count the number x of defectives found; if $x \le c$, where c is the maximum number of defects acceptable in the sample, the entire lot is accepted, otherwise it is rejected.

If c is set too high, many lots that contain more than the maximum allowable number of defectives will be accepted. The probability of this happening is called the *consumer's risk*, β (also known as the *type II error*). If c is set too low, then many lots that are clearly within the acceptable limit of defectives will be rejected. The probability of this happening is called the *producer's risk*, α (also known as the *type I error*). The aim is to choose a *sampling plan*, defined by values n and c, which keeps both these risks reasonably low. With any given sampling plan, we can associate a corresponding *operating characteristic* or *OC curve*, as depicted in the figure below. It shows the probability that a lot will be accepted or rejected for a given (assumed or actual) level of defectives in the lot.

Appropriate values for c and n that satisfy quality agreements between producers and customers can be selected either from tables, such as the military standard 105D (MIL STD 105D), or from comparing OC curves for different pairs [n,c].

N. Georgantzas

References

Montgomery, D.C. (1995) *Introduction to Statistical Quality Control*, 3rd edn, New York: Wiley.
Schilling, E.G. (1982) *Acceptance Sampling in Quality Control*, New York: Marcel Dekker.

Web site

Engineering statistics handbook: www.itl.nist.gov/div898/handbook/pmc/section2/pmc2.htm

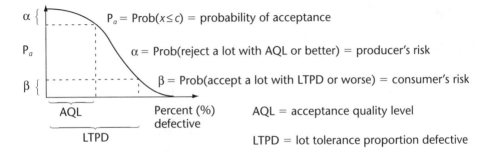

P_a = Prob($x \leq c$) = probability of acceptance

α = Prob(reject a lot with AQL or better) = producer's risk

β = Prob(accept a lot with LTPD or worse) = consumer's risk

AQL = acceptance quality level

LTPD = lot tolerance proportion defective

Figure: **Operating characteristic (OC) curve**

Action learning

Action learning is an approach to improve organizational functioning by focusing on teaching people to work more effectively in their current settings. Concentrating on real-world problems drawn from the participants' work environments, action learning uses a facilitator and a group of colleagues, or so-called action set, as probers and supporters of the learning.

One of a number of approaches, among them **action science**, emerging from **action research**, action learning is strongly associated with Reg Revans who coined the term and has given the practices their general contours.

Action learning aims to achieve important behavioural change through facilitated reflections in a group setting. The group grapples with real problems brought to it directly from the members' work lives. Dealing with participants drawn from the same general organizational setting, it probes their behaviour and develops a variety of (not necessarily mutually consistent) recommendations for action, in the hope of making them more 'natural' or deployable to the learner.

The facilitator's role is limited to keeping the process on track, managing the probing of the selected learner's actions and issues and helping to organize the results of the reflections into specific recommendations for the learner to apply. The facilitator does some work with the action set to assist the group in becoming a better vehicle for its members' learning. Though facilitation philosophies and practices vary from strongly non-interventionist to more directive, rarely does the facilitator reach for deeper levels of analysis of the problems presented, such as the psychological profile of the problem owner, the political economy of the organization in question, and the like. Rather, problems are taken on at the level on which they are presented and worked with through group brainstorming and various other group process techniques.

Although action learning shares some features with action science, the latter moves the interventions to a deeper and more confrontational level than does action learning.   *D.J. Greenwood*

References

McGill, I. and Beaty, L. (1992) *Action Learning: A Guide for Professional, Management and Educational Development*, London: Kogan Page.

Raelin, J.A. (1997) 'Action learning and action science: are they different?', *Organizational Dynamics*, 26(1):21–34.

Revans, R. (1990) *Action Learning*, London: Blond & Briggs.

Revans, R. (1982) *The Origins and Growth of Action Learning*, Bromley: Chartwell Bratt.

Senge, P.M. (1990) *The Fifth Discipline: The Art and Practice of the Learning Organization*, London: Random House.

Action research

Action research has its beginnings in the work of Kurt Lewis in the mid-1940s. He recognized the difficulties of studying complex social and psychological processes by traditional experimental methods of trying to isolate individual aspects. Instead, he advocated to do the research by becoming actively involved in practical, real-world problems. An inevitable consequence of this is that the researcher does not remain an 'objective' outside observer, but becomes part of the problem by his or her very presence, affecting both its processes and outcomes. Rather than see this as a serious disadvantage that could invalidate the research findings, action researchers make a virtue of this and use it to actively intervene and steer the process in directions that will lead to learning and hence greater insights into the processes. It is therefore important that the researcher enters the situation with a sufficiently well-formed set of hypotheses and methodology, which will allow them to reflect during and at the end of their involvement on both the theory and methodology and the learning achieved in the process. Usually the same or a further evolved theory, methodology and practice are applied to several different cases.

The process can be described as a self-reflective cycle of planning a change, acting and observing the process and consequences of the change, reflecting on these processes and consequences, then replanning, acting, observing, reflecting, and so on.

It is in this way that action research contributes to the advancement of theory, methodology and practice. However, it is clear that action research cannot confirm or refute hypotheses in a statistical sense. The conclusions drawn are much more qualitative and by necessity also more tentative.

Several of the **problem structuring methods** that make up **soft operational research** have evolved and continue being refined through action research, Checkland's **soft systems methodology** being the prime example. *H.G. Daellenbach*

References

Checkland, P. and Holwell, S. (1998) *Information, Systems and Information Systems*, Chichester: Wiley.

Lewin, K. (1946) 'Action research and minority problems', *Journal of Social Issues*, 2(4):34–46.

Action science

Action science is an approach to change in organizations and interpersonal relations based on improving people's ability to cease behaving defensively when challenged. It uses a combination of collaborative problem-solving tools in a group setting with a facilitator and a structured method of inquiry into the causes of unproductive individual and group behaviour. Beginning with Kurt Lewin's work on **action research** in the late 1940s, it evolved from ideas and techniques developed mainly by Chris Argyris *et al.* (1985), Donald Schön (1983) and by Argyris and Schön in collaboration (1996).

According to action science, most individual and group behaviour manifests *Model I* or **single-loop learning** features, meaning a tendency to intensify the very behaviours that created the problems in the first place. This involves poor inferences as the basis for future behaviour, defensiveness about one's own role in creating group problems, and a general collusion to maintain the status quo. Interventions in action science aim to enhance the ability of individuals and organizations to behave in a *Model II* or **double-loop learning** mode, changing behaviour toward greater openness and ability to modify patterns of operation to be more adaptive. This involves systematic inquiry into the causes of the problems, reduced defensiveness, and recognition of the role of one's own behaviour in generating and sustaining problems. While no organizations ever achieve Model II completely, the increased presence of Model II makes it a learning organization, meaning an organization capable of changing its internal structures and operations in an adaptive and more healthy way (Argyris and Schön, 1996). Senge (1990) provides a neat summary of these ideas.

Action science seeks to change organizations toward a more healthy and liberated state in which the capacities of the individuals are more effectively expressed and in which their behaviour matches the positive normative objectives that most of us generally articulate.

The facilitator plays a very strong role in action science, confronting the behaviour of the collaborators attempting to solve problems in their organizational lives, coaching them, and modelling Model II behaviour for them to see and imitate. A variety of structured techniques are used to assist in this facilitation, among them the 'left-column/right-column' exercises and the ladder of inference (Argyris, Putnam and McLain Smith, 1985).

Action science interventions can be quite deep and purposely destabilizing to organizational routines, as they probe the underlying causes of defensiveness and collusion. In this regard, it is a more confrontational form of intervention than **action learning**. Action science is also a more probing approach to learning and change compared to action learning and further distinguishes it from the more instrumental approach of the latter. *D.J. Greenwood*

References

Argyris, C., Putnam, R. and McClain-Smith, D. (1985) *Action Science: Concepts, Methods, and Skills for Research and Intervention*, San Francisco: Jossey-Bass.

Argyris, C. and Schön, D. (1996) *Organizational Learning II: Theory, Method, and Practice*, 2nd edn, Reading, MA: Addison-Wesley.

Lewin, K. (1946) 'Action research and minority problems', *Journal of Social Issues*, 2(4):34–46.

Schön, D. (1983) *The Reflective Practitioner: How Professionals Think in Action*, New York: Basic Books.

Senge, P.M. (1990) *The Fifth Discipline: The Art and Practice of the Learning Organization*, London: Random House.

Activity or entity cycle diagrams

They are a device used mainly in **simulation** to depict the sequence of activities and **state** changes that a **system** component undergoes over time. Consider a doctor's surgery where patients arrive (randomly), wait for an interview with the nurse, have the interview, wait for the doctor, consult with the doctor, and depart. If no other patients are waiting and the nurse is idle, the arriving patient starts the interview right away and similarly for the consultation. When the nurse or the doctor dismisses a patient, and no other patients are waiting, they become idle. The patients, nurse and doctor(s) are entities that engage in activities, usually jointly with another entity of a different type. While patients are temporary entities, both the nurse and the doctor(s) are permanent entities. Each goes through a cycle of activities. This is captured in the joint activity cycle diagram below.

Queues are usually depicted by circles, activities by rectangles. Each entity follows its own sequence of arrows. Two or more arrows entering the same rectangle usually means that the corresponding entities engage jointly in this activity, i.e. they are bound to each other for the duration. So the patient and the nurse join for the interview and go their separate ways afterwards, as depicted by the arrows going to different destinations. However, several arrows leading into a queue has no such implication.

Activity cycle diagrams are highly instructive devices to show what happens to the various entities over time, particularly in **queueing**-type situations. The entities can be people, vehicles, pieces of equipment, or spaces that other entities can occupy temporarily. The diagrams are a good preparation for programming computer

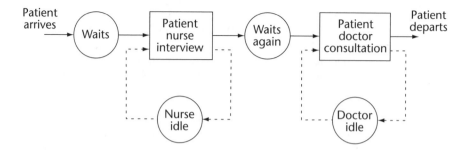

Figure: **Activity/entity cycle diagram**

simulation models. Several simulation software packages, such as Simul8, allow the input to be in the form of entity cycle diagrams. *H.G. Daellenbach*

References

Daellenbach, H.G. and McNickle, D.C. (2001) *Systems Thinking and OR/MS Methods*, Christchurch: REA, pp. 202–207.

Pidd, M. (1998) *Computer Simulation in Management Science*, 4th edn, Chichester: Wiley.

Adaptive processes

To evolutionary biologists, adaptation is the fundamental process of survival for living organisms. The fittest are those individuals that adopt the best strategies for their proliferation and ultimate survival. The concept has entered the realm of business. It is now common to state that 'in today's fiercely competitive global market only the fittest players (companies) will survive'. In this context 'fittest' is a synonym for strongest or most adaptable, but with a much sharper edge.

Defined generally, adaptation designates any process whereby a structure is progressively modified to give better performance in its environment – an environment that may be constantly evolving. So defined, adaptation is not confined to biology but has a critical role to play in fields as diverse as the social sciences, business, finance, economics, and computer science.

Two notable applications of this concept have relevance for MS/OR. In 1975 John Holland of the University of Michigan, starting from the premise that adaptations are fundamentally complex optimization processes, developed a radically new approach to optimization by search, which he aptly called **genetic algorithms**. Using a biological analogy, a set of points (genes) is subjected to a fitness evaluation employing a fitness function that measures how well each individual performs with respect to the task. The poor performers are gradually removed from the gene pool, leaving only the fittest. Genetic algorithms do usually not produce a **global optimal** solution but, as in nature, their primary goal is survival of the fittest, rather than simply optimization.

The second development occurred in **artificial intelligence** research. Considerable effort has gone into developing intelligent machines, e.g. robots that learn to improve their performance and adapt to new environments. Thus learning and adaptability become essential ingredients for intelligence. *Reinforcement learning* is the most widely used method, which has its roots in psychology. An action is reinforced (rewarded) if it results in a favourable response, otherwise it is punished. Other more elaborate schemes, such as hybrid, genetic and neural learning, and control and **fuzzy** behavioural control, have also been exploited in an effort to endow machines with adaptable behaviour. A number of modest successes have been realized, but an all-purpose adaptive machine still seems a long way off.

Systems that operate on the basis of its present, past and (anticipated) future states are called *anticipatory systems*. Surprisingly little research work has been done in this area. *A.D. Tsoularis*

References

Casti, J. (1991) *Searching for Certainty: What Scientists Can Know About the Future*, New York: Morrow.

Goldberg, D.E. (1989) *Genetic Algorithms in Search, Optimization, and Machine Learning*, Reading, MA: Addison-Wesley.

Rosen, R. (1985) *Anticipatory Systems*, Oxford: Pergamon.

Sutton, R.S. and Barto, A.G. (1998) *Reinforcement Learning*, Cambridge, MA: MIT Press.

Advanced distributed simulation

Simulation is a long-established tool of MS/OR for studying complex real-world **systems** by means of computer models which represent those features of interest in the system and how they change over time. A military combat operation, for example, may be simulated by representing the vehicles (aircraft, tanks, ships), component equipments (radars, guns, missiles), human combatants, environment (weather, navigational conditions), communications received from headquarters, and many other factors which determine the outcome of the battle. The traditional way of simulating such very complex systems required a very large amount of executable software and data residing on a single very large computer and took a long time to run.

Modern computer technology makes extensive use of *distributed systems*. These are computer network configurations, both local and remote, including the Web, where the same and different sets of data and software reside on separate computers, but can be accessed by all computers on the network. Advanced distributed simulation (ADS) is the application of this technology to simulation. Different components of a distributed simulation may run on different computers and communicate by sending data packets over the network. In the military example, the 'Blue' forces may run on one computer and the 'Red' forces on another computer at a distant site. Graphical user interfaces at each site may both depict all simulated entities, but only some entities will be fully simulated locally.

An important requirement in distributed simulations is that the simulated time in each component simulation is synchronized. This can be achieved by including the simulated time in the data packets sent over the network.

The component simulations may be developed and run by different groups of people as long as they agree on what data has to be sent and received. A 'Blue' ship may not need to receive data about 'Red' tanks, for example, but will need to know if a 'Red' missile is heading towards it. For very large distributed simulations it is important to restrict the amount of traffic on the network, as the processing of incoming data can slow down a simulation run. In an early form of ADS, termed *Distributed Interactive Simulation* (DIS), the network traffic was not limited in this way, but this problem has been overcome by the so-called *High Level Architecture* (HLA) developed by the US Department of Defense. In its most general form the components in an HLA simulation are not necessarily all simulations, but can include real equipment and humans as trainers as long as these components all send and receive data of the agreed kind.

It should be noted that the new technology of ADS is extending the traditional role of simulation as an analytical tool to include forms of training where the real-world systems involved are dangerous, expensive and/or geographically dispersed, such as military exercises, environmental protection and air traffic control. *R. Watson*

Reference

Lightner, M. and Dahmann, J. (eds) (1998,1999) Special Issues of *Simulation*, 71(6) and 73(5) on 'High Level Architecture', San Diego: Society for Computer Simulation International.

Web sites

Defense Modeling and Simulation Office: http://www.dmso.mil/

Simulation Interoperability Standards Organization: http://siso.sc.ist.ucf.edu/

Society for Modeling and Simulation International: http://www.scs.org

Agriculture, see applications of MS/OR in agriculture

AIMMS, see algebraic modelling languages

Algebraic modelling languages

Algebraic modelling languages are high-level languages specifically designed to formulate and prepare the input to commercial **mathematical programming** optimization software (solvers), such as MINOS, CPLEX or DICOPT, and then use the outputs from these solvers to prepare user reports in the style and format desired. They let the user formulate **linear, nonlinear** and **integer programming** problems using familiar algebraic modelling notation and concepts, such as sets of variables, equations and inequalities involving these sets, using summation notation, data tables, and commands such as 'minimize' and 'maximize', 'solve LP', 'display', etc. The languages turn this notation into a format that the solver can read. This allows the user to obtain and analyse solutions to his or her model without having to follow the often cumbersome and time-consuming input format of these solvers and without specifically knowing how the solution process works.

Prominent examples of algebraic modelling languages include *Advanced Integrated Multidimensional Modeling Software* (AIMMS), *A Mathematical Programming Language* (AMPL) and *General Algebraic Modeling System* (GAMS).

The problems can be described in a highly compact and natural way, using mathematical set notation, allowing similar parameter and decision variable definitions and constraints to be entered by single statements. Algebraic statements define the tables for the input data which is then entered into these tables, e.g. from a spreadsheet. This has the advantage that the model stays concise, even when the size of the problem (the data tables) grows. Also, the user can change the model or the data quickly and easily, and independently of each other.

The example below demonstrates the use of the AMPL syntax for a small **transportation problem**. Other languages use similar structures, but different syntax. The mathematical formulation is shown first, followed by its translation into the AMPL syntax:

Problem: Let $x_{s,d}$ = amount shipped from source $s = 1, 2$ to destination $d = $ A, B.

Minimize $10\,x_{1,A} + 15\,x_{1,B} + 20\,x_{2,A} + 18\,x_{2,B}$

subject to: $x_{1,A} + x_{1,B} \leq 50$ (supply 1 constraint)

$x_{2,A} + x_{2,B} \leq 50$ (supply 2 constraint)

$x_{1,A} + x_{2,A} \geq 60$ (demand A constraint)

$x_{1,B} + x_{2,B} \geq 40$ (demand B constraint)

$x_{s,d} \geq 0$, for all s, d.

AMPL representation:

Model:

Minimize cost:

      sum {s in SUPPLY_LOCATIONS, d in DEMAND_LOCATIONS}

            unit_cost[s,d] * x[s,d];

Subject to Supply_Constraints {s in SUPPLY_LOCATIONS}:

      sum {d in DEMAND_LOCATIONS} x[s,d] <= supply[s];

Subject to Demand_Constraints {d in DEMAND_LOCATIONS}:

      sum {s in SUPPLY_LOCATIONS} x[s,d] >= demand[d];

Data:

set SUPPLY_LOCATIONS	:=	1	2 ;
set DEMAND_LOCATIONS	:=	A	B ;

param unit_cost :   A   B   :=

               1   10   15

               2   20   18;

param supply   :=   1 50   2 50;

param demand   :=   A 60   B 40;

Additional features are easy access to the solver output for **sensitivity analysis,** the ability to export solutions to different files, the possibility of using looping and 'if-then-else' commands, and special modelling features, such as piece-wise linear functions.
M.C. Stewart

References

Bisschop, J. and Roelofs, M. (2001) *AIMMS: The User's Guide*, Haarlem, NL: Paragon Decision Technology B.V.

Brooke, A., Kendrick, D. and Meeraus, A. (1992) *GAMS: A User's Guide*, Danvers, MA: Boyd & Fraser, The Scientific Press Series.

Fourer, R., Gay, D.M. and Kernighan, B.W. (1993) *AMPL: A Modeling Language for Mathematical Programming*, Danvers, MA: Boyd & Fraser, The Scientific Press Series.

Algorithms, algorithmic solution methods

An algorithm is a set of logical steps and numerical operations performed repeatedly in a specified sequence, as depicted in the flow chart below. Each repetition is called an iteration. Most powerful solutions methods are algorithmic. An algorithm starts with an initial or incoming partial or complete solution, which may be the empty set (point 1 in chart). At each iteration, the incoming solution is changed in certain ways, using the rules of the algorithm (point 2). For an optimizing algorithm the change usually implies an improvement. The new solution so generated becomes the incoming solution for the next iteration (point 3). This process is repeated until certain conditions – referred to as *stopping rules* – are satisfied, such as 'an optimal or complete solution has been found,' or 'N iterations have been executed' or 'M minutes of computer time have elapsed' (point 4).

    For an algorithm to be a practical solution method, it has to have certain properties: (1) each successive solution should be an improvement over the preceding one; (2) successive solutions have to converge, i.e. get closer and closer to the optimal or complete solution; (3) convergence arbitrarily close to the optimal solution should occur in a reasonable number of iterations; and (4) the computation effort at each iteration has to be sufficiently small to remain economically acceptable.

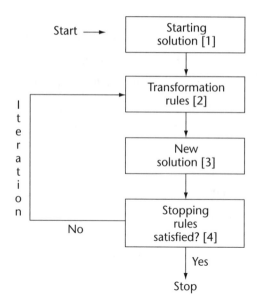

I
t
e
r
a
t
i
o
n

Figure: **Algorithm flow diagram**

A simple example is a *greedy algorithm* used, for instance, to allocate an initial amount of funds to a large number of potential investment projects. Assume each project i requires an initial outlay of C_i and offers to return R_i after one year. The following algorithm will find the set of investments that will maximize the return on the funds (the numbers in square brackets refer to the figure):

Step 1: Compute the ratios of R_i/C_i for all i. The initial list L of projects without allocation consists of all projects [1].

Step 2: In list L, let project k be the one with the largest ratio [2].

Step 3: If the amount of funds F left to allocate is larger then C_k, allocate that amount to project k. Remove project k from list L. The amount of fund left becomes $F - C_k$ [3]. Go back to step 2. If $F \leq C_k$, then allocate F to project k. STOP [4].

Many of the general-purpose MS/OR techniques, such as **linear** and **nonlinear programming** techniques, use algorithms for finding the optimal solution, and so do most **search heuristics** methods. Becoming thoroughly familiar with these algorithms is the major emphasis of most MS/OR university courses.   *H.G. Daellenbach*

Alternative optimal solutions

When more than one **feasible solution** (i.e. a solution that satisfies all constraints) to an optimization problem yields the same optimal value for the objective function (i.e. a maximum or a minimum, whichever is relevant), they are referred to as alternative optimal solutions. They occur most frequently in **linear programming** (LP) problems and problems solved by **dynamic programming**. In LP it means that several **basic feasible solutions** share the same optimal value of the objective function. They can be identified by the fact that any one optimal solution has one or more so-called nonbasic variables (i.e. variables that are zero in the optimal solution) with a zero **reduced cost** coefficient. (The reduced cost coefficient indicates by how much the value of the objective function would change if the corresponding variable is given a value 1.)

Preemptive **goal programming** – an LP that incorporates multiple objectives – depends heavily on the presence of alternative optimal solutions for the higher priority objectives.

AMPL, see **algebraic modelling languages**

Analog models

An analogous model substitutes the properties, features and behaviour of what is modelled by alternative means, such that the model is able to mimic whatever aspect of the real thing is of interest to the modeller. For example, the constantly updated picture that an air traffic controller observes on a monitor is an analog of the air traffic in a given sector of the air space as captured by radar devices. A flight simulator is an analog model that simulates how an aircraft responds to control commands, and so are many computer games or the game machines in a game parlour.

Analogy

Analogy means similarity, but not identicalness, of a feature or aspect to another feature with which it is compared. Analogies are used as a form of reasoning, known as *analogical reasoning*, as a teaching tool, as a method of enquiry into an unknown phenomenon, and as a creative way of communicating. Analogical reasoning involves a transfer of some explicit relationships from one context to another. In the question, 'Boy relates to man, as girl relates to . . . ?', analogical reasoning suggests that the answer is 'woman'. Analogies are regularly used as a teaching tool. The heart is described as a pump; electrical current is visualized as a liquid flow; and a set of mathematical functions are said to belong to a family. As a method of inquiry, researchers use analogies to gain insight into similar phenomena; they may suggest hypotheses to be tested. Psychology explores human behaviour through its resemblance to a rat's behaviour; **living systems theory** studies organizations as if they are living systems; and economies are studied as hydraulic systems with specific flows and leakages. Finally, analogies have been used as a creative form of communication for centuries, e.g. life is expressed as a journey, and love is said to be blind.

In MS/OR, analogy refers to the entity with which a given **problem situation** is compared. Drawing an analogy may highlight some feature of the situation that helps in the discovery of potential improvements. In MS/OR, analogies play all four roles described above. They are used in reasoning through the complex set of facts available in any problem situation. Many MS/OR methods are taught through the use of analogies, e.g. the analogy of machines breaking down with a **waiting line** situation, i.e. machines waiting for service. Some of the highly successful **heuristics**, such as **simulated annealing** and **genetic algorithms**, are based on chemical and biological analogies. Analogies also provide the key insights for discovering creative solutions. For example, the analogy between a problem situation and a drama might highlight the fact that the participants in the problem situation are simply enacting their roles according to some outdated script. Changing the script may be the key for improving the situation. Similarly, the analogy between the **knapsack problem** (which items to put into a backpack) and an allocation of limited funds to potential projects immediately suggests a possible solution technique. Communications with MS/OR analysts and clients also benefits from the use of analogies, such as **rich pictures** or **mind maps** to depict a problem situation.   *D.P. Dash*

References

Flood, R.L. and Jackson, M.C. (1991) *Creative Problem Solving: Total Systems Intervention*, Chichester: Wiley.

Keane, M.T. (1988) *Analogical Problem Solving*, Chichester: Ellis Horwood.

Analytic hierarchy process or AHP

The analytic hierarchy process, invented by T.L. Saaty in 1978, is a **multicriteria decision method** for identifying the alternative from a group of predefined possibilities that offers the best compromise over several conflicting objectives.

AHP consists of four major steps. Firstly, the objectives and alternatives are structured into a **hierarchy**. Then, using a numeric scale, the decision maker compares (a) each pair of the lowest-level objectives, and (b) each pair of alternatives under each lowest-level objective. Doing this signifies how important an objective is relative to the others, and how good each alternative is towards reaching an objective. The third step uses the resulting matrices to determine importance weights for the objectives, and achievement scores for the alternatives under each objective. The final step then ranks the potential alternatives.

Consider the following simplified example: A student has to decide which major subject to study. His overall objective is to find 'his best subject'. He decides the subject should be both enjoyable and offer good career opportunities – his second-level objectives. Subject enjoyment can, in turn, be broken down into third-level objectives of 'interest in subject' and 'difficulty', and for career opportunities he sees 'starting salary' and 'prestige of subject' as important. He has five potential major subjects. The figure below shows the **hierarchy diagram** for this situation.

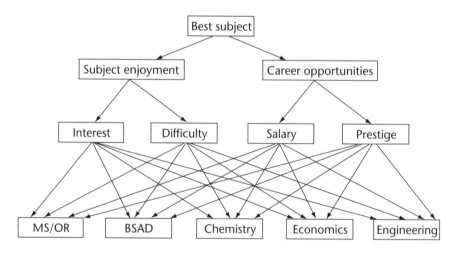

Figure: **Hierarchy diagram for subject selection**

Table 1 shows the result of the second step of AHP; the pairwise comparisons of third-level objectives on a scale from 1 (implying the two factors are equally important) to 9 (representing an extreme preference of the row factor over the column factor). For instance, 'interest in the subject' is strongly preferred over 'difficulty', hence the cell entry is 5. The reverse preference of the column factor over the row factor is indicated by the corresponding ratio 1/5. Given these preferences, a mathematical method, called *eigenvector analysis*, is then used to find the importance weights for the objectives, as listed in the last column of Table 1, and the scores for the alternatives under each objective.

Table 1: **Pairwise comparisons between third-level objectives**

	Interest	**Difficulty**	**Salary**	**Prestige**	**Importance**
Interest	1	5	4	5	0.575
Difficulty	⅕	1	3	3	0.226
Salary	¼	⅓	1	1	0.096
Prestige	⅕	⅓	1	1	0.103

Table 2 contains the final step of ranking the alternatives. The main body of the table shows the scores computed for the alternatives under each objective. A weighted average score is then computed for each alternative using the importance weights for the objectives, yielding a final ranking of the alternatives.

Table 2: **Overall ranking of subject alternatives**

Importance weights	0.575	0.226	0.096	0.103		
Alternative scores	Interest	Difficulty	Salary	Prestige	Weighted average	**Rank**
MS/OR	0.521	0.094	0.282	0.151	0.363	**1**
BSAD	0.057	0.165	0.138	0.098	0.093	**5**
Chemistry	0.057	0.057	0.204	0.287	0.095	**4**
Economics	0.124	0.457	0.174	0.168	0.209	**3**
Engineering	0.241	0.227	0.202	0.296	0.240	**2**

AHP has the following strengths: It can be used for problems with both quantitative and qualitative factors; the comparisons a decision maker has to make are of a relatively simple nature; and it easily can handle multiple decision makers by adding extra levels to the hierarchy.

Saaty's original version of AHP has undergone a number of changes to overcome several shortcomings, such as *rank reversal* caused by adding or deleting alternatives.

Applications of AHP to a variety of problems have been reported, such as selecting safety strategies for fire prevention, deciding between candidates when filling a job vacancy, and determining nuclear power plant locations in Finland.   *M.C. Stewart*

Reference

Saaty, T.L. and Vargo, L.G. (1991) *Prediction, Projection, and Forecasting: Applications of the Analytic Hierarchy Process in Economics, Finance, Politics, Games, and Sports*, Boston: Kluwer Academic.

Annuity, see **discounted cash flows**

Ant system or ant colony optimization

This is a **heuristic search** method used for finding solutions to combinatorial optimization problems. It was developed from the 1980s onwards and has proven successful in finding good solutions rapidly for a number of problem types, particularly those which can be represented as **networks** or **graphs**. Because it is a heuristic method, there is no guarantee that the optimal solution will be found, but many studies have shown that the solutions generally converge to the optimal one. It can be used successfully for problems with several optima, and will in many instances even then converge to the **global optimum**.

The concept underlying this method is the imitation of the behaviour of colonies of ants. When foraging, ants create odour trails for others to follow. Successful foraging leads to a strong trail, which creates a bias so that large numbers of ants from the colony follow the same route. Poor routes will have an odour which fades with time.

The search method provides a mechanism for randomly despatching artificial ants to look for feasible solutions to the problem. These solutions are represented by trails, which the artificial ants follow, making decisions at various points along their path. The ants return to the start, and the trail that they have followed is recorded. The better the trail, the stronger the signal (or bias) will be for later ants. In addition, the strength of the signal associated with the trail will decay with time. The best trails will acquire strong signals which do not fade, because they will be used over and over again. As time progresses, the members of the colony will tend to favour the best trails, from which a solution to the original problem can be found.

The method has been applied to problems of finding routes in networks, such as the **vehicle routing problem**. It is also possible to investigate some **scheduling problems**, by representing the order of jobs in a workshop as a trail for the ants to follow.　*D.K. Smith*

Reference

Bonabeau, E. and Théraulaz, G. (2000) 'Swarm smarts', *Scientific American*, March: 72–79.

Applications of MS/OR in agriculture

There have been many varied applications of management science to agriculture. Even though some aspects of farm management can be carefully controlled, most farming is subject to the **random** effects of the climate and spatial variation across the farm. Therefore the models which are used either include random features (e.g. **simulation**) or approximate the randomness. From the earliest days of statistical science, statisticians have helped farmers to design and analyse experiments about the treatment of crops and the husbandry of animals. This means that there are well-proven methods of experimental design which allow for experiments that balance the effects of several kinds of treatment on the yield of crops, while minimizing the effects of spatial variation.

A partial list of applications of MS/OR models and techniques includes:

Crop planning and management:

- **linear programming** to suggest the best use of fields in a farm, so as to maximize the average profit given the resources of space and labour available to the farmer;
- **forecasting** models to determine at which stage in crop production to use irrigation or to apply pesticides and fertilizers;
- **decision support systems** to maintain optimal plans during the growing season given weather conditions;
- investment and **replacement** models for farm equipment;
- scheduling the harvesting of crops to meet the capacity constraints on processing plants;
- control of the environment in glasshouses to accelerate or delay production so as to meet seasonal demands.

Livestock farming:

- **dynamic programming** to determine when members of a herd should be sent to market;
- models for optimal breeding programmes of cattle;
- optimal disposal of animal manure;
- **multicriteria decision models** for herd management.

Industrial applications related to farming:

- developing optimal feed-mixes for animals at different stages in their lives using linear and **integer programming**, as well as **nonlinear programming** (see also **diet problem**);
- optimal **vehicle routing** for milk collection, made complex by the seasonal variation of milk production, and need to keep collections from different farms in separate compartments;
- optimal maintenance of food processing equipment during the harvest season;
- forecasting models for agricultural commodities.

Some models incorporate the effect of farm decisions on a wider environment, such as the consequences of water use. Reports of applications can be found in *Agricultural Systems* and *Journal of Dairy Management.* *D.K. Smith*

Applications of MS/OR to banking and financial markets

Banks and more recently financial markets have seen numerous applications of MS/OR. One of the most visible applications in banking is the use of a single queue where customers wait to approach the first teller that becomes free. This is the result of a **queueing** study. Banks also routinely use credit rating or assessment models for all sorts of loan applications. These models are usually based on an extensive **regression** analysis of past loan performance data, periodically updated to reflect demographic and economic trends. More recently, **neural networks** have been successfully applied to this problem. **Facility location** models have been used both for the siting of branches and of automatic teller machines.

The relative economic performance and operating efficiency of individual bank branches is being assessed by **data envelopment analysis**. A late 1990s project dealt with the design of a bank transaction clearing and payment system for China.

A large area of application deals with portfolio and security management and valuation. **Portfolio selection** models are routinely used for the management of various investment funds managed by banks. **System dynamics** models help in determining threshold points for when to buy and when to sell securities in response to price trends. **Linear programming** is used to control financial risk by portfolio immunization, i.e. strategies that match portfolio composition in terms of return and risk and sensitivity of securities to shifts in interest rates with those of the liabilities they have to fund. Pricing of derivatives and various options is done with the help of **Monte-Carlo simulation**, neural networks, linear programming, and the famous **Black–Scholes formula** for European call options.

Foreign exchange traders exploit arbitrage opportunities via **networks** and **expert systems**. **Dynamic programming** is used to find the best sequence of splitting large trades into a sequence of smaller trades so as to reduce the effect on security prices. Currency options are valued using similar models as for the pricing of derivatives and options.

Stochastic dynamic programming and **network** models have been proposed for the combined management of bank cash balances and short-term treasury bills. *H.G. Daellenbach*

References

Das, S. (1998) *Risk Management and Financial Derivatives: A Guide to the Mathematics*, New York: McGraw-Hill.

Jarrow, R., Maksimovic, M. and Ziemba, W.T. (eds) (1994) *Handbooks in Operations Research and Management Science: Finance*, Amsterdam: North-Holland.

Mulvey, J.M. (1994) 'An asset liability investment system', *Interfaces*, 24(3):22–33.

Zenios, S.A. (ed) (1993) *Financial Optimization*, Cambridge: Cambridge University Press.

Ziemba, W.T. and Mulvey, J.M. (1998) *World-wide Asset and Liability Modeling*, Cambridge University Press (UK).

Applications of MS/OR to crew scheduling and manpower planning

The efficient utilization of human resources is one of the most important aspects of effective management in many areas of business and industry, both manufacturing and service industries, such as military, health and transportation. Crew scheduling problems (where 'crew' stands for all forms of staff) can vary in both size and complexity and also in the value of the 'solution'. Simple problems occur where there is a predictable and constant crew demand (e.g. in manufacturing) but when demand varies significantly over time (e.g. in transport operations) or demand can be difficult to predict with certainty (e.g. for call-centres, casinos, and airport customs and immigration) the problems are generally very much more complex. It is possible to identify four phases of the crew scheduling problem, each varying in relative importance depending on the application area:

- *Manpower planning:* This phase is concerned with forecasting future crewing requirements and identifying promotion paths and crewing training schedules. This is particularly important in military planning and in the airline industry where there are many skill or qualification levels amongst crew.
- *Work requirements:* This involves the determination of the number of crew of each particular skill level who will be required during each period of time. In situations where workload is highly variable (e.g. casinos), work requirements might be determined as a function of customer demand or by **simulation** (e.g. airport customs and immigration). In the transportation sector, they are usually defined directly by the timetable or flight schedule. Work requirements can often be represented in a graph of the number of crew required in each time period.
- *Shift or duty schedule construction:* For a given work requirement or a given set of tasks (e.g. a timetable of scheduled airline flights), a schedule of shifts or duties has to be constructed to provide the required number of crew in each period or to ensure that all tasks are covered, satisfying various rules and possible restrictions, such as shift duration, part-time shifts, start times, and regularity. Further restrictions may include the number of shifts or duties allocated to particular crew bases. In the airline industry, this process usually constructs tours of duty (often called pairings). Each tour of duty starts and ends at the same crew base and includes a sequence of duty periods and rest periods over many days. In this phase, no account is taken of individual crew members. The problem is usually modelled as a *generalized set partitioning optimization model* (a special form of zero–one **integer program**) with an objective of minimizing the operating cost of the duties produced.
- *Rostering:* This involves assigning individual crew members to the shift schedule produced, including provision for 'days off'. It takes into account the qualifications of the individual crew member and the content of their previous roster assignment. As with duty construction, there are usually many rules which must be satisfied. The problem is modelled as a generalized set partitioning optimization model with an objective of producing a roster which satisfies as many individual crew requests or preferences as possible. Often when the work requirements are stable, it is possible to construct a cyclic roster in which crew members move from one roster line to the next in a cyclic pattern.

Although the integer programming models for the duty construction and for the rostering problem can involve thousands of constraints and many millions of variables, they can be solved efficiently by clever exploitation of their structure and special **branch-and-bound methods.** *D.M. Ryan*

References

Butchers, E.R., Day, P.R., Goldie, A.P., Miller, S., Meyer, J.A., Ryan, D.M., Scott, A.C. and Wallace, C.A. (2001) 'Optimised crew scheduling at Air New Zealand', *Interfaces* 31(1):30–56. (Finalist paper in the 2000 Franz Edelman Award.)

Collins, R.W., Gass, S.I. and Rosendahl, E.E. (1983) 'The ASCAR model for evaluating military manpower policy', *Interfaces*, 13(3):44–53.

Mason, A.J., Ryan, D.M. and Panton, D.M. (1998) 'Integrated simulation, heuristic and optimisation approaches to staff scheduling', *Operations Research*, 46(2):161–175.

Ryan, D.M. (2000) 'The development of crew scheduling systems for Air New Zealand', *OR/MS Today*, 27(2): 26–30.

Applications of MS/OR in electricity sector

The electricity sector has a long tradition of MS/OR application, with a significant literature on **optimization**, parallelling the mainstream MS/OR literature from its earliest until the present time, and may well provide the single largest demonstration of successful MS/OR in action. This success stems, in part, from the economic importance of electricity generation, and from the fact that, from an early date, the sector has been managed from sophisticated control centres which coordinate activities on a large scale, thus providing an environment in which real-time data, and computer processing power, are available.

Electricity sector organizations face all the same problems as organizations in other sectors, and so may apply a wide range of MS/OR techniques to deal with areas such as **inventory control**, or vehicle **replacement** policy, for example. The electricity sector has also featured prominently in more general energy sector modelling efforts. But distinctive traditional applications areas have included:

- forecasting load and hydrological inflows;
- planning investment in generation and transmission systems;
- planning the optimal use of water, or other stored 'fuels';
- optimizing unit commitment, to determine which generation units will be on-line for each dispatch period over some **planning horizon**;
- optimizing unit dispatch to determine how much to generate from each unit within a dispatch period;
- use of *Optimal Power Flow* (OPF) models to determine a dispatch which minimizes the cost of generation, after accounting for both congestion and losses in the transmission system.

These applications range in time frame from decades down to real time, with a typical dispatch period being as short as five minutes. Large-scale **linear** or **nonlinear programming** models are often employed, with integer variables typically required to represent commitment, or investment variables. Reservoir management problems have proved a particularly fertile area for the application of new techniques, such as **dynamic programming** and **stochastic mathematical programming**. The OPF application is noteworthy, in that it involves nonlinear optimization of the real (active) and imaginary (reactive) parts of the complex variables representing each generator in an AC network. Solving such problems for realistic-sized systems, with 10 000 network nodes (or 'buses'), was a major success for *sparse matrix techniques*.

Traditionally, the sector has been dominated by (near) monopoly enterprises which has facilitated development of *large-scale optimization*. But, more recently, the electricity sector has been subject to major change, as deregulation has become popular around the globe. By breaking up the old monopolies, this movement has threatened some established MS/OR applications, but also created a raft of new challenges and opportunities. One major challenge is the development of **game theory** or **complementarity** models to predict the behaviour of participants in markets with imperfect competition. But, most importantly, the electricity supply system in many regions is now coordinated via a market, and these markets are typically cleared by optimization models which determine an optimal generation dispatch, based on offers received from generators, for each dispatch period, as frequently as every five minutes in some markets, with the results typically communicated via real-time control systems. All electricity is then bought and sold at the **shadow prices**, or **dual prices**, produced by these models, giving them a central role in a financially critical sector. *E.G. Read*

Application of MS/OR to emergency services

Emergency services include the police, fire services, air and road ambulance services, air, sea and land search-and-rescue services, civil defence, and accident-and-emergency clinics.

Applications in police services address staffing levels, allocation of patrol units to various precincts, and design of patrol beats to meet minimum response standards.

Facility location and **simulation** models are routinely used to determine the best location or relocation of fire and ambulance services. The problem is to find locations such that all areas to be serviced can be reached within a given response time with a given (high) probability. **Multicriteria decision models** attempt to balance the type of equipment each location should have and how many vehicles or crews to dispatch for various classes of fires. Other applications look at prediction of fire risks using simulation, and building evacuation strategies using **transshipment models** to identify bottlenecks. Forest fire fighting strategies have been explored using **dynamic programming** and **risk analysis**.

Search theory and **multiattribute utility analysis** have been applied to develop search strategies for search-and-rescue services.

Simulation and **operational gaming** are techniques applied for civil defence and disaster planning and training. Risk assessment models are important tools for understanding risks associated with emergencies, such as aircraft collisions at airports, oil spills and forest fires.

Simulation and **queueing** models help find best staffing levels and treatment priorities for accident-and-emergency clinics. *H.G. Daellenbach*

Reference

Pollock, S.M., Rothkopf, M.H. and Barnett, A. (eds) (1994) *Handbooks in Operations Research and Management Science, 6: Operations Research in the Public Sector*, Amsterdam: North-Holland. Several chapters deal with emergency services.

Applications of MS/OR in energy modelling

MS/OR has played a significant role in the development of most energy sectors, including oil and gas, electricity, and coal. But energy modelling, as a distinctive application area, emerged to prominence in response to the oil crisis of the 1970s, with the Energy Modelling Forum at Stanford University playing a major role in facilitating comparison between models, and fostering debate among modellers, both with respect to modelling techniques and to real-world predictions. Typically, energy models plan the mix of energy supply from various renewable sources to meet the demand for energy from various economic sectors of a nation, or even of the entire world, over a horizon of decades. *Energy/economy models* extend the concept by modelling the interaction of the energy system with the economy. These models are surveyed by Manne *et al.* (1979).

In smaller countries with a tradition of centralized planning such models may be used in a prescriptive mode, to plan developments. But in larger, decentralized, economies such models have always been descriptive policy models. Conceptually, we might expect prescriptive models to involve top-down optimization, and descriptive models to involve simulation of market interactions between independent decision makers. But many policy models, such as the famous Stanford PILOT, developed by Dantzig in 1978, employ large-scale optimization on the assumption that a market economy will actually produce something like the theoretical optimum. Conversely, many prescriptive models have employed price-driven **decomposition** as a computational device to break models into manageable units, and to link sub-models of different types. Such models not only pushed the computational limits, but demanded major organizational investment, on a national and international scale.

Perhaps the largest single modelling effort undertaken was the *Project Independence Evaluation System* (PIES), developed by the Energy Information Agency of the US government. This modelling system involved several very large **linear programming** models, each describing an energy supply or demand sector, along with an econometrically derived model of the US economy linked by an iterative price decomposition mechanism described by Hogan (1975). Murphy *et al.* (1988) describe a modern successor, the *Mid-term Future Energy Forecasting System* (MFEFS). *The Market Analysis modelling system* (MARKAL) (Fishbone and Abilock, 1981) is another major example which played an important role as the focus of a major international effort, in which versions of the MARKAL model were produced for many countries.

These models mark an era of intense developments in OR theory and application. Recent years have seen the focus of modelling efforts follow the public policy shift away from the concept of an 'energy crisis' towards establishment of competitive energy markets, coupled with a broader concern about environmental issues, such as global warming (Weyant, 1998). But more traditional energy/economy models continue to play an important role in the developing world where central agencies still play a strong role.

E.G. Read

References

Fishbone, L.G. and Abilock, H. (1981) 'MARKAL, a linear programming model for energy systems analysis', *International Journal of Energy Research*, 5:353–375.

Hogan, W.W. (1975) 'Energy policy models for Project Independence', *Computers and Operations Research*, 2:251–271.

Manne, A.S., Richels, R.G. and Weyant, J.P. (1979) 'Energy policy modeling: a survey', *Operations Research*, 27(1):1–36.

Murphy, F.H., Conti, J.J., Sanders, R. and Shaw, S.H (1988) 'Modelling and forecasting energy markets with the Intermediate Future Forecasting System', *Operations Research*, 36(3):406–420.

Weyant, J. (1994) 'Energy policy applications of operations research', in Pollock, S., Rothkopf, M.H. and Barnett, A. (eds) *Handbooks in Operations Research and Management Science 6: Operations Research and the Public Sector*, Amsterdam: North-Holland.

Weyant, J. (1998) *Energy and Environmental Policy Modeling*, Boston: Kluwer Academic.

Applications of MS/OR to environmental issues and natural resources

As early as the 1960s, MS/OR began to address the management and exploitation of natural resources, such as forests (see **applications of MS/OR in forestry**), wildlife, water usage, and fisheries, and problems of solid waste and air pollution management.

Linear and **dynamic programming**, including **multicriteria decision-making** models, are used for the study of water management for irrigation, flood control, power generation, and recreation, covering whole river basins. In fact, the *bicriterion method* for linear programming was developed specifically to deal with such problems. Water pollution and solid waste management models use the same set of tools. **Multiobjective mathematical programming** models help in finding suitable locations for solid waste disposal sites. **Data envelopment analysis** is being used to assess air pollution emission policies of power plants.

Multiobjective programming models have been used to develop management plans for vast tracts of lands. These models attempt to strike a balance between land use for recreation (hunting, tramping, and off-road vehicle use), grazing, mineral exploration, and wilderness and habitat protection. The US Bureau of Land Management has been at the forefront of these efforts.

Analysis of disposal sites for hazardous waste has been done using **multiattribute utility analysis**. **Cost–benefit analysis** is routinely used to assess environmental impacts of a vast range of public projects, such as the siting of motorways, river flood control and urban storm sewer works, open-pit mines, and so on.

The risk of depletion of the world fisheries resources has forced many countries to impose catch limits. The difficulty in fisheries is that much of the catch by any boat, particularly within the 200-mile national zones, is a mixture of species. Accounting for such aspects has led to **nonlinear programming** models. MS/OR has also been used to establish systems of tradable quotas used by a number of countries.

Without doubt the most famous study addressing the future of Earth has been the 'Limits to growth' project by Meadows and Meadows in the early 1970s. They built a **system dynamics** model that explored the consequences of natural resource use and depletion, pollution, population growth and technological development on the economic and ecological well-being of this planet for various **scenarios** of high, medium and low natural resource usage. Their predictions are rather sobering, but also raised a lot of methodological questions about the modelling approach, its underlying assumptions, the validity of projections into the distant future, and the model itself.

More recent efforts have addressed the greenhouse effect and ozone depletion through the use of dynamic global energy–economy–environment aggregate **general equilibrium models** and multi-sector econometric models. *H.G. Daellenbach*

References

Bjorndal, T. and Munro, G. (2000) *The Economics of Fisheries*, Amsterdam: Harwood Academic Publishers.

Greenberg, H. (1995) 'Mathematical programming models for environmental quality control', *Operations Research*, 43:578–622.

Meadows, D.H. *et al.* (1972) *Limits of Growth*, Washington: Signet Books.

Pollock, S., Barnett, A. and Rothkopf, M. (eds) (1994) *Handbooks in Operations Research and Management Science, 6: Operations Research and the Public Sector*, Amsterdam: North-Holland. Contains articles by G. Golden and E.A. Wasil on fisheries, C. ReVelle and J.H. Ellis on air and water quality, and P.R. Kleindorfer and H. Kunreuther on siting of hazardous facilities.

ReVelle, C. and McGarity, A. (eds) *Design and Operation of Civil and Environmental Engineering Systems*, New York: Wiley.

Applications of MS/OR in forestry

Operations research techniques have been applied to practically all aspects associated with forestry, from tree planting to wood processing. Some of the earliest applications, used by large wood processing firms, dealt with the conversion of cut trees into finished products. Based on statistical surveys, logs at each forest location are classified into numerous types according to quality and dimension. For each log type, yield tables are compiled for conversion into various finished products at the processing plants (sawmills, wood panel plants, pulp mills) operated by the firm – sawn timber, plywood and other aggregate wood panels, and pulp for paper production. The firm wants to determine the optimal allocation of logs to its processing plants so as to maximize the difference between the value of the finished products and the total cost incurred for log transportation and processing, subject to the capacities of the processing plants and finished product marketing requirements. The problems are formulated as large **linear programs** of several thousand constraints and tens of thousands of variables. They cover a single period, e.g. a month, or several periods, allowing goods to be temporarily stored. Since they cover the entire operation of a firm, they are often referred to as company-wide or corporate models.

Other applications deal with the following:

- *Silvicultural regimes:* Tree planting patterns, tree thinning and pruning scheduling, and tree harvesting planning over time. Different regimes yield different wood quality and yields. The objective is to maximize the **net present value** of the trees at harvest. The models used are linear programs and **simulation**. The latter caters for random aspects.
- *Long-term tree crop rotation planning* to meet long-term demand forecasts, again to maximize the net present value from a given tract of land, usually formulated as large linear programs.
- *Tree bucking:* Felled trees are cut in the forest into more easily transported lengths, called logs. Cutting into incorrect lengths may substantially reduce the conversion value of the logs. The aim is to develop bucking schedules so as to maximize the value of the trees for the planned conversion into finished products. The problem is usually solved by **dynamic programming**. Whole trees transported to the sawmill can be X-rayed to assess quality and the information processed on computer for determining optimal bucking patterns in real time.
- Assessment of the stumping price of a forest: One of the approaches used here is the tree bucking model, applied to a sample data from the forest.
- Planning of forest road construction by **network** models.
- Planning for forest fire protection and fighting, using **multiobjective programming models**.
- Data collection strategies for forest yield estimation and stumping value assessment, needed as input into planning models.
- Planning of forest machinery maintenance and **replacement**.
- Forest management with multiple objective programming model, to cover financial returns from logging, erosion control, recreational value, and wildlife management. *H.G. Daellenbach*

References

Bare, B.B. *et al.* (1984) 'A survey of systems analysis models in forestry and forest products industries', *EJOR*, 18:1–18.

Davis, L.S. and Johnson, K.N. (1987) *Forest Management*, 3rd edn, New York: McGraw-Hill.

Epstein, R. *et al.* (1999) 'Use of OR systems in the Chilean forest industries', *Interfaces*, 29(1):7–29.

Applications of MS/OR in health services

Health services are major industries in all developed countries and have been the subject of many MS/OR investigations. MS/OR approaches used include optimization methods (when objectives are clear and problems are well structured), 'what if . . . ?' or **sensitivity analysis** models (when problems are well structured, but objectives are unclear or competing), statistical methods (when understanding and interpretation of performance indicator information are the main focus), and **soft operational research** methods (when gaining a common understanding of issues faced is the major requirement). Below are some typical examples.

Health services, like many other industries, have their own versions of resource blending problems where the concern is to maximize outputs subject to resource constraints, or minimize costs subject to achieving output targets. For example, **linear programming** has been used to attempt to optimize whole health systems subject to cost constraints, to maximize the effectiveness of immunization programmes subject to resource constraints, and to minimize the cost of hospital meals subject to nutritional and aesthetic constraints (the **diet problem**).

Disease control, e.g. AIDS, is a very important area where 'what if . . . ?' modelling methods, such as **simulation**, can be used to predict the likely consequences of possible health services interventions. When good quality data on the effects of interventions is available the models are predominantly quantitative; in other cases studies rely much more on expert opinion and become more exercises of **problem structuring** and **scenario analysis**. Workforce planning is another area where a similar style of modelling is appropriate. 'What if . . . ?' models can be used to predict the short-term consequences of training extra nurses, but will need to incorporate expert judgement about the proportion that will be lost due to aggressive recruitment by overseas health providers (see **applications of MS/OR to crew scheduling**).

Health service managers, like managers in many public sector organizations, are concerned that resources (e.g. hospitals, teams of health professionals) are operating efficiently. Whilst many performance indicators exist (e.g. hospital occupancy rate, patient throughput), their interpretation is often ambiguous and dependent on the mix of patients treated and the range of treatments offered. Statistical methods such as **regression**, and more recently **data envelopment analysis**, have been used to compare performance levels and to identify benchmarks for good practice.

Many health services' activities can be viewed as **queueing systems**, e.g. patients waiting in an A&E department or outpatient clinic, patients on a waiting list, X-rays to be read, pathology tests to be performed, ambulance patients to be transported. The basic question in all these systems is what level of server resources to provide so that the 'customers' do not have to wait too long. Queueing and simulation models can often be used to help answer these questions.

One of the most successful applications has been the use of **multicriteria decision models** to the management of individual blood banks and blood product distribution centres, where initial outdating rates of old blood of over 20 per cent were reduced to below 5 per cent while keeping shortage levels down to 1 to 2 per cent.

D. Worthington

References

Boldy, D. (ed.) (1981) *Operational Research Applied to Health Services*, London: Croom Helm.

Cropper, S. and Forte, P. (eds) (1997) *Enhancing Health Services Management: The Role of Decision Support Systems*, Buckingham: Open University Press.

Pierskalla, W.P. and Brailer, D. (1994) 'Applications of operations research in health care delivery,' in Pollock, S., Rothkopf, M. and Barnett, A. (eds) *Handbooks in Operations Research and Management Science, 6: Operations research and the Public Sector*, Amsterdam: North Holland.

Applications of MS/OR in marketing, product sales and distribution

MS/OR models in marketing fall into three main groups: models for market analysis, models of the marketing mix, and decision models for marketing strategy.

Some of the first, albeit rather crude MS/OR models deal with brand loyalty and brand switching, usually formulated as **Markov chains**. Based on consumer purchasing behaviour, predictions can be made as to the ultimate market share of each brand. More recent studies use high-powered econometric and statistical methods, such as conjoint analysis to measure consumer satisfaction.

The output of market analysis models serves as inputs into decision models, such as J.D.C. Little's 1975 'BRANDAID' model that considers the entire marketing mix of product or service offered, pricing, advertising and distribution.

Marketing mix models cover actual and potential applications for sales effort allocation, using **dynamic programming**, sales force deployment – the **travelling salesperson problem** is an example of this – and product distribution, the latter as part of **logistics**, **supply chain management**. Design of advertising and promotion strategies and funds allocation for promotion efforts have been formulated as **mathematical programs**, and also analysed by **Bayesian decision analysis**, by **risk analysis** and by market simulation games. Marketing **decision support systems** have gained importance in the light of the ever-shortening product life cycles of many products. Globalization, e-business and e-commerce offer new opportunities for the use of MS/OR. *H.G. Daellenbach*

Reference

Eliashberg, J. and Lilien, G.L. (eds) (1993) *Handbooks in Operations Research and Management Science, 5: Marketing*, Amsterdam: North-Holland.

Applications of MS/OR in medicine

Surprising as it may seem, the first applications of MS/OR in the medical field occurred as early as 1943 (R. Dorfman, 'The detection of defective members in large populations', *Ann. Math. Stat.*). Since the 1970s, with the increasing availability and power of **expert systems** and **artificial intelligence**, research and applications of MS/OR techniques have flourished in medicine. Medical diagnoses and subsequent treatment, analysis and comparison of medical procedures, and the dynamics of pathology of various diseases have been fruitful areas for the use of MS/OR methodology and techniques. Below follows a small sample of applications.

The largest effort so far has gone into diagnosis. One of the earliest applications has been the use of artificial intelligence for pattern and image recognition for detecting abnormalities in cells, e.g. cancerous cells. Special-purpose expert systems have been developed for the diagnoses of certain families of diseases. **Decision trees, Bayesian decision analysis** and **risk analysis** are being used to assess the results of diagnostic tests in view of reducing the rate of both *false positive* and *false negative* conclusions. **System dynamics models** of disease pathology, based on **differential equations**, help in the early detection of diabetes. Statistical techniques and time series analysis studied the relationship between various diseases and the heart rate variability.

Evaluation and monitoring of surgical performance was used to estimate the risks of perioperative mortality. Stochastic modelling compared the effectiveness of two different treatment therapies for HIV infection. A combination of techniques was used to show that pooled testing of donated blood for HIV is more cost-effective than individual testing without affecting reliability. **Simulation** helps in the evaluation of clinical trials. **Dynamic programming** and **sensitivity analysis** have been used to estimate bounds on the failure profile for a new type of prostheses to be cost-effective. Risk analysis has been applied for DNA testing. **Data envelopment analysis** is seen as a valuable tool to evaluate physicians' relative decision-making efficiency. There is a potential for its use to compare different treatment plans. Simulation helps in planning the treatment and services for patients with coronary artery disease. **Mathematical programming** approaches have been proposed for the detection, prognosis and treatment of a number of diseases.

There is little doubt that medicine offers exciting opportunities for the use of MS/OR. *H.G. Daellenbach*

Reference

Gass, S.I. and Harris, C.M. (2000) *Encyclopedia of Operations Research and Management Science*, Norwell, MA: Kluwer Academic. Contains an extensive bibliography on pages 506–508.

Applications of MS/OR in military services

Science has been applied to warfare for hundreds of years, but especially starting with WWII. Crucial scientific developments in WWII include the invention of long-range sensors such as radar and sonar, the proximity fuse, the digital computer, and, of course, the atomic bomb. Partly because of the complicated tactics needed to employ these new weapons, WWII also saw the widespread application of the **scientific method** to the conduct of warfare, an activity that was called *Operational Research* in the UK and *Operations Research* in the USA (the slight difference in names still persists). The main features of this discipline are collection of operational data, construction of abstract models that are quantified by the data, and use of the models to find better ways of operating. The teams that applied this method generally included scientists of various specialities, as well as military officers. Their successes were impressive enough that the same ideas and methods were applied to industrial problems after the war, and curricula in OR were instituted at several universities. The first book on OR (Morse and Kimball, 1951) is entirely devoted to military problems of the type encountered in WWII. Typical OR questions of the time were: 'Where should search for enemy submarines be conducted? How large should convoys be? Large convoys of ships tend to be slow and overwhelm port facilities, but are easier to defend from submarine attack. Is it better for bombing crews to bomb the enemy or practise bombing?' The answer was sometimes the latter, since practice bombing permits feedback to the crew about accuracy in a way that is not possible in combat. Exchange rates for fighter aircraft depend strongly on the type of aircraft involved. In order to guide the development of new models, what are the important aircraft characteristics? A more recent book in the same tactical spirit is Wagner, Mylander and Sanders (1999).

Although OR is no longer a primarily military activity, it is still widely employed by the military. A great deal of modern military OR effort is spent in preparing for future conflicts, with computer **simulations** of combat often replacing the tense reality of WWII. The modern goal may be to develop tactics for use in changing circumstances, to predict the level of logistical support required, to determine the value of new equipment, or to spend a budget 'optimally'. The Military Operations Research Society publishes several books on the subject, including a reprinting of Morse and Kimball (see also Washburn, 1994). Military OR is for the most part strongly adapted to dealing with a secretive enemy whose decisions must be allowed for, but not all problems faced by military organizations are of this type. Some problems are similar to those faced by large civilian organizations, examples being **logistics**, communications, **manpower planning**, and the allocation of scarce resources. Solutions follow similar lines.   *A. Washburn*

References

Bracken, J., Kress, M. and Rosenthal, R. (1995) *Warfare Modeling*, Military Operations Research Society.

Hughes, W. (ed.) (1997) *Military Modeling for Decision Making*, Military Operations Research Society.

Morse, P. and Kimball, G. (1951) *Methods of Operations Research*, New York: Wiley.

Wagner, D., Mylander, W. and Sanders, T. (1999) *Naval Operations Analysis*, 3rd edn, Naval Institute Press.

Washburn, A. (1994) Chapter 4 in Pollock, S.M., Rothkopf, M.H. and Barnett, A. (eds), *Handbooks in Operations Research and Management Science, 6: Operations Research in the Public Sector*, Amsterdam: North-Holland.

Web-site

Military Operations Research Society (MORS): http://www.mors.org

Applications of MS/OR in the oil industry

The petroleum and oil industries have seen some of the first large-scale applications of MS/OR techniques, starting from the mid-1950s, when the first **linear programming** (LP) commercial computer codes became available. Some companies, such as Standard Oil of California, Shell Oil, Exxon and BP had research groups that were actively involved in advancing MS/OR methods and theory. For example,

Standard Oil teamed up with The RAND Corporation to develop their own LP code which included additional statements and functions built into the FORTRAN programming language that allowed specification of constraints, objective function, and **upper bounding** of variables, as well as facilities for *separable programming*, 25 years before the implementation of similar concepts in the **algebraic modelling languages**, such as GAMS, AMPL or AIMS. By the early 1960s, most oil companies had highly active MS/OR groups. MS/OR activity pervaded all aspects of operations. Major applications include:

- gasoline blending from various blending stocks to meet chemical specifications at minimum cost, formulated as LPs, using separable programming features to deal with nonlinearities;
- LP refinery optimization models for converting various types of crude oils into gasolines, diesel and fuel oils, and by-products;
- oil field production operations to determine the number of bores, pumping rates, and injection strategies to maximize extraction rates or volumes, by mixed **integer** and **nonlinear programming** and by **simulation**;
- oil field lease bidding models;
- offshore platform and marine terminal design, using simulation;
- company-wide models for the optimal coordination of all activities, from crude field production, pipeline and tanker usage, multiple refinery operations, to distribution to customers, again formulated as large LPs;
- optimal pipeline planning and operations using **networks** and simulation;
- crude oil tanker transportation scheduling, using simulation;
- maintenance planning using **critical path** scheduling;
- inventory control for motor oils and greases;
- **logistics** planning for product distribution, including depots location;
- real-time road tanker loading and **vehicle routing** systems;
- siting of service stations;
- determining spatial configuration of service stations, including number of pumps, and waiting space, using **queueing** models. *H.G. Daellenbach*

References

Bodington, C.E. and Baker, T.E. (1990) 'A history of mathematical programming in the petroleum industry', *Interfaces*, 20(4):117–127.

Bodington, C.E. (1995) *Planning, Scheduling and Control Integration in Process Industries*, New York: McGraw-Hill.

Applications of MS/OR in sport

It should come as no surprise that MS/OR finds many applications in both team and individual sport, such as association football, baseball and cricket, tennis, squash, sailing, and so on, dealing with decisions made by the players themselves, by administrators of the game, and by analysts who study measures of performance of the players and teams. The MS/OR techniques used include **linear programming**, **integer programming**, **dynamic programming**, **decision analysis**, **simulation**, **Markov chains**, **heuristics** and **statistical analysis**.

Decision making by players and coaches during the game involves **uncertainty** about the outcomes and actions of opponents, and decisions need to be made as the game progresses. For example, baseball teams must consider the probabilities of the outcomes when making strategy choices at critical times in the game. The decision to use a 'pitch hitter' with only one 'out' left has great possible returns, but is high risk. The safer strategy may give a more certain outcome, but with less potential reward. Probabilistic models, including Markov chains, have been used to determine good strategies. Similarly, optimal serving strategies for tennis players have also been formulated as Markov chains.

Sports administrators have to schedule tournaments that require teams to play each other a certain number of times and at specific locations. These problems can be formulated as integer and linear programs, or may be solved using heuristics. Schemes for drafting of players between teams, which is usually done to get more balanced teams, have been formulated as Markov chains.

Analysts try to rank teams, before and during the playing season, so as to determine the chance that a given team will finish in a certain position in a points table, such as making the playoffs or being relegated to a lower division. They are faced with large numbers of uncertain variables and information, including last year's performance, recent performance, current table position, and remaining schedule. Decision analysis and statistical **regression** have proved to be effective tools.

Surely one of the most well-known uses of MS/OR in sport is the Duckworth/Lewis method for cricket. This method, accepted by the International Cricket Council, sets a revised target number of runs for the team batting second, when a limited-overs cricket match has been shortened (usually due to rain) after it has commenced. It is designed so that neither team benefits or suffers from the interruption. The revised target is based on the recognition that teams have two resources – overs remaining and batsmen left – with which they can score runs. Targets are calculated for all possible combinations of these two resources. Analysis of past data has shown the relationship between total number of runs and these two factors to be an exponential model. *S. Batstone and M.C. Stewart*

References

Duckworth, F. and Lewis, T. (1998) 'A fair method of resetting the target in interrupted one- day cricket matches', *JORS*, 49(3):220–227.

Gass, S. and Harris, X. (1998) *Encyclopedia of Operations Research and Management Science*, Hingham, MA: Kluwer. Pages 777–780 contain an extensive list of journal articles.

Gerchak, Y. (1994) 'Operations research in sports', in Pollock, S.M., Rothkopf, M.H. and Barnett, A. (eds), *Handbooks in Operations Research and Management Science, 6: Operations Research in the Public Sector*, Amsterdam: North-Holland.

Machol, R.E., Ladany, S.P. and Morrison, D.G. (eds) (1976) *Management Science in Sports*, Amsterdam: North-Holland/TIMS.

Applications of MS/OR in transport industries

Air, sea, rail and road transport have seen numerous applications of MS/OR for investment planning, maintenance and **replacement** decisions, day-to-day scheduling of all sorts, **vehicle route scheduling**, and real-time dispatch. There are practically no planning and operational aspects that have not seen the hand of MS/OR. The list below summarizes some of the more important applications.

Air transport: In today's competitive air travel and airfreight market, an operator's economic viability depends on efficient scheduling. It is therefore not surprising that the most spectacular modelling efforts have gone into the combined scheduling of aircraft to routes, crews to aircraft, and aircraft maintenance. These models even determine where to refuel the aircraft to take advantage of fuel price differences. The models are huge mixed **integer programs**. Their efficient solution, even with the massive computational capacity of today's computers, has necessitated the development of new techniques. (See also **applications of MS/OR to crew scheduling**.) Stochastic models are used to determine cost-effective over-booking limits to make up for 'no-shows', i.e. ticket holders that do not show up or cancel shortly prior to the flight.

Sea transport: Scheduling of cargo pick-up and drop-off for a fleet of ships of different capacities, cargo space configuration, and speed, is without doubt the most important problem for a sea transport operator. For coastal shipping, the effect of the tides also needs to be taken into account. **Simulation** models have been used for ship scheduling since the late 1950s. Today, such models are used dynamically to reassign and reroute ships as the demand for cargo space varies. **Network models** are also used. Associated with sea transport is the efficient operation of both container and conventional cargo ports, both in terms of equipment and facilities planning (usually involving **queueing** models) and in terms of day-to-day unloading and loading.

Rail transport: Efficient use of rail wagons and engines, and crew assignments, are the important issues. Scheduling of engine usage poses similar problems as the scheduling of aircraft. Crew assignments have to meet maximum working hours as well as bringing crew back to their home base, avoiding *dead runs* (i.e. trips where the crew are passengers) as much as possible. Both give rise to large network models. Efficient use and redistribution of rail wagons is explored by network models, **linear programming** and simulation and more recently by continuous tracking of wagons via automatic recording devices situated strategically over the entire rail network and fed into a central computer.

Public road passenger transport: The problems are similar to the scheduling of engines and drivers for rail transport with the addition of catering for the early morning and evening rush hours. In contrast to rail, maintaining bus schedules can be hampered by the effects of other road user traffic. This calls for dynamic reassignments of both buses and drivers.

Road freight transport: In addition to finding optimal routes for individual or fleets of vehicles making pick-ups or drop-offs over a given geographical area, usually via **heuristic methods**, more recent applications have considered the dynamic scheduling of fleets of vehicles nationwide, including the opportunities for back-hauling.

Cost–benefit analyses have also been done in most countries on various aspects related to the transport industries, such as the construction or upgrading of airports, rail tracks, and roads. Several studies analysed the effects of the introduction of certain types of vehicles on road deterioration, accident costs and environmental impacts.   *H.G. Daellenbach*

Reference
Odoni, A. and Wilson, N.H.M. (1994) 'Transportation', in Pollock, S.M., Rothkopf, M.H. and Barnett, A. (eds), *Handbooks in Operations Research and Management Science, 6: Operations Research in the Public Sector*, Amsterdam: North-Holland.

ARIMA processes

ARIMA processes are mathematical models of time series used for **forecasting**. ARIMA is an acronym for 'AutoRegressive, Integrated, Moving Average'. Each of these phrases describes a different part of the mathematical model. ARIMA processes have been studied extensively and are a major part of *time series analysis*. They were popularized by George Box and Gwilym Jenkins in the early 1970s; as a result, they are also known as **Box–Jenkins models**. Box and Jenkins (1970) effectively put together in a comprehensive manner the relevant information required to understand and use ARIMA processes.

The ARIMA approach to forecasting is based on the following ideas:

1. The forecasts are based on linear functions of the sample observations.
2. The aim is to find the simplest models that provide an adequate description of the observed data. This is sometimes known as the principle of *parsimony*.

Each ARIMA process has three parts: the *autoregressive (or AR) part*, p; the *integrated (or I) part*, d; and the *moving average (or MA) part*, q, written in shorthand as ARIMA(p,d,q).

AR: This part of the model describes how each observation is a function of the previous p observations. For example, if $p = 1$, then each observation is a function of only one previous observation. That is,

$$Y_t = c + \phi_1 Y_{t-1} + e_t$$

where Y_t represents the observed value at time t, Y_{t-1} represents the previous observed value at time $t-1$, e_t represents some random error and c and ϕ_1 are both coefficients. Other observed values of the series can be included in the right-hand side of the equation if $p > 1$:

$$Y_t = c + \phi_1 Y_{t-1} + \phi_2 Y_{t-2} + \dots + \phi_p Y_{t-p} + e_t$$

I: This part of the model determines whether the observed values are modelled directly, or whether the differences between consecutive observations are modelled instead. If $d = 0$, the observations are modelled directly. If $d = 1$, the differences between consecutive observations are modelled. If $d = 2$, the differences of the differences are modelled. In practice, d is rarely more than 2.

MA: This part of the model describes how each observation is a function of the previous q errors. For example, if $q = 1$, then each observation is a function of only one previous error. That is,

$$Y_t = c + \theta_1 e_{t-1} + e_t$$

Here e_t represents the random error at time t, e_{t-1} represents the previous random error at time $t-1$, and c and θ are both coefficients. Other errors can be included in the right-hand side of the equation if $q > 1$.

Combining these three parts gives the diverse range of ARIMA models. There are also ARIMA processes designed to handle seasonal time series, and vector ARIMA processes designed to model multivariate time series. Other variations allow the inclusion of explanatory variables.

ARIMA processes have been a popular method of forecasting because they have a well-developed mathematical structure from which it is possible to calculate various model features such as prediction intervals. These are a very important feature of forecasting as they enable forecast uncertainty to be quantified.   *R. Hyndman*

References

Box, G.E.P. and Jenkins, G.M. (1970) *Time Series Analysis: Forecasting and control*, San Francisco: Holden-Day.

Makridakis, S., Wheelwright, S.C. and Hyndman, R.J. (1998) *Forecasting: Methods and applications*, New York: Wiley.

Pankratz, A. (1983) *Forecasting with Univariate Box-Jenkins Models: Concepts and cases*, New York: Wiley.

Art gallery problem

Imagine an art gallery in which it is necessary to watch every wall and all the open spaces in all rooms. Where should the guards (or security sensors) be placed? People and electronic devices 'see' in straight lines. The limit of their vision from their location is the first wall in any direction. It is assumed that they can see in any direction. If the gallery is on one floor, then the problem is one of geometry in two dimensions. In the simplest case, with all the walls as straight lines, the problem can be solved by counting the number of corners (say N) and dividing by 3, rounding down if necessary. This gives the minimum number of guards. It has been shown that there are examples which need this number.

Even with two dimensions, the problem becomes much harder when there are obstacles inside the gallery, such as lift-shafts or large pieces of art that block the view. In many cases, a good answer can be found by dividing the space into triangles, with the corners of the triangles at the corners of the gallery and the obstacles. The corners of the triangles are coloured using three colours, so that each triangle has one corner of each colour. Then a guard is placed at all the corners of the least-used colour; they can each see the whole of their triangle(s), and so together they can see the whole space. The figure below demonstrates this. Forming the smallest number of triangles indicates that Y is the least used colour in both cases. Guards placed there can survey the entire gallery for both cases.

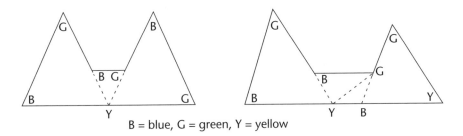

B = blue, G = green, Y = yellow

Figure: **Art gallery problem**

When the height of the gallery becomes important, as would be the case where the problem was one of lighting, then methods of combinatorial optimization have been applied for finding the solution, with varied success.

The problem is also referred to as Chvatal's problem. *D.K. Smith*

Reference
O'Rourke, J. (1987) *Art Gallery Theorems and Algorithms*, New York: Oxford University Press.

Artificial intelligence or AI

Artificial intelligence is the study of methods and ideas that will enable the computer to perform tasks that would normally require human intelligence. Its two main aspects are methods for knowledge representation – the counterpart to human memory – and general mechanisms for problem solving. In the 1950s two distinct AI camps emerged. The first endeavoured to build models of intelligence by drawing inspiration from biology and the second one by creative programming. In 1955, Allen Newell and Herbert Simon, the recipient of the Nobel prize in economics in 1978, demonstrated that it was in principle possible that the computer's language of strings of bits could be used to represent features of the real world, which could be subsequently manipulated internally via formal rules encoded in specialized software. These formal rules had to capture common-sense understanding and reasoning if the computer was to simulate important aspects of human intelligence. Thus the potential of the computer as an information-processing machine could serve as a vehicle for an artificial mind.

Research in AI gained momentum in the 1960s. Marvin Minsky, one of the key figures in the AI movement, claimed that within a generation the problem of AI would be largely solved, but by the 1980s the whole AI movement had run into grave difficulties. Although systems for subject-specific knowledge representation could be built, no general theory of problem solving, i.e. reasoning and common sense, nor a coherent framework for its realization had been developed. Minsky, in a complete reversal of his position, was openly conceding that the AI problem was one of the hardest ever undertaken. An ambitious project by the Japanese Ministry of Trade and Industry also ended in defeat in the early 1990s.

Probably the most successful commercial applications of AI are **expert systems** (ES). An ES is a sophisticated program with the ability to make logical deductions by utilizing expertise in a narrow domain of human knowledge encoded in the system's knowledge base. A widely cited example is the expert diagnosis of a particular medical condition whose well-documented symptoms reside within the ES knowledge base. Critics of AI point out correctly that ES are limited by their narrow domain of expertise and their inability to extrapolate beyond the boundaries of their knowledge base.

Considerable success has been reported in applying conventional ES to routine financial decision-making operations, such as approval of credit lines, mortgage underwriting, financial planning, and the underwriting of complex insurance policies. All these activities involve some form of risk assessment. It is the view of financial experts that AI technology holds great promise for enhancing a multitude of tasks performed in the financial industries, and particularly in investment portfolio management activities.   *A.D. Tsoularis*

References

Genesereth, M. and Nilsson, N. (1987) *The Logical Foundations of Artificial Intelligence*, Los Altos, CA: Morgan-Kaufmann.

Gottinger, H.W. and Weimann, H.P. (1990) *AI: A Tool for Industry and Management*, New York: Ellis Horwood.

Trippi, R.R. and Lee, J.K. (1996) *AI in Finance and Investing*, Chicago, IL: Irwin Professional Publishing.

Assignment problem

Consider the problem of assigning n candidates with different skills and aptitudes to perform n different jobs (of roughly equal amount of work), each requiring different skills and abilities. Let c_{ij} denote the suitability or the penalty or cost of assigning candidate i to do job j. You wish to assign candidates to jobs so as to achieve maximum overall suitability or minimize total penalties. A feasible solution is one where each candidate does exactly one job and all jobs are covered. Problems of this nature are known as assignment problems. They may be assigning lawyers to court cases, jobs to machines, vehicles to deliveries, sales personnel to sales territories, contracts to bidders, and so on.

The assignment problem is a special case of the **transportation problem** of transporting goods from m sources to n destinations so as to minimize the total transportation cost. In the assignment problem $m = n$, and each candidate represents a source with capacity $a_i = 1$ and each job is a destination with demand $d_j = 1$. The decision variables $x_{ij} = 1$ if candidate i is assigned to job j and 0 otherwise. If the c_{ij}'s are penalties the assignment problem becomes

$$\text{Minimize} \sum_{i=1}^{n} \sum_{j=1}^{n} c_{ij} x_{ij},$$

subject to

$$\sum_{j=1}^{n} x_{ij} = 1, \text{ for } i = 1, \dots, n$$

$$\sum_{i=1}^{n} x_{ij} = 1, \text{ for } j = 1, \dots, n$$

and $x_{ij} \geq 0$ and 1, for all i and j.

The first set of constraints specifies that each candidate is assigned, while the second set requires that each job is done. Although we do not specify that the x_{ij} need to be integer, i.e. 0 or 1, the special structure of this **linear program** – as for the transportation problem – will always guarantee that the optimal solution is integer. Since the right-hand sides of all constraints are equal to 1 this implies that each candidate is assigned to one and only one job.

The above formulation can easily be extended to cater for the number of candidates not being equal to the number of jobs, for the objective function being maximized, and for the presence of forbidden pairs of candidate–job assignments.

The optimal solution can be computed using a linear programming or transportation problem solution algorithm. In 1955, when computers could only handle small linear programs, H.W. Kuhn developed a simple and efficient algorithm, known as the **Hungarian method**, for solving the assignment problem. In today's world of high-speed PCs this method has become an interesting historic relic.   *B. Chen*

References

Hillier, F.S. and Lieberman, G.J. (2001) *Introduction to Operations Research*, 7th edn, New York: MacGraw-Hill, ch. 8.
Chvátal, V. (1983) *Linear Programming*, New York: Freeman, ch. 20.

Attributes

Attributes are characteristic properties of entities of interest to the analyst. In statistics, observations have numeric and qualitative or categorical attributes, e.g. gender, age, income, and residence of an individual in a random sample. (In statistics a reference to 'attribute' data is usually interpreted to mean that the attribute is categorical, e.g. gender, yes/no, on/off, type, etc.)

System components, **simulation** entities, system activities and interactions have numeric and/or categorical attributes that are measured by **state variables**. For example, in a **queueing** system the status of a server is an attribute; it is either 'idle' or 'busy', where idle may be assigned the number 0 and busy the number 1. The number of people waiting in a queue is an attribute of the entity 'queue'. The state of a system at any given point in time is given by the values of all state variables, i.e. the collection of all attributes.

Assessing how well a desired objective has been achieved requires that the objective is expressed by a measurable attribute, such as cost. Sometimes more than one attribute contributes toward the objective function. For example, the objective 'clean air' is rather vague. It has to be translated into specific aspects that can be measured – so-called *surrogate* attributes, such as the number of dust particles and the concentration of certain gases, e.g. nitrogen oxides. These are the attributes used to judge how clean the air is. These in turn may be combined into a single measure through a **multiattribute utility function**. As the name itself indicates, it is a utility function that captures the intrinsic worth of an outcome that consists of the attribute values of several aspects. To use again the clean air example, each possible control measure will result in given levels of the various attributes mentioned. The overall **effectiveness** of each measure is then judged on the basis of the multiattribute utility function.

Autopoiesis

Autopoiesis (the theory of *self-producing systems*) is an ambitious theory aiming to explain both what is particular about **living systems** and how human thought and language arises from our biological foundations (Maturana and Varela, 1980, 1987). It does so in a very coherent and consistent manner and has generated significant interest and debate in fields as diverse as organization theory, business strategy, **information systems**, law, and sociology (Mingers, 1995).

The very concept of a self-producing or self-constructing system has interesting and radical implications. Traditionally, systems theory has dealt with **open systems** that process or transform inputs from its environment into outputs. Such a view can be applied to, say, a company quite easily – resources are taken in, undergo production processes, and result in products and services. Societies could also be seen as complex adaptive systems which used internal **feedback** processes to change their structures to better survive in a turbulent and changing environment.

An autopoietic system, however, is quite different – it does not transform inputs into outputs, instead it transforms itself into itself. What is meant by this is that the outputs of the system, that which it produces, are its own internal components, and the inputs it uses are again its own components. (Note, however, that it does always require some elements from the environment, and it does excrete waste.) It is thus in a continual dynamic state of self-production. A classical example of a physical autopoietic system is the biological cell. Of more relevance for management is the question of whether social or organizational systems might be autopoietic (Luhmann, 1986; Mingers, 1992). If they are, then there are several important consequences:

- It can explain how social systems can survive for long periods despite changing their structure and components. We can see many groupings – families, companies, religions, cultures and societies – that exhibit tremendous long-term stability and persistence despite enormous changes in their environment, and their own internal membership and structure.
- The self-producing nature of the system means that we do not have to look for external inputs and outputs, nor do we have to see the system as functionally dependent on other systems. Its 'purpose' is simply its own continual self-production.

The origin of change and development is placed firmly within the system rather than from the environment, whilst the concept of *structural coupling* shows how, nevertheless, systems and their environments can mutually shape each other.   *J. Mingers*

References

Luhmann, N. (1986) 'The autopoiesis of social systems,' in Geyer, F. and van der Zouwen, J. (eds), *Sociocybernetic Paradoxes*, London: SAGE Publications.

Maturana, H. and Varela, F. (1980) *Autopoiesis and Cognition: The Realization of the Living*, Dordrecht: Reidel.

Maturana, H. and Varela, F. (1987) *The Tree of Knowledge*, Boston: Shambhala.

Mingers, J. (1992) 'The problems of social autopoiesis', *Int. J. Gen. Sys*, 21(2):229–236.

Mingers, J. (1995) *Self-Producing Systems: Implications and Applications of Autopoiesis*, New York: Plenum Press.

B

Backwards induction, see **decision trees**

Bargaining

The terms bargaining and **negotiation** are sometimes used interchangeably, but there are some subtle distinctions between the two. Bargaining is the system of activities, or a subprocess, within the broader process of negotiation. It refers to the stage in negotiations when the parties seek to attain their own goals, when some or all of those goals are in basic conflict with the goals of the other party. For example, if we said someone 'bargained hard', we would mean that they made relatively few concessions, or that the negotiation process was particularly long and difficult. The two leading writers in the field of bargaining theory are Walton and McKersie (1965). Their work, *A Behavioural Theory of Labor Relations*, is the seminal work on bargaining. Although it relates specifically to labour relations, the concepts which they developed have been applied to many bargaining contexts.

Walton and McKersie distinguish four different bargaining subprocesses. *Distributive bargaining* refers to those activities which we would normally associate with the term bargaining. It refers to the situation where one party's gain is the other's loss: it is fixed-sum bargaining, usually over the allocation of resources such as money, power, or status. For example, most labour relations bargaining falls into this category, as do a lot of negotiations over sale and purchases. *Integrative bargaining* involves very different tactics and refers to the achievement of objectives which are not in conflict with those of the other party. These objectives are more in the nature of a common problem which can be resolved by joint problem-solving behaviour. This is the so-called *win–win* approach to bargaining made famous by Fisher and Ury (1981).

The other subprocesses identified by Walton and McKersie apply more specifically to labour relations bargaining. The first is *attitudinal restructuring*, which is the medium- to long-term influencing of the relationship between the parties that takes place both during bargaining and at other times. *Intra-organizational bargaining* refers to those negotiations which take place within each bargaining team or side: it is the process of getting everyone on your side in alignment. It recognizes that people may bring different agendas to the table, even though they are on the same side. Sometimes it can be as difficult to reconcile these agendas as it is to deal with the other party. Taken together, these four parallel subprocesses comprise a significant part of the negotiation process.   *I. Brooks*

References
Fisher, R. and Ury, W. (1981) *Getting to Yes*, Boston MA: Houghton-Mifflin.
Walton, R. and McKersie, R. (1965) *A Behavioural Theory of Labor Relations*, Ithaca, NY: ILR Press.

Barrier methods, penalty methods, sequential unconstrained maximization technique (SUMT)

In general, it is easier to optimize a nonlinear objective function in several variables if there are no constraints imposed on the variables than if they are subject to constraints. Barrier and penalty methods are two approaches that provide a means of reformulating a constrained **nonlinear programming problem** into an unconstrained problem. Both incorporate the constraints into the objective function in the form of *barrier functions* or *penalty functions* that make the region close to or beyond the boundary of the **feasible region** unattractive. The resulting unconstrained problem is then solved by any one of several ordinary unconstrained optimization techniques, such as a *steepest ascent method* or **gradient method**.

The boundary of the feasible region can be thought of as a fence or wall. One side is feasible, the other infeasible. Barrier functions prevent you from walking smack up to the wall coming from the feasible region. Penalty functions force you back over the fence if you stray outside the feasible region.

Penalty functions add a penalty to the value of the objective function on any solution that is outside the feasible region, while solutions inside do not incur a penalty. Consider the following simple nonlinear programming problem:

$$\text{Maximize } f(x) = 10x - x^2$$

$$\text{subject to} \quad x \le 2$$

The unconstrained optimum is $x = 5$. Penalty methods replace the constraint on x by a penalty function which is incorporated into the objective function. The following is a possible form of a penalty:

$$P(x) = -(\text{maximum } [\, 0, x - 2\,])^2$$

The severity of the penalty is controlled by multiplying it by λ. Added to $f(x)$, we get an unconstrained optimization problem of the form

$$\text{Maximize } C(x) = f(x) + \lambda P(x) = 10x - x^2 - \lambda (\text{maximum } [\, 0, x - 2\,])^2$$

This unconstrained problem is now solved for x. Note that for small values of λ, the optimal x is larger than 2, violating the constraint. Letting λ sequentially become larger and larger, the optimal x approaches its constrained optimum of 2 closer and closer.

Barrier methods work in a similar way. They are particularly useful when it is not possible to calculate the objective function outside the feasible region. By adding a barrier function to the objective function, feasible solutions very close to the edges of the feasible region become unattractive, while for solutions away from the edges the objective function value is not affected significantly. As for penalty methods, the optimization with barrier functions is performed sequentially, using different barriers as the algorithm proceeds, with the finishing point of one iteration becoming the starting point of the next.

With the availability of ever more powerful nonlinear optimization software and high-level **algebraic modelling languages**, such as AIMMS, AMPL and GAMS, these methods have lost much of their importance and appeal.   *D.K. Smith*

Reference
Nocedal, J. and Wright, S.J. (1999) *Numerical Optimization*, New York: Springer Verlag, ch. 17.

Basic feasible solutions in linear programming (LP), basic variables, basis

The **simplex method** of LP works with so-called *basic solutions*. If a problem has m constraints and a total of n variables (including **slack and surplus variables**), then a basic solution is one where exactly $n - m$ variables are set to zero, called the *non-basic variables*, while the remaining m variables may assume nonnegative values, called the *basic variables*. The basic variables form the so-called *basis*. (There is the added assumption that none of the constraints can be expressed as a linear combination of some of the other constraints.) If such a basic solution meets all constraints and all variables also satisfy the nonnegativity conditions, then it is also a **feasible solution**. A basic feasible solution corresponds to a corner point or vertex of the feasible region. At least one optimal solution to an LP is always at a vertex.

Consider the following set of constraints:

$$x_1 - x_2 + x_3 \quad\; = 6$$
$$2x_1 + x_2 \quad\; + x_4 = 18$$

Since there are two equations, a basis has two basic variables. If x_1 and x_2 are the non-basic variables at zero value, then x_3 and x_4 are the basic variables. Their solution can be read off the constraints as $x_3 = 6$, and $x_4 = 18$. In this example, there are a total of six bases. Four of the bases have all variables nonnegative, hence they are feasible basic solutions, while two have one variable assume a negative value.

Bayesian decision analysis

Bayesian decision analysis is a variation of statistical **decision analysis**. Both involve finding the best decision or strategy (i.e. a sequence of conditional decisions, where subsequent decisions depend on which one of the possible future states of the world eventuates) in the face of an uncertain future. The decision **criterion** used is the maximization of expected benefits or the minimization of expected costs. **Decision trees** are an effective way to do this analysis.

Bayesian decision analysis argues that in many instances it may be advantageous to purchase or obtain additional information that may give more foreknowledge about what the future holds, usually in the form of 'predictions' about certain aspects of the future. This is particularly so if traditional decision analysis shows that the value of *perfect information* is considerably higher than the cost of acquiring more information. 'Predictions' about possible futures may be obtained by various means, such as from a small-scale experiment, a small market survey, a **Delphi study**, or from expert opinion. In most cases, these predictions will not be perfect; they could be in error. If we can infer, say from past experience or from educated guesses, what the probabilities of various errors are, it is possible, with the help of *Bayes' Theorem*, to adjust the original probability distribution over the future states for each type of prediction. (Bayes was an English clergyman and mathematician in the 18th century.) The original probabilities are referred to as the *prior probabilities* and the updated probabilities as the *posterior probabilities*.

However, before committing funds to acquire additional information, the decision maker will want to know if it is, in fact, worthwhile to acquire the additional information or not. A new decision choice is thus added to the original problem, i.e. 'acquire information' which is penalized by the cost of doing so, and 'do not' which has no direct cost and leads to the original problem. The 'acquire information' branch of the expanded problem is evaluated on the basis of the posterior probabilities, the original branch on the basis of the prior probabilities. The decision as to whether or not to acquire additional information is decided on the basis of the expected value of the two branches.

Bayesian decision analysis has also been extended to the case where the future states are expressed by a continuous probability distribution.

Bayesian decision analysis has been applied to problems dealing with the placement of nuclear power stations, hazardous waste disposal decisions, research and development projects, marketing decisions, medical diagnoses, and so on. Decision analysis software has options to deal with Bayesian analysis. *H.G. Daellenbach*

Reference

Hillier, F.S. and Lieberman, G. (2001) *Introduction to Operations Research*, 7th edn, New York: McGraw-Hill, ch. 15.

Beer game

The beer game is a manual **simulation** game designed to demonstrate the dynamic effects of a **system's** structure on its behaviour. It was originally conceived in the 1960s by Jay Forrester, the originator of **system dynamics**, at MIT. It simulates a production–distribution system for the supply of a single brand of beer. Four independent teams manage the system: a retailer, a wholesaler, a distributor and a factory. The game simulates about 40 weeks of operation, with a facilitator keeping competing teams operating in parallel. Financial penalties of $1 a case are imposed on stock held in inventory and $2 per case for stock that cannot be supplied. This penalty structure makes players averse to stock-outs and to carrying high levels of stocks. Not unlike most supply chains, this one also suffers from poor communication between **stakeholders**, compounded by delays in the delivery of orders and the receipt of stocks of beer. Consequently, when such a system is subjected to a sudden external shock, for example a one-time increase in customer demand, a ripple effect is set in motion. Retailers see their stocks decrease and place extra orders to the wholesalers, who in turn see their stock dwindle suddenly. Expecting a possible further rise in orders from the retailers, they in turn increase their own orders to the distributors, and so on until the factory. This results in the propagation of a wave of orders down through the supply chain with increasing amplitude as risk-averse decision makers struggle with poor information. Over-ordering at the retail level now causes orders to the wholesalers to fall drastically, who in turn suddenly see their stock reach unprecedented levels, as more supplies ordered earlier still arrive from the distributors, with the latter facing the same situation, except of even greater magnitude. As retailers see their stocks return to acceptable levels, they start ordering again and the whole process repeats itself down the chain, resulting in renewed wild swings in inventories at all levels.

In supply chain terms this phenomenon is known as the *bullwhip effect*, or the *Forrester effect*. Forrester used it as one possible explanation for the occurrence of short-term business cycles. The game can be used as a basis for discussing issues as diverse as **information systems** in supply chains, the effect of using improved decision rules, **business re-engineering processes**, and decision making under time constraints. Computer models of the beer game have been used to study chaotic behaviour of systems. *J. Barton*

References

Senge, P.M. (1990) *The Fifth Discipline*, New York: Doubleday, ch. 3.

Sterman, J.D. (2000) *Business Dynamics*, Boston: Irwin McGraw-Hill.

Van Ackere, L., Reimer, E. and Morecroft, J.D.W. (1993) 'Systems thinking and business process redesign: an application of the beer game', *European Management Journal*, 11(4) (December):412–423.

Bender's decomposition, see decomposition

Benefits and costs, relevant

Which costs and benefits should a decision maker take into account when evaluating the monetary effect of alternative decision choices? Any cost or benefit item now or in the future that for the **wider system of interest** as a whole changes as a consequence of any of the decision choices is relevant. Any costs or benefits that remain the same in total terms, regardless of the decision choice, are irrelevant and may be ignored.

Note that the total effect on the wider system of interest (e.g. the firm or organization as a whole) must be considered, not simply those on the **narrow system of interest** (e.g. the activity or department of the firm or organization primarily involved). The reason is that a decision choice may result in cost and benefit changes in the narrow system which may be partially or fully offset in the wider system. The decision choice should be based on their total effect, not just the effect on the narrow system. For example, the use of cheaper raw materials in the manufacture of a product will reduce the costs for the production department, but may result in higher guarantee costs for the sales department. The total effect may be detrimental. Another example is given by the allocation of so-called overhead costs (administrative and management costs incurred by the firm as a whole) to the various activities of the firm. A decision choice may affect the level of overhead allocation to an activity that could be misinterpreted as an increase in the cost of that activity, whereas in fact the total overhead cost of the firm has not changed at all. Unless a decision choice affects such overhead costs for the firm as a whole, overheads are not a relevant cost item and should be ignored. Note that cost accounting figures often include arbitrary allocation of overheads. *H.G. Daellenbach*

Reference
Daellenbach, H.G. (1994) *Systems and Decision Making*, New York: Wiley, ch. 9.

Big-M method

The big-M method is used to find a feasible solution to a **linear program** (LP), when an initial **feasible solution** cannot be found by inspection. If all m constraints are of the less-than-or-equal form, an initial feasible solution can be found by inspection by setting the m **slack variables** equal to the right-hand side of the corresponding constraints and all original decision variables equal to zero. (The slack variables are the difference between the right-hand side and the left-hand side of these constraints.) A feasible solution that has exactly m variables nonnegative and the remaining ones zero is called a **basic solution** and the basic variables form a so-called basis. The **simplex method** works with basic feasible solution.

If some of the constraints are equalities or of the larger-than-or-equal type, other methods have to be used. One of them is the big-M method. Its idea is simple. We use the trick of augmenting the problem by adding to each equality and larger-than-or-equal constraint a so-called *artificial variable*. For example, if the original constraint is

$$5x_1 - 2x_2 = 10$$

we change it to $\qquad 5x_1 - 2x_2 + x_3 = 10$

where x_3 is an artificial variable which has no physical meaning. However, an initial basic solution for this augmented problem can now be found by inspection.

The next step is to rid the basis of the artificial variables. If we succeed in doing this, then we have found a basic feasible solution to the original problem. The second trick is to penalize the artificial variables in the objective function with a big cost, M, hence the name big-M method.

The simplex method is now applied to the augmented problem with its new objective function in the usual way. If the original problem has indeed a feasible solution (i.e. a basic feasible solution without any artificial variables), the simplex method will find it. In terms of the actual computational steps, there is little difference between what is known as the *two-phase method* and the big-M method.   *H.G. Daellenbach*

Reference

Winston, W.L. (1994) *Operations Research: Applications and Algorithms*, 3rd edn, Belmont, CA: Duxbury, pp. 164–169.

Bin packing and container packing

Practical applications of the bin-packing problem arise in the context of both packing and cutting problems. Packing problems require a set of objects to be fitted within a given space or area, whilst cutting problems require a given set of 'pieces' to be cut from a given resource. In both instances the problem can be modelled as that of fitting a collection of objects into well-defined regions so that they do not overlap.

Bin-packing problems are characterized by the dimension of the problem, the number and shape(s) of the objects and of the regions used for packing. For example, the one-dimensional problem involves minimizing the number of 'bins' of a given capacity required to fit a given collection of objects. A typical application might be minimizing the number of steel bars of length L required to produce a set of n smaller bars of known lengths l_1 to l_n.

The two-dimensional case has several variants. The direct counterpart of the one-dimensional problem involves minimizing the number of rectangles of dimension L × W required to fit a set of n rectangles of dimensions $l_1 \times w_1$ to $l_n \times w_n$. Other commonly encountered formulations involve minimizing the length required to fit all the pieces into a rectangle of given width, or maximizing the area (or value) of a subset of the pieces fitted into a single rectangle of given dimensions. As well as the obvious applications in cutting stock material so as to minimize the cost of meeting a given set of orders, or to maximize the value obtained from a given sheet, the two-dimensional case is appropriate for packing problems in which rectangular boxes of a common height are to be packed in layers. If all the boxes are identical then the problem is known as the pallet-loading problem. Further variation occurs if the pieces or the packing area are not rectangular, for example loading circular drums, or cutting garment pieces from a strip of material.

All of the above variants can be extended to a problem in three dimensions. However, as the most common application areas are in the packing of rectangular objects, only the rectangular variants have been widely studied.

Even the one-dimensional problem is **NP-hard**. Thus **heuristic solution methods** are popular. The simplest order the pieces according to a set of rules (e.g. largest first) and pack them in sequence according to a set of placement rules appropriate to the objective. More recently, techniques such as **simulated annealing** and **genetic algorithms** have been applied. There has also been some success with exact solution methods for small to moderate-sized problems in one and two dimensions. These methods are usually based on **integer programming** techniques.   *K.A. & W.B. Dowsland*

References

Dowsland, K.A. and Dowsland, W.B. (1992) 'Packing problems', *EJOR*, 56:2–14.
Dyckhoff, H. and Finke, U. (1992) *Cutting and Packing in Production and Distribution*, Berlin: Springer-Verlag.

Black boxes, systems as black boxes

The complexity of real life may be such that we have no or only incomplete knowledge of the inner workings of a **system**. Clearly, this is the case for human brain functions. So, there is no full understanding of how humans learn. Similarly, in spite of the enormous progress made in meteorology and computer processing capacity, weather systems are only partially understood. Computers or other machinery fail for a myriad of reasons. It may be impractical to keep track of individual causes, nor may it be important for determining the number of service personnel needed. So only aggregate records are kept.

In each of these examples, the inside of the corresponding system is left largely empty. All we know, or keep track of, are the inputs into and the outputs from the system. For the lay observer it looks like one of those black control boxes, with lots of wires into and out of the box, but no way of knowing what is under the cover. If our aim is to predict the output of such a system in response to various inputs, we may indeed not have to know the details of its inner workings, even if this were possible. In such instances, all we need to discover is the form of the functional relationship between inputs and outputs, i.e. a few functions and equations that mimic the system transformation process. For example, **regression analysis** may show how the output changes to various configurations of input controls.

In other situations the transformation process could be modelled in all its detail. However, we know from past experience that certain transformation rules produce desirable and/or cost-effective combinations of outputs. Hence, rather than represent the transformation process in full detail, it may be adequate to view the inner working as a black box and simply express the sequence of activities by a single functional relationship, one for each transformation rule. Applications of this approach are intricate multistage industrial processes, like an oil refinery or a sawmill, where the model allows the transformation process to choose among several alternative yield tables to convert the inputs – crude oil or sawn logs – into finished products. Subsystems which receive inputs from and provide outputs to other components of the system are often modelled in this way.

Black–Scholes call option pricing model

In 1973 Fischer Black and Myron Scholes published a paper in which they derived a formula, using stochastic **differential equations**, which can be used to determine the value of a *European call option*. An option is the right to buy or sell another asset (such as a share) at a given price within a given period of time. A 'call' option is the right to buy something, frequently a fixed number of shares in a company, and 'European' indicates an option that can only be exercised at the end of the time period. The formula can be written as:

$$C = S\,N(d_1) - Ke^{-rt}\,N(d_2),$$

$$\text{where} \quad d_1 = (\ln(S/K) + (r + \sigma^2/2)t)/\sigma\sqrt{t}, \quad d_2 = d_1 - \sigma\sqrt{t}$$

Here C = theoretical call premium, S = current stock price, t = time until option expires, K = option strike price, r = risk-free interest rate, σ = standard deviation of stock returns (volatility), and $N(\ )$ is the cumulative normal distribution function.

An interpretation of the formula is that the first term represents the benefit that one would expect to gain from actually purchasing the stock, while the second term is the present value associated with paying the exercise price when the option expires (hence the continuous discounting factor e^{-rt}). The normal distributions account for the uncertainty about the final price, when compared with a risk-free investment (usually taken to be government bonds).

The model as stated requires a number of fairly restrictive assumptions: no dividends, the ability to estimate the volatility, European options rather than, for example, American options which can be exercised at any time, and a lognormal distribution for the forward return on the asset. Several extensions and modifications to the original model have been proposed, which allow many of these assumptions to be relaxed, although often at the expense of simplicity and ease of computation. However, the basic model has remained extremely popular and successful, as reflected in the Nobel Prize awarded to Scholes in 1997. It is logically simple, computationally efficient and very **robust**.

D.C. McNickle

References

Black, F. and Scholes, M. (1973) 'The pricing of options and corporate liabilities', *Journal of Political Economy*, 81:637–659.

Das, S. (ed.) (1998) *Risk Management and Financial Derivatives: A Guide to the Mathematics*, New York: McGraw-Hill.

Bootstrapping, see judgemental bootstrapping, statistical bootstrapping

Boundary

Boundary is one of the most important **systems concepts**, although its precise meaning differs from one systems theory to another. Below, two different and highly influential understandings of boundary are discussed. It is important to be aware that these two uses of the same word are quite distinct from one another and point to the existence of two different MS/OR paradigms (or sets of assumptions).

In **general systems theory**, and in the theories of **open systems** and **living systems**, a boundary divides the system from its environment (e.g. to take the case of a human being, the skin is its boundary). All of these theories see systems as actually existing in the world, rather than being metaphors or theoretical constructs through which reality is interpreted. In an open system, or a living system, the boundary is permeable: it allows inputs into the system and outputs to leave it. This view of boundaries has been embodied in several MS/OR methodologies – most notably socio-technical **systems thinking**, where organizations are seen as open systems adapting to, and interacting with, their environments.

The second definition of boundaries is that they are social or personal constructs that define the limits of the knowledge that is to be taken as pertinent in an analysis. All analyses are inevitably partial – they assume boundaries – and where these are set will depend on the value judgements made by the analyst and/or those he or she consults. This is the case whether or not the analyst actively considers what the boundaries should be. Hence, making boundary judgements explicit and subjecting them to critique widens the possibilities for choice about what to include or exclude from analysis.

When setting boundaries, the analyst and other participants may choose a conventional system boundary, i.e. one that demarcates an organization or other social system from its environment. However, other boundary judgements are also possible, e.g. a boundary demarcating a set of relatively diverse people or things that a person or group currently sees as non-interacting, but which they believe ought to interact to bring about a social change.

Three authors who have worked on the idea of making boundary judgements in MS/OR are C. West Churchman, Werner Ulrich and Gerald Midgley. Churchman's primary focus is on the analyst's personal process of reflection concerning the right boundary to choose. Ulrich stresses the rational justification of value and boundary judgements, and argues for dialogue (where possible) between those involved in, and affected by, an MS/OR project so that they can jointly determine the appropriate boundaries. Midgley discusses what happens when some **stakeholders** and issues become marginalized (or devalued) in analyses, and how MS/OR projects can take account of processes of marginalization.

Ulrich and Midgley both use the term **boundary critique** to refer to the process of considering different possible boundaries. Boundary critique is central to the understanding of critical awareness in both **critical systems thinking** and **critical systems heuristics**. *G. Midgley*

References

Churchman, C.W. (1970) 'Operations research as a profession', *Management Science*, 17:B37–53.

Churchman, C.W. (1979) *The Systems Approach and its Enemies*, New York: Basic Books.

Midgley, G. (2000) *Systemic Intervention: Philosophy, Methodology, and Practice*, New York: Kluwer/Plenum.

Midgley, G., Munlo, I. and Brown, M. (1998) 'The theory and practice of boundary critique: Developing housing services for older people', *JORS*, 49:467–478.

Ulrich, W. (1983) *Critical Heuristics of Social Planning: A New Approach to Practical Philosophy*, Berne: Haupt. Reprinted 1994, Chichester: Wiley.

Ulrich, W. (1996) *A Primer to Critical Systems Heuristics for Action Researchers*, Hull: Centre for Systems Studies.

Boundary critique

Boundary critique is the methodological core idea of **critical systems heuristics** (CSH) (Ulrich, 1983). Increasingly, it is also recognized as a central concept of **critical systems thinking** and of critical professional practice in general. In the terms of CSH, the idea is that both the meaning and the validity of professional propositions always depend on *boundary judgements* as to what facts (observations) and norms (valuation standards) are to be considered relevant and what others are to be left out or considered less important. Such boundary judgements are constitutive of the reference systems to which refer all our claims to knowledge or rationality, in professional practice as well as in everyday life.

Systems thinking – the effort to consider the 'whole relevant system' (Churchman, 1970) – cannot alter the fact that all our claims remain 'partial' (Ulrich, 1983), in the double sense of being selective with respect to relevant facts and norms and of benefiting some parties more than others. This is what boundary critique (Ulrich, 1996, 2000; Midgley *et al.*, 1998) is all about; it aims at disclosing this inevitable partiality. A systematic process of boundary critique needs, first, to identify the sources of selectivity, by surfacing the underpinning boundary judgements. Second, it needs to question these boundary judgements with respect to their practical and ethical implications and to surface options, through discussions with all concerned **stakeholders** (note that their selection in turn represents a boundary judgement in need of critique). As a third and last consequence, based on these two critical efforts it may then become necessary to challenge unqualified claims to knowledge or rationality by compelling argumentation, through the emancipatory use of boundary critique. CSH offers a conceptual framework for all three tasks.

Basic to the entire process is grasping the ways in which a specific claim is conditioned by boundary judgements. CSH explains this by means of the 'eternal triangle' of *reference system*, *facts* and *values*, as depicted in the figure below: Whenever we propose a problem definition or solution, we cannot help but assert the relevance of some facts and norms as distinguished from others. Which facts and norms we should consider depends on how we bound the reference system, and vice versa; as soon as we modify our boundary judgements, relevant facts and norms are likely to change, too.

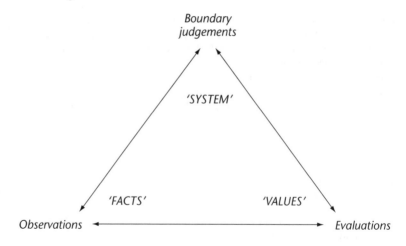

Figure: **The 'eternal triangle' of boundary judgements, observations, and evaluations** (*Source*: Ulrich, 2000, p. 252)

Thinking through the triangle means to consider each of its corners in the light of the other two. For example, what new facts become relevant if we expand the boundaries of the reference system or modify our value judgements? How do our valuations look if we consider new facts that refer to a modified reference system? In what way may our reference system fail to do justice to the perspective of different stakeholder groups? Any claim that does not reflect on the underpinning 'triangle' of boundary judgements, judgements of facts, and value judgements, risks claiming too much, by not disclosing its built-in selectivity.

Once the selectivity of the reference system in question has thus been grasped in terms of underpinning boundary judgements, systematic boundary critique then means exploring its implications for all the parties concerned, regardless of whether or not their concerns have been included in the underpinning reference system. CSH conceives of this larger context as the *context of application* of a professional proposition, as opposed to the *primary system of concern*. The context of application considers all the effects that a professional claim may impose on third parties, including stakeholders whose concerns are not represented by the primary system of concern. Both the primary system of concern and the context of application can be examined systematically by means of CSH's boundary questions (see also **critical systems heuristics**). *W. Ulrich*

References

Churchman, C.W. (1970) 'Operations research as a profession', *Management Science*, 17(2):B37–53.

Midgley, G., Munlo, I. and Brown, M. (1998) 'The theory and practice of boundary critique: developing housing services for older people', *JORS*, 49(5):467–478.

Ulrich, W. (1983) *Critical Heuristics of Social Planning: A New Approach to Practical Philosophy*, Bern: Haupt. Reprint edition 1994, Chichester: Wiley.

Ulrich, W. (1996) *A Primer to Critical Systems Heuristics for Action Researchers*, Hull: Centre for Systems Studies, University of Hull.

Ulrich, W. (2000) 'Reflective practice in the civil society', *Reflective Practice*, 1(2):247–268.

Box–Jenkins modelling

The Box–Jenkins approach to modelling **ARIMA processes** was described in a highly influential book by statisticians George Box and Gwilym Jenkins in 1970. An ARIMA process is a mathematical model used for **forecasting**. Box–Jenkins modelling involves identifying an appropriate ARIMA process, fitting it to the data, and then using the fitted model for forecasting. One of the attractive features of the Box–Jenkins approach to forecasting is that ARIMA processes are a very rich class of possible models and it is usually possible to find a process which provides an adequate description to the data.

The original Box–Jenkins modelling procedure involved an iterative three-stage process of *model selection, parameter estimation* and *model checking*. Recent explanations of the process (e.g. Makridakis, Wheelwright and Hyndman, 1998) often add a preliminary stage of *data preparation* and a final stage of *model application* (or forecasting).

1. Data preparation involves transformations and differencing. Transformations of the data (such as square roots or logarithms) can help stabilize the variance in a series where the variation changes with the level. This often happens with business and economic data. Then the data are differenced until there are no obvious patterns such as trend or seasonality left in the data. Differencing means taking the difference between consecutive observations, or between observations a year apart. The differenced data are often easier to model than the original data.
2. Model selection in the Box–Jenkins framework uses various graphs based on the transformed and differenced data to try to identify potential ARIMA processes which might provide a good fit to the data. Later developments have led to other model selection tools such as *Akaike's Information Criterion*.
3. Parameter estimation means finding the values of the model coefficients which give the best fit to the data. There are sophisticated computational algorithms designed to do this.
4. Model checking involves testing the assumptions of the model to identify any areas where the model is inadequate. If the model is found to be inadequate, it is necessary to go back to Step 2 and try to identify a better model.
5. Model application of forecasting is what the whole procedure is designed to accomplish. Once the model has been selected, estimated and checked, it is usually a straightforward task to compute forecasts. Of course, this is done by computer.

Although originally designed for modelling time series with ARIMA processes, the underlying strategy of Box and Jenkins is applicable to a wide variety of statistical modelling situations. It provides a convenient framework which allows an analyst to think about the data, and to find an appropriate statistical model which can be used to help answer relevant questions about the data. *R. Hyndman*

References

Box, G.E.P. and Jenkins, G.M. (1970) *Time Series Analysis: Forecasting and control*, San Francisco: Holden-Day.

Makridakis, S., Wheelwright, S.C. and Hyndman, R.J. (1998) *Forecasting: Methods and applications*, New York: Wiley.

Pankratz, A. (1983) *Forecasting with Univariate Box-Jenkins Models: Concepts and cases*, New York: Wiley.

Branch-and-bound methods

In most real-world optimization problems, the number of **feasible solutions** is finite. It seems natural to use enumeration for finding an optimal solution. Unfortunately, this finite number can be, and usually is, very large and in practice complete enumeration is impossible. Therefore, it is imperative that any enumeration procedure, if used at all, be cleverly structured so that only a tiny fraction of the feasible solutions actually need to be examined. Branch-and-bound method is one of such approaches.

The basic concept underlying branch-and-bound techniques is to 'divide and conquer'. Since the original problem is too large to be solved directly, it is divided sequentially into smaller and smaller subproblems, called *branches*, until all these subproblems can be conquered, i.e. ruled out for further consideration. This is referred to as *fathoming*. It is this fathoming that avoids complete enumeration.

The general procedure for branch-and-bound algorithms with a maximizing objective is as follows. We start with an initial feasible solution – possibly chosen arbitrarily – as the first *incumbent*, i.e. a feasible solution of the highest objective function value found so far. The problem is then divided into two subproblems or branches for fathoming. For any unfathomed branch, we determine whether it has a feasible solution to the original problem. If not, that branch is ruled out from further consideration, i.e. is fathomed. If the branch has at least one feasible solution, we obtain (relatively easily) an *upper bound* on the best possible objective function value for all feasible solutions in that branch. If it is not better than the current incumbent that branch is fathomed. If it is better and a feasible solution is found with an objective function value equal to the upper bound of the branch, then that feasible solution replaces the current incumbent, whose branch is now fathomed in turn. If no feasible solution with an objective function value equal to the upper bound of the branch is found, that branch is divided into two new branches. The procedure of branching and fathoming continues until all branches are fathomed. The incumbent at that point is the optimal solution to the original problem.

Integer and mixed **integer programming** problems are solved by branch-and-bound algorithms. Branch subproblems are relaxed to regular linear programs by removing the integer restrictions on the variables. New branches are introduced whenever the optimal solution to that branch has fractional variable values by adding constraints that force one of the fractional variables to the next higher or next lower integer value. Upper bounds on branches are based on the optimal linear programming solution to that branch. If a branch produces an optimal solution that is integer, it can be fathomed. *B. Chen*

References

Hillier, F.S. and Lieberman, G.J. (2001) *Introduction to Operations Research*, 7th edn, New York: McGraw-Hill, ch. 13.

Papadimitriou, C.H. and Steiglitz, K. (1982) *Combinatorial Optimization: Algorithms and Complexity*, Englewood Cliffs, NJ: Prentice-Hall, ch. 18.

Brand switching, see Markov chains

Break-even analysis

Break-even analysis determines the output level $Q = B$, called the *break-even point*, of a production or service activity for which the total cost of the output $C(Q)$ is just covered by the total revenue $R(Q)$ generated by the activity, i.e. at the output level $Q = B$, $C(B) = R(B)$. For any $Q < B$, total cost $C(Q) >$ total revenue $R(Q)$, i.e. the activity results in a loss, while for any $Q > B$, $C(Q) < R(Q)$, i.e. the activity starts producing a profit.

If total revenue is proportional to the output level, i.e. $R(Q) = pQ$, where p is the unit revenue (or net selling price), and the total cost is the sum of a **fixed cost** F for any positive output Q and a total **variable cost** that increases proportionally with the output level, i.e. $C(Q) = F + vQ$, where v is the unit cost, the break-even point occurs at the level B, where $F + vB = pB$. Solving this expression for B yields the simple formula:

$$\text{Break-even point} = B = \frac{F}{p - v} \text{ for all } p > v.$$

For $p \leq v$, no break-even point exists. Cost and revenue functions that are nonlinear or consist of several segments may have to be solved numerically.

Consider the following example: Assume that the fixed costs (administration, editing, cover design, index production, type-setting, plates, advertising, etc.) associated with producing this book amount to $F = £15,372$, while the variable production and distribution cost per unit, including royalties, amounts to $v = £17.25$. (Note though that some components of the unit variable cost are likely to decrease with Q.) If the book is sold for a net price of $p = £26.40$, then the number of copies that must be sold to break even is

$$\text{Break-even point} = B = \frac{15{,}372}{26.40 - 17.25} = 1\ 680 \text{ copies}$$

For this cost and revenue structure, at least 1 680 copies must be sold before this book will begin to generate a profit for the publisher. If fewer copies are sold, the publisher will make a loss. Not all fixed costs will then be recovered.

Most uses of break-even analysis assume linear cost and revenue functions. Furthermore, the method gives no measure of the risk involved since it assumes that costs and net unit revenue are known with certainty, whereas in reality either or both are random variables. It is therefore a somewhat crude tool.   H.G. Daellenbach

Reference

Daellenbach, H.G. and McNickle, D.C. (2001) *Systems Thinking and OR/MS Methods*, Christchurch: REA, pp. 17–20.

Business process re-engineering

Business process re-engineering (BPR) came to prominence in 1993 through Hammer and Champy's book *Reengineering the Corporation: A manifesto for business revolution*. Re-engineering is defined as:

'the fundamental rethinking and radical redesign of business processes to achieve dramatic improvements in critical, contemporary measures of performance, such as cost, quality, service and speed.' (Hammer and Champy, 1993, p. 32)

BPR was presented as counterbalancing the incremental improvement approach of **total quality management** (TQM) through its emphasis on demolishing what exists and building something new. Re-engineering

- first determines what a company must do, then how to do it;
- ignores what is and concentrates on what should be;
- is about business reinvention and achieving quantum leaps in performance;
- focuses on the broad process, not the component tasks;
- is led by a senior level management champion; and
- is ambitious, rule-breaking and makes creative use of information technology.

Re-engineering uses many of the basic tools of TQM, but applies them in a different context. While TQM asks initially 'How can processes be done better?' BPR starts out with 'What processes should be done?' BPR is sometimes described as a 'strategy of last resort' to counteract extended periods of complacency and neglect where organizational processes have not been subject to hard scrutiny. Re-engineering is also distinguished by the importance placed on the use of information technology to enable the achievement of new goals. Information technology is promoted as providing creative solutions to problems that may not yet be known to exist. As with TQM, BPR is a team-based approach to organizational problem solving, but by contrast, it is a top-down strategy actively driven by senior management, rather than delegated to an empowered workforce.

Re-engineering was adopted as the next solution to the problem of low competitiveness of American companies in the global marketplace. By the mid-1990s, Champy was writing that 'Reengineering is in trouble'. The problem was not with the re-engineering process, but with management. Management attitudes and practices had not been re-engineered: BPR was seen as the solution to organizational problems rather than as a technical part of the solution. It is noteworthy that the concerns being aired about BPR echo those being expressed in the mid-1980s about **quality control circles**. The underlying message is remarkably similar to the messages of Imai, Juran and Deming: process improvement cannot reach its full potential without change in management values, attitudes and behaviours.

From the mid-1990s on, debate has focused on the potential for complementary use of BPR and TQM/Continuous Quality Improvement (CQI). BPR, to achieve major breakthroughs, should be followed by TQM/CQI to maintain and improve on the gains from re-engineering. In this way BPR and TQM can both contribute to the development of an organizational culture of ongoing improvement.   *D.J. Houston*

References

Champy, J. (1995) *Re-engineering Management: The mandate for new leadership*, New York: Harper Collins.

Hammer, M. and Champy, J. (1993) *Re-engineering the Corporation: A manifesto for business revolution*, New York: Harper Collins.

C

CAD/CAM, computer-aided design/computer-aided manufacturing

The design process for new products and components traditionally has been an iterative one in which the product specifications are refined in successive stages based upon the designer's experience, computations, sketches, and drawings. Significant advances in computing technology over the last few decades has enabled the present-day designers to use a tool called *computer-aided design* (CAD) which uses computational and graphics software that enhances the design productivity substantially. The geometry of the components can be graphically displayed and manipulated easily on video monitors. Extensive databases are developed that allow an engineer to try different designs and test their performance, thereby eliminating some of the time and expense of physical mock-ups, models and prototypes. Sometimes, an already existing design in the database may be found that is suitable for a new product, eliminating the duplication of effort. The information stored in the CAD database can be directly transmitted to manufacturing and other departments.

Computer-aided manufacturing (CAM) is a broad term that encompasses the direct use of computers in process control. CAM systems control the machine tools on the shop floor. The machines typically perform a variety of operations; the machine receives its instructions from a computer about the sequence and specifications of its operations. The computer programs can be stored in the manufacturing database, retrieved, updated and revised as components are added or redesigned, and transmitted electronically within a factory location or to different factory locations across the world. The benefits to production from CAM are many: more reliable instructions to machines for operations, more consistent product quality, closer tolerances for the finished parts, and lower labour costs. However, CAM systems are very expensive to install.

CAD/CAM, or CADICAM, is the integration of CAD and CAM. The ultimate goal is a system that will take the engineer's design (developed using CAD) directly to the manufacturing stage (manufactured using CAM). In effect, a single push of a button would make a picture become a real object. A large number of organizations use CAD/CAM technology for product design and manufacturing. Despite a high installation cost these systems repay the initial investment in less than three years in many cases.

While organizations enjoy great benefits by having CADICAM, too much CAM can be costly. Errors in a computer program can result in a huge number of erroneous parts being produced, albeit efficiently. However, with proper human intervention such errors can be minimized. Therefore, human involvement is not eliminated by CAD/CAM; it is being deployed in new ways.   *R. Sridharan*

References

Adam, E.E. and Ebert, R.J. (1996) *Production and Operations Management,* 5th edn, Englewood Cliffs: Prentice-Hall.

McClain, J.O., Thomas, L.J. and Mazzola, J.B. (1992) *Operations Management,* 3rd edn, Englewood Cliffs: Prentice-Hall.

Call options, see Black–Scholes call option pricing

Capacity planning

Capacity is built to satisfy demand for a product or a service. Its traditional definition is the maximum output rate of an operation. It is always measured in units of output per time period. For example, a soft drink bottling firm may build a facility to fill 20 000 bottles per day, and each teller in a bank may, on average, be able to serve 150 customers per day.

Capacity usually cannot be adjusted continuously, but comes in discrete chunks of adding or withdrawal of individual pieces of equipment, each with its own fixed capacity and cost. The demand for a product (or service) is a significant factor influencing the best amount of capacity to acquire. It is a matter of balancing the cost of excess capacity and capacity shortages. Since future demand is uncertain both are **random variables**. A major concern in capacity planning is the ability to cope with such variability. Firms use several creative approaches to improve capacity planning, such as increase **forecasting** accuracy, influence the level of future demand, and offer incentives to customers that will reduce fluctuations in demand, e.g. by offering price discounts for seasonal products during low demand periods or by finding new products to make use of seasonal excess capacity, and last but not least the traditional use of inventories as buffers to smooth out production levels.

Long-term capacity planning typically covers two to ten or more years and involves considerable investments. For instance, in the power industry it could be in excess of ten, whereas for supermarkets it may be as short as two. It involves changes in facilities and equipment, as well as technology, taking into account the long-term demands for the products and the current phase of their life cycle(s). Medium-term capacity planning extends from three months to two years. Its concerns are planning overall activity levels and its consequences on securing resources, such as raw material sources, appropriate permanent and temporary employee levels, outsourcing and subcontracting needs, and overall inventory levels, as well as routine replacement, overhaul and maintenance of equipment. Short-term capacity typically covers the coming three months. Short-term capacity can only be varied within narrow ranges, with overtime being the major means for temporarily increasing capacity. The needs come mainly from existing commitment, such as customer orders on the books and short-term forecasts. Hence, short-term capacity planning is concerned with detailed day-to-day work scheduling, including overtime, and procurement and management of raw materials and supplies (see **logistics, supply chain management**).

MS/OR approaches are used for all forms of capacity planning. Long-term capacity planning situations can be modelled and analysed using **integer programming, risk analysis**, and **scenario planning** via simulation. Medium-term capacity planning situations can be modelled using **linear programming, dynamic programming, simulation**, stochastic **inventory models** and **line balancing**. Short-term planning has the largest degree of complexity. It involves detailed **scheduling** and **sequencing** of activities or jobs to make the most of limited daily capacity. These are notoriously the most difficult problems for which good optimization techniques are actively explored.

R. Sridharan

References

Schmenner, R.W. (1993) *Production/Operations Management*, 5th edn, New York: Macmillan.
Shroeder, R. (1999) *Operations Management: Contemporary concepts and cases*, Boston: McGraw-Hill.

Capital asset pricing model or CAPM

The capital asset pricing model is a generalization of the Markowitz **portfolio selection** model to capital assets pricing (i.e. marketable securities) for the market as a whole. It earned W.F. Sharpe, together with H.M. Markowitz and M.H. Miller, the 1990 Nobel Prize for Economics. It starts out from the premise that through *diversification* the investor will only invest in portfolios that lie on the **efficient frontier** of all possible portfolios. Efficient portfolios diversify away all unsystematic risk, leaving only the market risk arising from market-wide economic factors (such as changes in interest rates). Furthermore, by also investing in a riskless security (i.e. treasury bills) or borrowing, the investor is able to maximize her or his **utility**. This is depicted in the figure below. The straight line from R_f through M is known as the *capital market line* (CML). M is the *market portfolio*, where all securities are represented in proportion to their market value. The riskless security has a return of R_f and the market portfolio of R_m and a risk of σ_m (measured by its standard deviation). A *risk-averse* investor with a utility function parallel to U_1 will select a mixture P of the riskless security and portfolio M, while a *risk-seeking* investor with a utility function parallel to U_2 will select portfolio Q, which consists of borrowing at the riskless rate R_f and investing the personal wealth plus the amount borrowed in M.

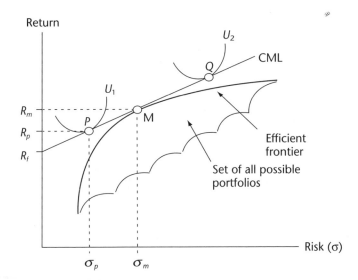

Figure: **Capital market line**

The CML has a slope equal to the ratio $[R_m - R_f]/\sigma_m$. It measures the additional return in excess of the risk-free rate that the investor expects per unit increase in risk. The expected return on portfolio P can thus be expressed as

$$E(R_p) = \text{riskfree return} + \text{risk premium} = R_f + \sigma_p \frac{R_m - R_f}{\sigma_m}$$

where σ_p is the standard deviation of P. However, since efficient portfolios diversify away any unsystematic risk, only the systematic risk is left, which for any security i is measured by the so-called *beta coefficient, β_i*:

$$\beta_i = \frac{\text{Cov}(R_i, R_m)}{\sigma_m^2} = \frac{\rho_{i,m}\sigma_i\sigma_m}{\sigma_m^2} = \frac{\rho_{i,m}\sigma_i}{\sigma_m}$$

$\text{COV}(R_i, R_m)$ is the covariance between portfolio i and the market portfolio, σ_i its standard deviation, and $\rho_{i,m}$ the correlation coefficient between the two. By definition, the market beta coefficient, $\beta_m = 1$. A security with a beta coefficient of less than 1 is called a *defensive security*. It experiences lower than average returns in a rising market, but also smaller than average losses in a declining market. On the other hand, a security with a beta coefficient larger than 1 is called an *aggressive security*. It shows above average gains in a rising market, but also greater than average losses in a declining market. The beta coefficient of a portfolio of securities is simply equal to the weighted average of the individual betas.

The expected return on security i can now be expressed as a function of its beta coefficient as follows:

$$E(R_i) = R_f + [R_m - R_f]\beta_i$$

This is referred to as the *capital asset pricing model*. It predicts a linear relationship between the beta coefficient and the return of a security. Its line is called the *security market line*. Beta coefficients are regularly used by investment analysts.

A. Gunasekarage

Catastrophe theory

In the late 1980s, late 1990s and 2000/01, stock markets all over the world experienced sudden and large price drops, followed by periods of consolidation and then gradual price increases. One possible explanation for such behaviour is that one or more important underlying aspects or variables reach critical levels that trigger such disruptions. The mathematical study of such abrupt changes has given rise to *bifurcation theory* and *catastrophe theory*.

The classical example of such behaviour is what happens when some structure, such as a bridge or platform, is gradually subjected to greater and greater loads. Initially there will be an almost imperceptible change in the shape of the structure, but when the load reaches some critical value, the structure collapses abruptly. Although such phenomena have been recognized for millennia, as indicated by the proverb 'the straw that broke the camel's back', it is only since the 1960s and 1970s that mathematicians have been able to model this behaviour with mathematics.

As mentioned above, stock market crashes could conceivably also be subject to such behaviour. For instance, a continued deterioration of gross national product below a critical point could suddenly cause share investors to divest their holdings in favour of safer investments, such as bonds, causing a dramatic drop in share prices.

Although catastrophe theory offers a coherent mathematical framework for analysing such discontinuous behaviour, its power is mainly descriptive and its ability to make quantitative predictions doubtful. It has been used to explain behaviour in biology, population dynamics and sociology, as well as physics and engineering, but no real successes have been recorded in the field of economics.

A.D. Tsoularis

Reference
Williams, G. (1997) *Chaos Theory Tamed*, Washington: Joseph Henry Press. Uses little mathematics; excellent source of information, with glossary.

Causal loop diagrams

Causal loop diagrams depict cause-and-effect relationships between various aspects, entities or variables. If item X affects item Y, this causes one or more **attributes** or properties of item Y to change, such as its numeric value or its status. This is shown by connecting the two with an arrow. A change in Y may in turn become the cause of a change in Z, and so on, resulting in a *cause-and-effect chain*.

A positive sign (or the letter 's', standing for 'same direction') attached to the arrow head means that a strengthening of item X causes a strengthening of Y, while a negative sign (or the letter 'o' for 'opposite') indicates that an increase in X results in a decrease in Y. An arrow from an item farther down in the cause-and-effect chain to an earlier item is a **feedback loop**. If the number of negative signs attached to the arrows on the entire loop is odd, the loop is negative, i.e. positive and negative effects fully or partially cancel each other out, thereby dampening the cumulative effect of the causes. An even number of negative signs indicates positive feedback. The cumulative effects get reinforced along the chain, causing the system either to explode or collapse. Such destabilizing behaviour is usually undesirable.

The causal loop diagram below depicts a production/inventory system. 'Orders received' from customers are added to the 'order book' (the backlog of orders received but not yet shipped). 'Orders received' affects the 'order pattern observed' which in turn influences the 'sales forecast'. Increases in the 'order book' or the 'sales forecast' both cause an increase in 'production'. However, the response in 'production' is delayed by the **lead time** it takes to initiate an increase and the time to produce the finished goods, labelled 'production lead time'. A rise in 'finished goods stocks' has a dampening effect on 'production'. Obviously, 'production' increases 'finished goods stocks'.

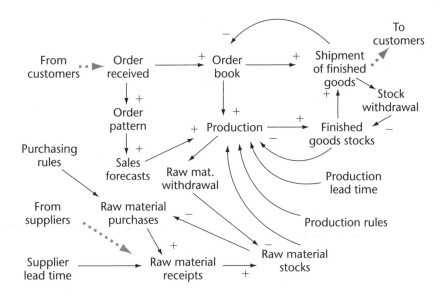

Figure: **Causal loop diagram for production/inventory system**

Note that some arrows have no signs attached. They are either constant inputs, such as the current production lead time, or they constrain an item or variable. For example, 'production' can only happen if sufficient 'raw material stocks' are available. The diagram contains two simple and one more complex feedback loops, all of them negative, as desired.

Causal loop diagrams show the transformation process in a **system** and highlight complexity in **systemic** relationships between systems variables, and in particular the presence of feedback loops and their nature. They are also the starting point of a **system dynamics** analysis and other simulation modelling.

However, they can also be used to depict qualitative cause-and-effect relationships and are then sometimes referred to as *cause-and-effect diagrams.*   *H.G. Daellenbach*

References

Maani, K.E. and Cavana, R.Y. (2000) *Systems Thinking and Modeling – Understanding Change and Complexity*, Auckland: Pearson Education NZ.

Martin, J. (1984) *Block I Introduction Workbook*, Technology course T301: Complexity, Management and Change: Applying a Systems Approach, Buckingham: The Open University Press.

Cause-and-effect thinking, reductionist thinking

Russell Ackoff (1973) – philosopher, operations researcher and systems thinker – said that the intellectual foundations of the traditional scientific model of thought are based on two major ideas. The first is *reductionism*: the belief that everything in the world and every experience of it can be reduced, decomposed or disassembled into ultimately simple indivisible parts. Explaining the behaviour of these parts and then aggregating these partial explanations is assumed to be sufficient to allow us to understand and explain the behaviour of the whole. So disassembling a clock into its components and explaining the function of each should be all that is needed to explain that the clock measures time. Wrong, all it can explain is how the two arms sweep around in a circle. The property of measuring time goes beyond the explanation of the parts. It is an **emergent property**, something that none of its parts or subset of parts have by themselves; something that only the whole has.

Applying reductionism to problem solving translates into breaking a problem into a set of simpler subproblems or **subsystems**, solving each of these individually and then assembling their solutions into an overall solution for the whole. However, it is obvious that even if each is operated at its highest **efficiency**, the sum of the individual solutions does not necessarily produce an overall solution that is best for the system as a whole. The individual solutions ignore the interactions, interdependence and synergies between the subsystems. What is best for one subsystem may cause a serious loss of efficiency in another.

The second basic idea is that all phenomena are explainable by (usually) unidirectional cause-and-effect relationships. A thing X is taken to be the cause of Y, if X is both necessary and sufficient for Y to happen. Hence, 'cause X' is all that is needed to explain 'effect Y'. The causes could be the means to achieve some end – the effect of the cause. The end itself becomes the means to achieve some other end, creating a so-called *means–ends chain*. Many human activities and decision processes are based on this idea.

Causal relationships may not be just one-way. There could be mutual causality between two things, i.e. X affects Y, but is in turn affected by Y. The two are interdependent, giving rise to **feedback loops**. Dealing with one alone, while ignoring the other, may not achieve the desired results. Unfortunately, linear cause-and-effect thinking all too often ignores this. For example, most public policy deals separately with poverty and with health. But poverty tends to cause poor health, which in turn causes poverty. Dealing with both simultaneously, rather than just with each individually, is likely to be much more **effective** in improving both.

This is, though, not to say that reductionism and cause-and-effect thinking should be discarded. It only means that they must be complemented with a systemic mode of thinking, i.e. **systems thinking**. *H.G. Daellenbach*

References

Ackoff, R.L. (1973) 'Science in the Systems Age: Beyond IE, OR, and MS', *Operations Research*, (May–June): 661–671.

Jackson, M. (2000) *Systems Approaches to Management*, New York: Kluwer/Plenum.

Central limit theorem

Many projects in MS/OR or statistics involve **random** phenomena that consist of a large number of individual independent events. For example, a milk-processing plant collects milk from 126 dairy farms. The daily milk production of each farm is a random variable that depends on the number of cows. The daily total milk collected is the sum of these 126 random variables. For planning purposes, the manager needs to know the form of the probability distribution of this total.

If the individual components all follow a certain known theoretical distribution, the distribution of the sum can be found analytically. For instance, the sum of *Poisson* variables is also Poisson. More often, however, no simple analytic solution exists. It is precisely for this reason that the central limit theorem and its many versions are so vitally important. In its simplest form the central limit theorem states that the distribution of the sum of n independent and identically distributed random variables with means of μ and a variance of σ^2 approaches a normal distribution with a mean of $n\mu$ and a variance of $n\sigma^2$, as n gets larger and larger, regardless of the shape of the distribution of the individual variables. The theorem does not explicitly state how large the number of components must be for the approximation to be sufficiently good. If the individual random variables have a distribution that is fairly symmetric, then an n as small as 15 to 20 may be enough. For highly skewed distributions, 30 to 50 or even more may be needed.

Other forms of the central limit theorem state that the average of n random variables also approaches a normal distribution as n increases, with a mean of μ/n and a variance of σ^2/n, regardless of the population distribution. This property is extensively used in statistical inference and hypothesis testing.

The results of the central limit theorem are also applied to the case where the individual random variables have different distributions. If n is sufficiently large, the normal distribution may still serve as a good approximation with a mean equal to the sum of the means and a variance equal to the sum of the variances. The approximation will be worst at the far tails. Obviously, the more similar the individual distributions are in terms of mean and variance, the smaller n can be.

Although the normal distribution assumes a continuous random variable, it is also used as an approximation for discrete variables, if its mean and variances are relatively large, i.e. in the hundreds or thousands. *A.D. Tsoularis*

Reference

Levine, D.M., Krehbiel, T.C. and Berenson, M.L. (2000) *Business Statistics: A First Course*, Englewood Cliffs, NJ: Prentice Hall. Good introductory text with business examples.

Certainty equivalent, see utility functions

Chance-constrained programming

Chance-constrained programming is a form of decision making under uncertainty. Problems are formulated as **linear programs** with the addition of constraints that are allowed to be violated with a certain probability. These constraints involve **random** parameters whose values are unknown at the time of decision and will only be revealed after the decision has been made. A decision is feasible if constraint i is satisfied for at least a certain fraction, $0 \leq \alpha_i \geq 1$, of the possible outcomes for all i. The α_i-values are set arbitrarily by the decision maker. Uncertainty may be present in the right-hand side (RHS) of the constraint (e.g. the amount of resource available) or in the left-hand-side coefficients (LHS) (e.g. the unit resource use for each activity). For each random parameter a probability distribution of possible values must be given.

Each constraint that must individually hold with a certain probability is called a *single-chance constraint*. A single-chance constraint might be that the water in Lake A on a given day has to be at least 16 000 million cubic metres with a probability of 95 per cent.

A set of constraints which must hold simultaneously with a certain probability are called *joint-chance constraints*. Joint-chance constraints might be that the water in Lake A is at least 16 000 million cubic metres and the water in Lake B is at least 15 000 million cubic metres simultaneously with a probability of 90 per cent.

A single-chance constraint with uncertainty solely in the RHS may simply be replaced by an equivalent **deterministic** constraint. The new RHS is found by inverting the probability distribution. When uncertainty is also present in the constraint's LHS coefficients, the constraint might still be converted to an equivalent deterministic constraint. For joint constraints the **feasible region** may not remain well behaved, i.e. it may be **non-convex**.

Solution methods for dealing with general chance constraints rely heavily on the type of distributions the random variables are drawn from. In light of this, care must be taken when modelling chance constraints. Given the rather simplistic manner in how uncertainty is modelled, chance-constrained programming has found few real-life practical applications.

The alternative to modelling chance constraints explicitly is to allow the constraints to be violated and to apply a penalty cost on the amount of violation (e.g. $100 for each cubic metre of water below the 16 000 million level). Increasing the penalty will tend to reduce the probability that the constraint is violated. **Sensitivity analysis** can be used to determine the appropriate level of the penalty.　　　*S. Dye*

Reference
Prékopa, A. (1995) *Stochastic Programming*, New York: Kluwer Academic Publishers.

Chaos theory

Chaos is the disorderly, but deterministic, i.e. not **random**, evolution or behaviour of a nonlinear, dynamic **system** over a long time. 'Deterministic' describes any system behaviour that can be completely specified (i.e. determined) by one or more nonlinear equations and a starting point or initial **state of the system**. Hence, chaos results from a deterministic process, but to the casual observer the behaviour may look random and unpredictable. Furthermore, chaos only happens in systems where the behaviour is described by nonlinear functions. Chaos theory is the body of mathematical knowledge used to analyse situations where chaos arises.

The easiest way to see how something changes over time is to graph it. The plot below shows the share price for a given company. Although the fluctuations look highly irregular, they are not necessarily chaotic. Only a detailed analysis of data will reveal whether the price fluctuations are random or, alternatively, chaotic. Roughly speaking, if the long-term prediction of the share price is impossible, the price follows a chaotic pattern. It could follow a very simple nonlinear equation, i.e. complex behaviour does not necessarily have a complex origin, or it could need a set of many polynomial and trigonometric equations in many variables to represent it.

Share price $

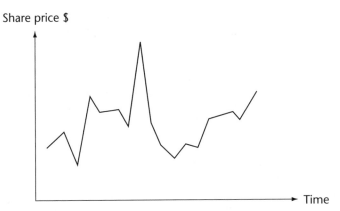

Time

Figure: **Share price fluctuations over time**

There is the temptation to label any haphazard-looking process as chaotic. At present, chaos turns up in computer **simulations**, but has not been conclusively found in real-world data.

It was inevitable that the concept of chaos and its ramifications would attract the attention of economic theorists, as the performance of the financial, commodity and foreign exchange markets is notoriously volatile and nearly impossible to predict. There has been an explosion in recent years of empirical work searching for possible explanations based on chaos theory for various types of economic and financial time series, but again no conclusive evidence in favour of chaos has been found.
A.D. Tsoularis

References
Gleick, J. (1987) *Chaos: making a new science*, New York: Penguin. This best-seller brought chaos into the limelight. Interesting examples and historical facts.
Williams, G. (1997) *Chaos Theory Tamed*, Washington: Joseph Henry Press. Uses little mathematics to explain complexity. Excellent source of information, with glossary.

Chinese postman problem, see delivery person problem

Closed systems, open systems

The father of **general systems theory**, Ludwig von Bertalanffy, introduced the concepts of closed and open systems. A closed system has no interactions with any environment. No inputs, no output (see **systems concepts**). In fact, it has no environment. In contrast, open systems interact with the environment, by receiving inputs from it and providing outputs to it.

In real life there exist no truly closed systems, except maybe the entire universe, and even that depends on our religious beliefs. Any real-life system has an environment with which it interacts, even if only in a small way. So, the concept of a closed system is of theoretical interest. With no interactions with an environment, its behaviour is regulated entirely by the interactions among the components of the system and its initial or starting conditions. These determine to the last detail how the system behaves. Hence, a closed system must be **deterministic**.

In contrast, open systems can be either deterministic or **stochastic**. In fact, all stochastic systems are also open systems since the factors that introduce the randomness in the behaviour are the result of forces or events not included inside the system, usually because their causes of randomness are not fully understood. A stochastic system may exhibit an interesting long-run behaviour, in the sense that it approaches a sort of equilibrium or **steady-state** that is independent of its starting state. If disturbed it will tend to again return to this equilibrium after a sufficient length of time. Bertalanffy named this property *equifinality*.

Scientists, particularly in the biological or physical sciences, may try to create artificially closed systems that are as far as physically possible insulated from their environment. Their only inputs are initial starting conditions. But the starting condition are control inputs, so these systems are not truly closed. By providing different initial **states of the system** the analyst can observe how the system behaviour responds to different initial conditions.

The literature on **systems thinking** sometimes refers to a system that has minimal exchange with its environment as a closed system. Similarly, it might say that the way a given scientific reasoning looks at a phenomenon implies it is treated as a closed system. So reductionism, a mode of analysis that divides things into smaller parts and then explains the functioning of the whole in terms of the explanations of the parts, is said to treat these parts as if they were closed systems.

Systems defined for decision-making purposes are always open systems, since by definition the decisions or the decision-making rules are inputs into the system.

H.G. Daellenbach

References

Daellenbach, H.G. (2001) *Systems Thinking and Decision Making*, Christchurch: REA, ch. 3. Basic introduction.

The Open University, Walton Hall, Milton Keynes, MK7 6AA, UK, has a number of course books on systems that are excellent reading for beginners.

Clustering, cluster analysis, or segmentation analysis

Describing or predicting salient characteristics of multivariate data (e.g. observations on luxury car buyers that measure three variables: income, level of savings, and gender) is usually easier for homogeneous subgroups of observation than for the data as a whole. Hence, faced with multivariate data, the analyst may first wish to group observations into sets, such that the members of each set have similar characteristics, but the sets are distinctly different from each other.

Cluster analysis is a family of techniques of multivariate data analysis for doing this. All techniques aim to identify the number of these groups and assign their members in a systematic way according to some predetermined criterion of similarity. In the luxury car buyer example, similarity could be measured by the closeness of two observations in three-dimensional space. All techniques follow a similar process. The individual N observations in the data are at first all treated as separate groups. The two observations that are closest to each other in terms of the chosen criterion for similarity are joined together, forming a group of two and $N-1$ groups of one observation. At the second round, a new individual either joins the group of two or forms a new group with another individual, the criterion again being closest similarity. This process of joining individuals and/or groups continues until all are in the same group, or until it becomes unreasonable to join the remaining observations or groups of observations. For example, the process may stop when a pattern matches an a priori model from theory or previous data, or when the last two clusters which have been joined are much further apart than the previous two. There are, though, no firm rules about what 'much further apart' should mean. The final choice of what the best cluster pattern is remains largely a matter of experience.

If measures of similarity between any two individuals are relatively simple, similarity between groups or between a group and an individual is less straightforward. For example, when trying to join an individual to a group, is similarity measured with respect to the closest individual in the group, with respect to the individual in the group farthest away, or with some measure for the centre of gravity of the group, or even a statistical measure, such as the variance of the members? Each choice will lead to the formation of different groupings. Which criterion is best or most effective to form homogenous groups differs from problem to problem and may have to be explored by trying out several. Clustering is not an exact science, but data analysis experience will help.

Cluster analysis has been widely used for consumer research, such as categorizing customers according to their purchasing habits, and is one of the analysis methods used in **data mining**.   *D.K. Smith*

References
Hair, J.F. jr, Anderson, R.E., Tatham, R.L. and Black, W.C. (1998) *Multivariate Data Analysis*, Upper Saddle River, NJ: Prentice-Hall, ch. 9.

Krzanowski, W.J. (1990) *Principles of Multivariate Analysis: A User's Perspective*, Buckingham: Open University Press, ch. 3.

SAS/STAT® User's Guide, Version 8 (1999), 3 volumes, Cary, NC: SAS Institute Inc.

Web site
SAS/STAT: http://www.sas.com

Cognitive mapping

A cognitive map captures a person's view about a situation, represented in the form of a network of statements, expressing ideas, means and ends, linked together by arrows indicating the direction of connections between statements or *cause-and-effect* relationships. Since the way something is said is important and may have connotations associated with it that go beyond the words, whenever possible the person's own words are used.

Consider the following short conversation between career counsellor C and a student S:

C: Tell me how you see your future.

S: I am torn between wanting to take a year off [1], travelling a bit [2], or go for a job right away, settle down [3], start earning some money [4].

C: Do you have large debts?

S: No, but it's my dad's wish that I should not wait too long before getting a job [5].

C: Have you had any job interviews yet?

S: Yes, but I waited a bit too long. All the good jobs seem to be gone already [6].

C: You know what kind of job you would like?

S: Yes, with one of those international consulting firms [7].

C: Why?

S: Varied interesting project work; travel abroad [8]. No dull factory office job for me [7]!

C: Travelling seems really important. Why?

S: I want to see the world, meet interesting people [9].

A skilled analyst will create the map for the student's views while he talks. The resulting map is shown below (the brackets identify S's verbal statements).

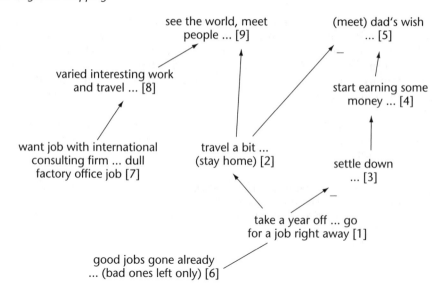

see the world, meet people ... [9]

(meet) dad's wish ... [5]

varied interesting work and travel ... [8]

start earning some money ... [4]

want job with international consulting firm ... dull factory office job [7]

travel a bit ... (stay home) [2]

settle down ... [3]

take a year off ... go for a job right away [1]

good jobs gone already ... (bad ones left only) [6]

Figure: **Cognitive map**

Cognitive maps consist of sets of concepts, called *constructs*. Each is shown with two poles – the idea expressed and its perceived opposite, either stated or inferred, separated by an ellipsis (...) and read as 'A rather than B'. The student's first answer gives rise to several constructs and are here numbered in the sequence they are made. Note that opposites are not logical opposites but perceived opposites. For instance, for construct (1), if the student does not take a job immediately, he will take a year off as opposed to other alternatives he could consider, such as getting a job in a few months. If the opposite pole is simply the negative, it is not shown. Arrows always connect first poles between constructs, unless a negative sign is attached to the arrow, which implies that a first pole connects to a second pole. Usually highest goals are shown on top, most detailed actions at the bottom. Although deceivingly simple-looking, they require considerable skill and experience to draw up.

Cognitive maps have similar purposes as **mind maps**, except that they more clearly reflect a person's thinking in his or her own words and perceived links. They reveal conflicts, ambiguities and contradictions that may need clarification and result in map changes. They help structure a **problem situation** and may suggest opportunities for new actions. In the workplace they have been used to develop a framework for detailed job descriptions and job training, activity planning, and long-term strategic planning.

Cognitive maps are the prime inputs into **strategic option development and analysis** or SODA, a **problem structuring method**. *J. Holt*

References

Eden, C. (1990) 'Using cognitive mapping for strategic options development and analysis (SODA)', in J. Rosenhead *Rational Analysis for a Problematic World*, Chichester: Wiley, pp. 21–42.

Eden, C. and Simpson, P. (1990) 'SODA and cognitive mapping in practice', in J. Rosenhead *Rational Analysis for a Problematic World*, Chichester: Wiley, pp. 43–70.

Eden, C., Jones, S. and Simms, D. (1983) *Messing About in Problems*, Oxford: Pergamon.

Web site

Banxia Decision Explorer Bibliography: http://www.banxia.com/dexplore/debiblio.html

Combinatorial optimization

Optimization problems seek a 'best' choice of decision variables (or a configuration, or a set of parameters) which will achieve some **objective**. Such problems can be conveniently divided into two categories: those whose decision variables are continuous (i.e. they can assume any number, as is the case in **linear programs**), and those whose decision variables are discrete (i.e. they can be chosen only from a large finite set, as is the case in **integer programs**). Optimization problems with discrete decision variables are often called *combinatorial optimization problems* (COPs), to indicate that distinct combinations of variables are being sought. More loosely, a COP is any optimization problem that has a finite (or countably infinite) number of **feasible solutions**.

There are many application areas in which COPs arise, often in connection with the management of scarce resources in order to increase **efficiency** and productivity. Some applications include: **supply chain management** and distribution, production **scheduling**, data analysis, capital budgeting, **facility location**, DNA specification, X-ray crystallography, **portfolio selection** and communication network design. Two of the most famous COPs can be described as follows.

1. The **travelling salesperson problem**: For a set of n locations, find a minimal length tour which starts at a given location, visits all other locations exactly once, and returns to the starting point.
2. The machine **scheduling** problem with due dates: Find the processing sequence for m jobs on a single machine, with given processing times and due dates of each job, such that the total tardiness (= 0 if job not late, completion time minus due date if job is late) of all jobs is minimized.

In order to 'solve' (1), each of the $n!$ possible tours needs to be considered in some way; for (2) there are $m!$ possible sequences. So why not simply list all the potential solutions for a given set of data, evaluate each one, and select the best? The answer lies in the huge number of solutions that practically sized COPs have; for example, if there are $n=25$ locations in the data for (1), and one trillion tours could have their length evaluated each second, then it would take almost 500 000 years to evaluate all routes! Such 'brute-force' approaches are clearly useless when an answer is needed in real time, so other methods have been devised – methods that guarantee to find the best solution for a 'small' problem or that construct a 'good' approximate answer for larger problems. Most of these techniques are based upon ideas from linear programming, recursion, (partial) enumeration, **heuristics**, or **Monte-Carlo simulation**. The choice of technique used depends on the amount of time available for computation, how the problem is formulated, the reliability of the data and the quality of the solution required. *J.W. Giffin*

References

Syslo, M.M., Deo, N. and Kowalik J.S. (1983) *Discrete Optimization Algorithms*, Englewood Cliffs, NJ: Prentice Hall.
Winston, W.L. (1994) *Operations Research Applications and Algorithms*, 3rd edn, Belmont, CA: Duxbury.

Community operational research

Community operational research (COR) is intervention in the service of community development: working for improvement by dealing with issues that have a perceived negative effect on either the whole of, or sections of, local communities.

Early COR writers and practitioners saw their main clients as being community groups: those organizing and/or campaigning at a grassroots community level on a largely unfunded basis. However, later practitioners broadened COR practice to include working on social issues more generally, without a clearly defined client. In these cases there is usually a need to involve a network of groups and organizations spanning the community, voluntary, public and private sectors, where the COR practitioner may have a pivotal role in facilitating relationships between key people and organizations.

While there are different views about the extent to which COR can legitimately engage with the agendas of statutory and business organizations, as opposed to the agendas that emerge solely from community groups and voluntary organizations, there is nevertheless a common commitment to a practice of meaningful community involvement. In other words, the agenda of a COR project should take seriously the views of people outside formal organizations – preferably through their direct participation, but at the very least through a genuinely open process of consultation. Because of this focus on participation, and the frequently experienced need to work with multiple **stakeholders** who have conflicting views, there is a tendency to draw more often on **soft operational research** methods supporting debate than 'hard' quantitative, methods – although the latter are used on occasion. There is also a strong emphasis on moderating the power of the 'expert' practitioner, and ensuring the meaningful inclusion (and if possible the empowerment) of marginalized stakeholders in COR projects.

Two examples will illustrate the diversity of COR practice. First, in the 1980s, Charles Ritchie conducted a now well-known intervention with tenants on a housing estate in a city in the North of England. These tenants took over the management of their estate from local government, and needed support in a variety of areas, such as negotiating with local government, strategic planning, and budgeting. He worked with the tenants in a participative manner over several years. The idea was not simply to provide analytical input (although this was needed on occasion), but to support the tenants in developing skills in the use of appropriate MS/OR techniques so they could eventually manage their affairs on their own.

In the above case, the analyst aligned himself closely with one group: the tenants. In contrast, in a project tasked with designing new services for runaway children living on the streets, the analysts involved the police, health services, welfare organizations, the education authority and charities. They used powerful stories gathered from children about their experiences on the street to make a compelling case for change. They then modelled the vicious cycles that both the children and the organizations were trapped in, and these models fed into the use of problem structuring methods to support both children and the agencies to design new services and new ways of working.   *G. Midgley*

References

Midgley, G. and Ochoa-Arias, A.E. (eds) (2002) *Community Operational Research: Systems Thinking for Community Development*, New York: Kluwer/Plenum.

Ritchie, C., Taket, A. and Bryant, J. (eds) (1994) *Community Works: 26 Case Studies showing Community Operational Research in Action*, Sheffield: Pavic Press.

Complementarity and complementary slackness

A variety of physical and economic phenomena are most naturally modelled by saying that certain pairs of inequality constraints must be complementary, meaning that at least one of the pair must hold with equality. These constraints may in principle be accompanied by an objective function, but are more commonly used to construct *complementarity problems* (CPs) for which a feasible solution is sought, much like solving a system of simultaneous equations.

For example, we know that in markets where supply of a good exceeds the demand for it, i.e. all buyers have completely satisfied their needs, no consumer is willing to buy more units of the good. With no buyers, the price of the good must fall to zero.

On the other hand, when all supply is bought by consumers (i.e. supply equals demand), additional units are valuable, and consumers may be willing to pay a positive price for them. This is equivalent to the complementarity statement:

$$\{0 \leq \text{price}\} \text{ complements } \{\text{supply} \geq \text{demand}\}$$

One of these conditions must hold at equality. Either supply equals demand and we have a positive price, or supply exceeds demand and the price is zero.

CPs are important in MS/OR **optimization**. For example, each **linear program** (LP) has associated with it another LP, known as the *dual LP*, while the original is the *primal LP*. With each constraint in one LP, we associate a variable in the other and vice versa. If we know one, we can infer the other. This is known as **duality** – a special case of complementarity. At the optimal solution it must be true that if a constraint in either the primal LP or the associated dual LP is not binding, then the corresponding variable in the other problem must be zero. This is known as the *complementary slackness condition of LP*.

In the above example, 'price' is actually the dual variable associated with the supply/demand constraint. More generally, the optimality conditions for a constrained **nonlinear programming problem**, i.e. the **Karesh–Kuhn–Tucker conditions**, form a CP. These conditions are a collection of equations and inequalities describing the constraints, the slope of the objective function, and nonnegativity conditions on the decision variables and the **Lagrange multipliers** (the equivalent of the dual variables). When combined with the complementary slackness conditions, the KKT conditions are a system of complementary equations.

Complementarity has applications in management science, chemistry, engineering, and large-scale economic models, such as **general equilibrium models**.

S. Batstone

References

Cottle, R.W., Pang, J.S. and Stone, R.E. (1992) *The Linear Complementarity Problem*, Boston: Academic Press.

Ferris, M.C. and Pang, J.S. (1997) 'Engineering and economic applications of complementarity problems', *SIAM Review*, 39:669–713.

Ferris, M.C. and Munson, T.X. (2000) *GAMS/PATH User Guide*, Version 4.3.

Complexity theory

Complexity theory, like **chaos theory**, abandons the quest of **general systems theory** to look for laws of regularity and stable equilibria or **steady-states** in order to explain the long-run behaviour of **systems**. Reasoning along different lines, the early chaos theory researchers showed in the 1960s that, as a result of complex **feedback loops**, the values of a function could exhibit complex and unpredictable variations even if the underlying structure was completely **deterministic**, i.e. that chaotic does not have to imply randomness. They also showed that even chaotic behaviour may have inherent order.

Complexity theory picks up similar themes and applies them to social systems as well as to natural and physical systems. According to Stacey (1992), the implications for management, and by extension for MS/OR activity, are that we need to accept that long-run behaviour, even of social systems, is inherently chaotic and hence unpredictable, and that it is evolving rather than approaching an equilibrium. Since long-term prediction is impossible, he claims that long-term planning and strong rigid organizational structures are useless, in contrast to short-term planning where predictions are not only more reliable, but where decision makers often can to a limited extent bear influence on the environment and hence may often bring about a more desirable immediate future. He suggests that since chaos is not disorder, it is more fruitful to unearth the hidden patterns, such as constancy in variation, consistent variability, regular irregularity, and patterns within disorder. Senge (1990) sounds a similar theme when he calls for the discovery of *system archetypes* which he claims all organizations tend to repeat, often in different forms, and urges managers to learn how small changes can create large effects that can break the destructive effects of dysfunctional systems archetypes.

Others call for organizations to adopt **double- and triple-loop learning**, to create flexible structures that are supportive of self-organization and therefore can adapt naturally and quickly to the evolving external environment and internal culture – rather self-evident conclusions.

In terms of practical guidelines for management or public policy in dealing with our increasingly turbulent environment, complexity theory has so far produced little of any concrete nature. It is mainly of theoretical interest. Even the claim that long-term planning is useless or even harmful ignores the fact that simply thinking about the distant future, say via **scenario analysis** or *contingency planning*, is already a valuable exercise since it will foster flexibility and the search for **robustness** and adaptability in planning. Long-term planning rarely commits a manager to more than implementation of the immediate actions, but may help in keeping options open.   *H.G. Daellenbach*

References

Jackson, M.C. (2000) *Systems Approaches to Management*, New York: Kluwer/Plenum, pp. 81–89, 190–201.

Senge, P.M. (1990) *The Fifth Discipline: the Art and Practice of the Learning Organization*, London: Random House.

Stacey, R.D. (1992) *Managing Chaos*, London: Sage.

Computational complexity, see NP-hard

Computer-aided design/computer-aided manufacturing, see CAD/CAM

Concave, see convex

Conceptual models, conceptualization

A conceptualization is something conceived in the mind, a mental picture or representation of a phenomenon, an entity, or a set of operations, that may of may not have actual counterparts in the real world. Its aim is to gain greater understanding or insight. Examples are: heaven and hell; the customer traffic in a bank seen as a **waiting line**; a flow diagram of how the components in an **information system** interact; the operation of an electric power system, viewed as a 'market' of autonomous power stations, each deciding on the amount of power to generate at any point in time in response to price cues issued by a central controller.

Conceptual models is a concept Checkland (1989) uses in his **soft systems methodology**. A conceptual model shows the **systemic** relationships of all activities needed to realize the transformation process of the *nominal system* (itself a conceptualization), defined for a given decision problem, including activities that monitor the performance of the system. It is usually expressed in the form of six to ten verbs, connected to each other in a logical influence or precedence sequence. As is true for the way Checkland views systems, conceptual models to not pretend to represent the real world or some ideal system. The purpose of the conceptual model is to allow comparisons with what is happening in the real world with a view to fostering a debate that may lead to change.

Reference
Checkland, P. (1989) 'Soft systems methodology', in J. Rosenhead (ed.), *Rational Analysis for a Problematic World*, Chichester: Wiley.

Constant returns to scale, see economies of scale

Consumer risk, see acceptance sampling

Container packing, see bin packing

Control theory

Control is the art of steering a dynamic process or a **system**, be that mechanical, electrical, biological or of economic nature, towards a desired **objective**. Since a process or a system will be invariably subjected to unpredictable disturbances originating in its surroundings, continuous monitoring is required to ensure that the designated goal is eventually reached or maintained. This kind of monitoring mechanism is known as **feedback control**. An everyday example of a feedback control system is the automatic cruise control in cars, which uses the difference between the actual and the preset speed to vary the fuel flow rate.

The mathematics of feedback control are the subject of control theory. Every control problem consists of the following elements:

1. An entity or phenomenon (e.g. a car; a manufacturing plant; a firm) that transforms inputs into outputs (e.g. fuel into car speed; raw materials into plant's finished products; promotion into firm's market shares) that is to be controlled over time.
2. A target or desired objective for the entity's operation (e.g. preset speed for car; target levels for finished product stocks; firm's desired share of market).
3. Controls that steer the actions or operation of the entity towards a desired objective (e.g. adjusting the opening of the fuel valve of the car; adjusting the input of raw materials into a manufacturing plant; adjusting the level of promotion to affect market share).
4. Means to monitor the actually achieved outputs and compare with the target (e.g. speedometer reading compared with preset speed; daily stock reports checked against target levels; monthly reported market share compared with target).
5. Exogenous uncontrollable aspects originating in the outside world (i.e. the system environment) that affect or influence the behaviour of the entity (e.g. inclines in the road; varying sales for finished products; promotional effort by competitors and/or changes in demand).

The process or operation of the entity to be controlled is represented by mathematical expressions that capture its behaviour over time and, in particular, its response to control inputs. The dynamic behaviour is given by the change in the **state of the system**, measured by state variables. The control problem then is to select an appropriate sequence of controls at discrete points in time or on a continuous basis to drive the system towards the desired goal. Such models can easily become highly complex such that the effect of controls may become difficult to trace numerically, i.e. the model becomes intractable. The analyst's task is to strike a balance between realism and model tractability.

Often a control problem includes a performance criterion which reflects monetary benefits or costs, energy consumption, or time elapsed to reach the target. The aim is to determine controls that reach the target while optimizing the performance criterion, such as maximizing benefits or minimizing costs. This is the province of the highly mathematical optimal control theory and the calculus of variations, usually involving a set of **differential equations.** *A.D. Tsoularis*

References

Kamien, M.I. and Schwartz, N.L. (2000) *Dynamic Optimization: The Calculus of Variations and Optimal Control in Economics and Management,* Amsterdam: North Holland.

Tapiero, C.S. (1988) *Applied Stochastic Models and Control in Management,* Amsterdam: North Holland. Chapter 1 outlines the uses of control theory in management with minimum mathematics.

Convex, convexity, concave, concavity

Of great interest to **optimization** is to know whether or not the objective function and/or the **feasible region** are well behaved. The former is the function to be maximized or minimized; the latter is the set of all combinations of the decision variables that satisfy all constraints. If the shape of both is 'well behaved', then finding a **global optimal** solution tends to be much easier than if this is not the case. Convex functions and convex feasible regions are well behaved.

A feasible region, or in fact any set of points in an *n*-dimensional space, is convex if we can draw a line segment between any two points in this collection, such that every point on the line segment is also in the set. This is depicted in the graphs in Figure 1.

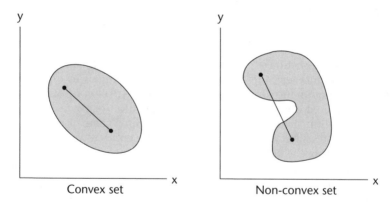

Figure 1: **Convex and non-convex sets**

In two-dimensional space, a function, $f(x)$, is convex if we are able to draw a line segment between any two points on $f(x)$, such that any point on the line segment is above the original function. This is depicted graphically in Figure 2. Every point on the line segment from point A to point B is above $f(x)$.

We can say that for any two values of x, say x_1 and x_2, $f(x)$ is convex if

$$f((1-\lambda)x_1 + \lambda x_2) \leq (1-\lambda) f(x_1) + \lambda f(x_2) \text{ for all } 0 \leq \lambda \leq 1$$

The left-hand side of this equation is the value of the function $f(x)$, while the right-hand side is the value of the line segment, expressed as the weighted sum of $f(x_1)$ and $f(x_2)$, with $(1-\lambda)$ and λ as the weights. For $f(x)$ to be convex, this equation must hold for every pair of x_1 and x_2 over which $f(x)$ is defined. Obviously, these definitions can be extended to more than two dimensions. In three dimensions, a convex function has the shape of a bowl.

A concave function is exactly the opposite of a convex function, i.e. $f(x)$ is concave, if the line between any two points on $f(x)$ is always below the original function.

From Figure 2 it follows that a convex function can have only one minimum, and by analogy a concave function can have only one maximum. Hence finding

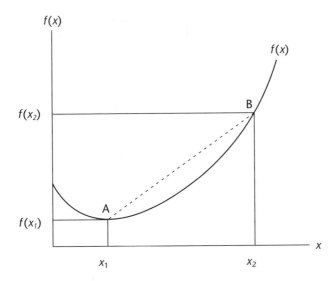

Figure 2: **Convex function**

a minimum to a convex function (or a maximum to a concave function) implies that this is the global optimum. This is not the case for functions that are neither convex nor concave, as shown in Figure 3. D is only a local minimum; E is the global minimum.

S. Batstone

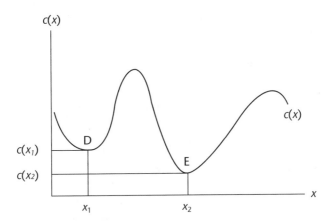

Figure 3: **A function that is neither convex nor concave**

Reference
Hillier, F.S. and Lieberman, G.J. (2001) *Introduction to Operations Research*, 7th edn, New York: McGraw-Hill.

Cooperative games

In 1950, A.W. Tucker created the famous game of *prisoners' dilemma*, while addressing a group of psychologists at Stanford University. The prisoners' dilemma is a two-person non-zero-sum game that involves two prisoners held separately, each offered two options: confess or remain silent. If each prisoner confesses, both will go to prison for 10 years. If neither confesses, then both will serve one year in prison on a minor offence. If one confesses and the other does not, the one who has confessed will go free, while the other will get 20 years. The penalty table is as follows:

Penalties in years of prison [prisoner 1; prisoner 2]		Prisoner 2 strategies	
		Confess	Do not confess
Prisoner 1 strategies	Confess	10; 10	0; 20
	do not confess	20; 0	1; 1

At first glance, it appears that the 'do not confess' option is the most beneficial for both, as they would have to spend only one year in prison. Since there is no communication between them, each one may be tempted to confess, hoping that his partner does not, and he will go free, or be afraid his partner will confess and he will get 20 years. However, if both confess there is no incentive for either one to change strategy as that would lead to a 20-year imprisonment. What has happened here is that the two prisoners have fallen into a strategy equilibrium that is clearly dominated by the 'do not confess' pair of strategies. Note though, even if they had agreed beforehand to be silent, both still would have had the incentive to defect, unless there were some way to enforce the agreement. However, this game demonstrates that communication and cooperation based on mutual trust, threat, or a binding agreement between the two parties would have converted this game into a 'win-win' situation.

Cooperative games are particularly important in economics. Here is a rather simple example that may illustrate the reason why. A Japanese and an US electronics company are both considering working on developing a superconductor. If both companies work on the superconductor, they will have to share the market, and each company will lose $10 billion. If only one company works on the superconductor, the company will earn $100 billion in profits. If neither company works on the superconductor, each company makes no profit. The profit table is as follows:

US / Japanese	Develop	Do not develop
Develop	−10; −10	100; 0
Do not develop	0: 100	0; 0

If we think of this as a non-cooperative game, it is much like a prisoners' dilemma, with the dominant strategy being the decision for both companies to work on the superconductors, thus losing $10 billion each. However, if the two companies agree to cooperate, the joint solution is that neither company works on the superconductor. An alternative scenario could see the Japanese government enter this game as a third player, offering the Japanese company a subsidy of $15 billion to work on the superconductor. The game now becomes:

US / Japanese	Develop	Do not develop
Develop	−10; 5	100; 0
Do not develop	0: 115	0; 0

In this case, the Japanese government and company form a coalition with the winning strategy to work on the superconductor regardless of the decision the US company makes. Obviously, the US and the Japanese companies could cooperate directly, going into partnership, and sharing the $100 billion. *A.D. Tsoularis*

References
Colman, A.M. (1995) *Game Theory and its Applications in the Social and Biological Sciences*, Oxford: Butterworth-Heinemann. A lucid account with a minimum of mathematics.
Poundstone, W. (1993) *Prisoners' Dilemma*, New York: Anchor Books, Doubleday. Interesting account of the role of game theory in the Cold War era.
Von Neumann, J. and Morgestern, O. (1944) *Theory of Games and Economic Behaviour*, Princeton, NJ: Princeton University Press.
Winston, W.L. (1994) *Operations Research: Applications and Algorithms*, Belmont, CA: Duxbury, ch. 15.

Cost–benefit analysis

Cost–benefit analysis or CBA is a formal procedure for assessing the desirability of public investment projects, such as river management schemes, motorway and airport construction, health, education and recreational services, all involving the use of public funds. A project is viewed as increasing *social welfare* if the sum of all net benefits accruing to all affected parties exceeds the totality of all costs, including non-traded goods, i.e. for which no market exists, such as *public goods* (e.g. recreational facilities; clean air) and *externalities* or spillover effects (e.g. pollution).

All costs and benefits associated with a project over its lifetime are assigned a monetary value. If the overall **net present value** (NPV) is positive or the ratio of NPV of all benefits and NPV of all costs is larger than 1, then the project is seen as potentially worthwhile from an economic welfare point of view.

Costs cover capital costs, and maintenance, operational and decommissioning expenditures associated with the project. Positive benefits include increases in existing economic outputs or new outputs (e.g. increased agricultural output due to irrigation), and improvement in existing values (e.g. higher property values due to better access; enhanced recreational fishery due to lower water pollution) and decreases in current costs (e.g. savings due to shorter travel time; decrease in health cost due to cleaner air). Negative benefits cover reduction in current values (e.g. noise pollution). Traded goods are valued at market prices (excluding taxes or subsidies). For other aspects, such as noise pollution, it may be possible to find *surrogate measures* that mimic the **opportunity cost** that a market mechanism would produce, such as the decrease in house prices. Redistribution of existing benefits and costs are excluded, since they do not change total welfare.

Although widely used, CBA has been much criticized for its debatable assumptions. Summing all costs and benefits implies linear **utilities** that are additive over all parties affected. The assumption that total welfare is increased if total net benefit exceeds total costs, regardless of who loses and who wins, ignores distributional effects and equity. Expected values are substituted for uncertainty. Putting dollar values on intangible aspects, such as enhancement, deterioration, or even loss of scenic beauty, health, life, ecological systems, a species, or an 11th-century Norman church (as in the third London Airport Study), is highly controversial. For this reason, it is recommended that no monetary value is put on such aspects, but that they be reported in qualitative terms, with the expectation that they are debated in the political process and in the public hearings usually associated with public projects. **Discounting** costs and irreversible benefit losses incurred by future generations (such as storing nuclear waste or the flooding of a wilderness) raises serious ethical issues. *H.G. Daellenbach*

References

Boardman, A.E. *et al.* (1996) *Cost–benefit analysis: Concepts and practice*, Upper Saddle River, NJ: Prentice-Hall.

Fuguitt, D. and Wilcox, S.J. (1999) *Cost–Benefit Analysis for Public Sector Decision Makers*, Westport, CN: Quorum Books.

Cost-effectiveness analysis

Cost-effectiveness analysis (CEA) is a procedure similar to **cost–benefit analysis** for assessing the desirability of public investment projects, particularly involving health, safety, military and environmental issues, where the benefits cannot be easily expressed in monetary terms. Examples are: reduction of breast-cancer deaths through mammography screening policies, reduction of accidents through road alignments, reduction of fishing by-kill of endangered species. CEA is a suitable approach if all options have the same type of benefits, but differ in both the costs involved and the size of the benefits.

All monetary costs associated with the project for all parties involved, including capital costs, maintenance, and operational costs, are **discounted** over the life of the project. Similarly, all benefits accruing to all parties involved are totalled. Options are compared and ranked according to either the ratio of total benefits to total discounted costs or the benefits achieved per, say, $1000 of costs incurred.

Since benefits are not expressed in monetary terms, CEA avoids the controversial aspects of how to value intangible aspects and benefits. However, its application is restricted to situations where there is only one type of benefit.

Since the early 1980s, CEA has found widespread application in the health sector and in medicine. Rather than simply look at lives saved, doctors have invented a benefit measure, called *QUALY*, that captures both the additional years of life saved and their quality. *H.G. Daellenbach*

Reference

Drummond, M.F. (1997) *Methods for the Economic Evaluation of Health Care Programmes*, Oxford: Oxford University Press.

Costs, see also **benefits and costs, relevant**

Costs, types of

- *Explicit costs* involve an 'out-of-pocket' transfer of funds between parties (examples: raw material cost paid to supplier; wages paid to workers).
- *Implicit costs* do not involve an 'out-of-pocket' transfer of funds. Some result from accounting practices (examples: depreciation of equipment; allocation of administrative and managerial fixed costs – so-called overheads – to various operations). Others are costs imposed on other activities, the wider system, or the environment (examples: productivity loss caused by disruptions in production; revenue foregone from lost sales due stock shortages; social cost of poverty).
- *Intangible costs* are implicit costs difficult to assess in monetary terms due to their nature and complexity (examples: 'loss of goodwill' due to lost sales; environmental cost of pollution; social cost of poverty). Intangible costs are never recorded in accounting systems.
- *Opportunity costs* are implicit losses incurred by foregoing the return of the best alternative opportunity (examples: return lost for using funds for activity A rather than B where B offers the highest return for the funds; profit lost from stock shortages). Opportunity costs are never recorded in accounting systems.
- *Variable costs* vary with the level of activity (examples: raw material cost and labour cost of machine operators varies proportionately with production level).

- *Fixed costs* are not affected by changes in activity level (example: overheads; start-up cost of an activity; acquisition cost of equipment used for activity). With some exceptions, fixed costs disappear when the activity ceases permanently. Unless fixed costs are different for different decision choices, they are not relevant for determining the best mode of operation of a system. They are, though, relevant for the separate decision on whether the operation should be undertaken or not.
- *Sunk costs* are costs incurred in the past which cannot be altered, regardless of the decision taken. Hence, they are irrelevant (example: equipment repair cost already spent should be ignored for a decision on whether to spend more on repairs or replace the equipment). Sunk costs are often erroneously included in the analysis.
- *Marginal cost* at a given level of activity x measure the increase in total costs incurred by increasing the level of activity by one unit to $x+1$ (example: say the printing cost of 4000 copies for this book is £13,120, while 4001 copies cost £13,124; the marginal printing cost of the 4 001th unit is £4). As the level of activity increases, the marginal cost often decreases initially due to greater **efficiency**, particularly in labour inputs (exhibiting increasing returns to scale), then becomes constant (constant returns to scale), and ultimately begins to increase due to manufacturing disruptions, emergence of production bottlenecks, and overtime cost (decreasing returns to scale). (See also **economies of scale**.)   H.G. Daellenbach

Reference
Daellenbach, H.G. (2001) *Systems Thinking and Decision Making*, New York: Wiley, ch. 9.

Counterintuitive outcomes

A counterintuitive outcome is a system output or system behaviour that seems to contradict intuition, common sense, or go against normal expectations, and cannot be readily explained by **reductionist** and/or **cause-and-effect thinking**. A simple example of a counterintuitive outcome is the chemical binding of hydrogen and oxygen – two gases – producing water, a liquid; or offering help to somebody in distress and getting abused for it.

Systems may exhibit counterintuitive outcomes. Often such outcomes go against long-established business rules. For example, it is a generally accepted business principle that a firm should push those products that offer the highest profit margin. Say a firm produces two products on the same assembly line as shown in the figure below. Both cost the same to produce, i.e. $90/unit, but product A has a profit margin of 50 per cent, while B only achieves 40 per cent. (The profit margin is [profit/selling price] × 100%.)

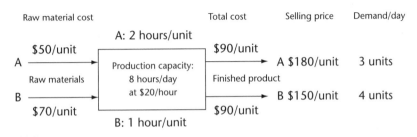

Figure: **Example of counterintuitive outcome**

According to the principle, the firm should produce as many of A as can be sold, i.e. four, and then use up the remaining production capacity of two hours to produce two units of B. The daily profit is then $3 \times \$90 + 2 \times \$60 = \$390$. Interestingly, a reversal of the above business principle produces a better result, namely the firm should produce as many as possible of the product with the lower profit margin and only then use the remaining production capacity to produce the one with the higher profit margin. The resulting output of four units of B plus two units of A has a total profit of $420 – higher by $30.

This is a counterintuitive result. Why does it happen? The answer is simple. The business principle ignores vital system interactions. In this case, the different use of production capacity of the two products. Every hour of capacity used by product B produces a profit of $60, while an hour of work on product A achieves only $45.

Outcomes that at first seem counterintuitive are usually not mysterious happenings. Most often, they can and should be explained by a sufficiently comprehensive **systems thinking** approach.   *H.G. Daellenbach*

Crew scheduling, see applications of MS/OR to crew scheduling

Criteria, objectives, objective function

Criterion, objective and goal are often used interchangeably. The *Webster Collegiate Dictionary* defines goal, in the sense used in MS/OR, as 'the end toward which effort is directed', while objective is defined as 'something toward which effort is directed, an aim, goal or end of action'. Objective and goal are thus the same. Objectives or goals are the end result of what most MS/OR effort aims to achieve, such as maximum profit, minimum cost, shortest distance, high environmental quality, equity, and so on. If the objective can be measured numerically, it is captured by the objective function. For example, in **linear programming**, the objective is expressed as

$$\text{maximize (or minimize) } z = \sum c_i x_i$$

where the x_i's are the decision variables, the c_i's their unit contribution towards the objective, and z the value of the objective function for a given combination of x_i values.

Criterion, however, is defined as a principle or 'a standard on which a judgement or decision may be based'. Both *principle* and *standard* imply a rule. So criterion is the rule used to judge how well an objective has been achieved, and this is its interpretation in MS/OR. The following example demonstrates the difference between objective and criterion. A small city wants to locate its only fire station so that the distance to all parts of the city is as short as possible. The objective or goal is 'shortest distance'. But how do you decide which location offers 'shortest distance'? We need a criterion for that. One criterion is to minimize 'the sum of the road distances' between all buildings and the fire station. This could result in most buildings being close, but some being very far away. A second criterion is to minimize 'the sum of the squared distances', thereby penalizing large distances more heavily. A third is to minimize 'the maximum distance' between any building and the fire station. Each of these is a criterion for judging the achievement level of the objective 'shortest distance', and each may result in quite a different 'best' solution.

In many situations, there is only one rule or criterion to judge how well an objective has been achieved. For instance, if the objective is 'highest profit', there is only one way to judge that, i.e. the size of the profit. The criterion coincides with the objective. It is this frequent coincidence that may be the cause why often criterion and objective are used (erroneously) interchangeably.   *H.G. Daellenbach*

Critical path method or CPM, Gantt charts

Developed by DuPont and Remington-Rand in 1956, the critical path method is a technique for planning and monitoring complex projects that consist of a number of tasks and activities, where certain tasks can only be started when certain other tasks have been completed. It calculates the minimum length of time in which a project can be completed and identifies those tasks, called *critical tasks*, which if delayed will delay completion of the project as a whole. The sequence of critical tasks form the *critical path*. The tasks on the critical path are the ones that must be monitored carefully to make sure their completion is not delayed.

Consider a project of building a house. The table below shows the data for this simplified illustration. Tasks A and B can be started at time 0. The earliest start time (ES) for tasks C and D is after two weeks, i.e. when A is finished. Task E has to wait for both B and C to finish. While C can be completed by the end of week 6, B takes 7 weeks. So ES for E is 7. Since D can be finished by week 5 and E by week 9, the whole project is completed end of week 9, as shown below. Tasks B and E form the critical path. Tasks A, C and D can be delayed somewhat, as shown in column 'Slack'. The latest start and finished times are shown in columns LS and LF. Figure 1 depicts the project in a diagram, where the nodes are the activities. The numbers above the circles are the EF times. The shaded tasks form the critical path.

Table: **The tasks for building a house**

Task Code	Task name	Expected time	Task precedence	ES	LS	EF	LF	Slack = LS−ES
A	Lay foundations, build exterior walls	2 weeks		0	1	2	3	1
B	Have windows, doors, etc. made	7		0	0	7	7	0
C	Build interior	4	A	2	3	6	7	1
D	Install roof, finish outside	3	A	2	6	5	9	4
E	Install windows, finish interior	2	B, C	7	7	9	9	0

where: ES = early start, LS = late start, EF = early finish, and LF = late finish.

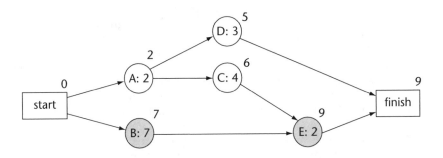

Figure 1: **Activity-on-node diagram for house project**

For actual monitoring and control of a project, these times are depicted as horizontal bars on the *Gantt chart* in Figure 2 (critical tasks shown shaded). For instance, from the panel (b) it follows that by the end of week 3, task A has to be finished to allow task C to start, while task B has to be 3/7 complete, or else project completion time will be delayed.   *N.C. Georgantzas*

References

Luttman, R.J., Laffel, G.L. and Pearson, S.D. (1995) 'Using PERT/CPM to design and manage clinical processes', *Quality Management in Health Care*, 3(2):1–12.

Plsek, P.E. (1993) 'Tutorial: Management and planning tools of TQM,' *Quality Management in Health Care*, 1(3):59–72.

Turner, J.R. (1993) *The Handbook of Project-based Management*, New York: McGraw-Hill.

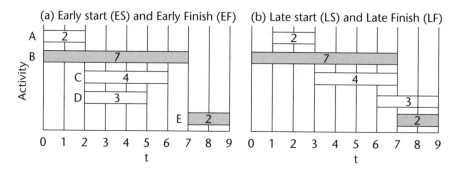

Figure 2: **Gantt charts for (a) ES & EF, and (b) LS & LF**

Web sites
Critical path analysis: http://www.mindtools.com/critpath.html

Critical systems heuristics

Critical systems heuristics (CSH) (Ulrich, 1983) represents the first systematic attempt at providing both a philosophical foundation and a practical framework for **critical systems thinking**. The Greek verb *'heurisk-ein'* means to find or to discover; heuristics is the art (or practice) of discovery. In management science and other applied disciplines, **heuristic** procedures serve to identify and explore relevant problem aspects, questions, or solution strategies, in distinction to *deductive* (algorithmic) procedures, which serve to solve problems that are logically and mathematically well defined. Professional practice cannot do without heuristics, as it usually starts from 'soft' (ill-defined, qualitative) issues such as what is the problem to be solved and what kind of change would represent an improvement.

A critical approach is required since there is no single right way to decide such issues; answers will depend on personal interests and views, value assumptions, and so on (see **Weltanschauung**). A critical approach does not yield any single right answer either; but it can support processes of reflection and debate about alternative assumptions. Sound professional practice is *critical practice*.

CSH aims to support critical professional practice through a critical employment of the systems idea. The methodological core idea is that all problem definitions, proposals for improvement and evaluations of outcomes depend on prior judgements about the relevant whole system to be looked at. Improvement, for instance, is an eminently systemic concept, for unless it is defined with reference to the entire relevant system, suboptimization will occur. CSH calls these underpinning judgements 'boundary judgements', as they define the **boundaries** of the reference system to which a proposition refers and for which it is valid.

Accordingly, the methodological core idea of CSH is to support systematic processes of **boundary critique**. To this end, CSH offers a framework of boundary concepts, as shown in the box below, that translates into a checklist of 12 critical boundary questions (Ulrich, 1987, 1996, 2000). (They are asked in both the 'who or what is(are)' and the 'who or what ought to be' modes.) They can be used, first, to identify boundary judgements systematically; second, to analyse alternative reference systems for defining a problem or assessing a solution proposal; and third, to challenge in a compelling way any claims to knowledge, rationality or 'improvement' that rely on hidden boundary judgements or take them for granted. The third application leads to an emancipatory employment of **systems thinking**; it offers both those involved in and those affected by professional practice a new critical

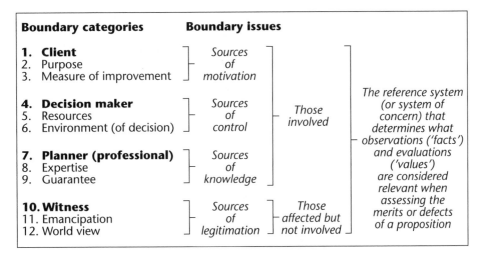

Boundary categories	Boundary issues		
1. Client 2. Purpose 3. Measure of improvement	*Sources of motivation*		*The reference system (or system of concern) that determines what observations ('facts') and evaluations ('values') are considered relevant when assessing the merits or defects of a proposition*
4. Decision maker 5. Resources 6. Environment (of decision)	*Sources of control*	*Those involved*	
7. Planner (professional) 8. Expertise 9. Guarantee	*Sources of knowledge*		
10. Witness 11. Emancipation 12. World view	*Sources of legitimation*	*Those affected but not involved*	

Figure: **Boundary categories of critical systems heuristics**
(*Source*: W. Ulrich, 1983, p. 258; 1996, p. 43; and 2000, p. 256)

competence, regardless of their theoretical knowledge or special expertise with respect to the problem in question.

In sum, CSH can be defined as a critical methodology for identifying and debating boundary judgements. Despite its emancipatory implications (the aspect for which it is best known), CSH should not be misunderstood and used as an emancipatory systems approach only; for its principle of systematic boundary critique is vital for sound professional practice in general, whatever importance may be attached to emancipatory issues. For the same reason, CSH does not aim to be a self-contained systems methodology, but is better understood as an approach that should inform all critical professional practice, whatever specific methodology is used.   *W. Ulrich*

References

Ulrich, W. (1983) *Critical Heuristics of Social Planning: A New Approach to Practical Philosophy*, Bern: Haupt. Reprint edition 1994, Chichester: Wiley.

Ulrich, W. (1987) 'Critical heuristics of social systems design', *EJOR*, 31(3):276–283.

Ulrich, W. (1996) *A Primer to Critical Systems Heuristics for Action Researchers*, Hull: Centre for Systems Studies, University of Hull.

Ulrich, W. (2000) 'Reflective practice in the civil society: the contribution of critically systemic thinking', *Reflective Practice* 1(2):247–268.

Critical systems thinking

Critical systems thinking (CST) is a systems perspective that first emerged in the late 1980s following criticisms of earlier ('hard' and 'soft') systems and MS/OR approaches. The latter were criticized on three grounds:

1. being overly *instrumental* in their approach, i.e., only exploring the means for reaching predefined ends rather than subjecting people's ends and the values informing them to analysis;
2. failing to take sufficient account of *power relationships* during intervention; and
3. getting involved in a *paradigm war* rather than appreciating that the 'hard' and 'soft' approaches are actually useful for different purposes, meaning that they should be seen as complementary rather than in competition with one another.

It is important to be aware that CST is not a methodology – although a number of CST writers have produced methodologies for operationalizing their ideas. It is perhaps more usefully viewed as a paradigm, or research tradition, where a community of practitioners with both common interests and different viewpoints debate the pros and cons of their positions in order to explore the implications for MS/OR theory and practice. Therefore, on a number of occasions, people writing under the banner of CST have argued for opposing ideas.

Having highlighted the fact that CST writings are quite diverse, it is nevertheless possible to identify key strands of thinking of interest to most people in the CST community. The three critiques of earlier work (above) are used below to structure a discussion of some of these key strands.

Instrumentality: There is a common concern that much MS/OR serves agendas that have been predefined by some **stakeholders** to the disadvantage of others. Some writers (e.g. Ulrich, Jackson, Mingers and Oliga) have related this to the ideas of Jürgen Habermas. Habermas argues that money and power are increasingly dominating community affairs, making it difficult for people to consider ends other than the ones 'given' to them in mainstream economic and political discourses. The antidote to this, according to Habermas, is the reconstitution of *civil society*, allowing more space for public debate about social ends. Similarly, in CST, those authors following Habermas have argued for participative methodologies that enable debate between stakeholders.

Recently, there has been a move away from Habermasian thinking on the grounds that continual reference to a single political theory limits MS/OR practice (e.g. the work of Gregory, Midgley and Jackson). However, the concern to challenge instrumentality still remains, and the reason is that opening debates about ends as well as means expands the possibilities for human choice and action.

Power relationships: Instrumentality is connected to the need to deal with power relations in MS/OR intervention. Some writers have talked about power that is experienced as one person or group constraining the actions of another, while others have seen power as linked with the kind of knowledge (often professional knowledge) that is used to order social affairs. Some of the most sophisticated work sees the two manifestations of power as interrelated. This thinking about power has given rise to many methodological and practical innovations. Examples include:

- Ulrich's **critical systems heuristics** (which is designed to enable communication between planners and ordinary citizens);
- Vega's methodological guidelines for evaluating social justice in health care based on expanding the possibilities for exploring alternative forms of knowledge; and
- Midgley's work on dealing with the marginalization of stakeholders and issues during intervention.

Paradigm war: The final critique of previous OR/MS ideas is that there is no need for a paradigm war between 'hard' and 'soft' OR practitioners – their methods can, and should, be viewed as complementary. The major justification for this complementarity – referred to as *methodological pluralism* or **multimethodology** – is that the use of a wide variety of methods enhances OR/MS practice. By understanding the strengths and weaknesses of different methods, they can be used together to address a wider set of purposes than any single method would be able to do.

Although there is widespread agreement on this practical rationale for methodological pluralism, there are differences on how this pluralism should be embodied in methodology. Some (e.g. Jackson) argue that different methods and methodological ideas should be drawn upon depending on the diagnosis of the problem situation – but no one set of methodological ideas should be given pride of place. Others (e.g. Ulrich and Midgley) argue that **boundary critique** needs to be placed up front in interventions, otherwise there is a danger that diagnoses of problem situations will be superficial and biased towards the interests of a narrow range of stakeholders. Boundary critique involves the consideration (by the analyst and stakeholders) of different possible boundaries for the inclusion of issues and people – and the values giving rise to boundary judgements are also up for analysis. Of course, boundary critique inevitably throws up tensions (and potentially conflicts) that need to be managed as part of the intervention process.

This and other issues with significant implications for intervention are being widely debated in the CST community, making CST a fruitful source of ideas for developing both OR/MS theory and practice. *G. Midgley*

References

Flood, R.L. and Jackson, M.C. (eds) (1991) *Critical Systems Thinking: Directed Readings*, Chichester: Wiley.

Flood, R.L. and Romm, N.R.A. (eds) (1996) *Critical Systems Thinking: Current Research and Practice*, New York: Plenum.

Jackson, M.C. (2000) *Systems Approaches to Management*, New York: Kluwer/Plenum, pp. 355–426.

Midgley, G. (2000) *Systemic Intervention: Philosophy, Methodology, and Practice*, New York: Kluwer/Plenum.

Ulrich, W. (1983) *Critical Heuristics of Social Planning: A New Approach to Practical Philosophy*, Berne: Haupt. Reprinted 1994, Chichester: Wiley.

Curse of dimensionality, see dynamic programming

Cutting plane algorithms

Most **integer programming** (IP) solution methods are based on solving a sequence of related *relaxed* **linear programming** (LP) problems until the optimal integer solution has been identified. An IP problem may be relaxed by removing the necessity that decision variables are integers. The relaxed LP has a **feasible region** that is larger than and completely contains that for the original IP. It can be solved efficiently with conventional LP solution techniques. If the optimal solution to the relaxed problem is not feasible for the integer problem, i.e. is not an integer solution, extra constraints are added to try to make the solution become integer. A *cutting plane* is such an added constraint. It 'cuts' away part of the feasible region of the relaxed LP without removing any feasible integer solutions.

A cutting plane algorithm repeatedly applies this device. It starts out with the original IP relaxed to an LP. If the optimal LP solution is integer, the **algorithm** terminates. If not, a new cut is added, cutting away part of the feasible region, and the new relaxed LP solved. This process is repeated, further and further restricting the feasible region of the relaxed LP until its optimal solution yields an integer solution. Given that the cuts are constructed never to remove an integer solution, the final LP solution found is also the optimal solution of the original IP problem.

The figure below demonstrates this process for the following IP problem:

$$\text{Maximize} \quad z = x_1 + 2x_2$$

$$\text{subject to} \quad -x_1 + 3x_2 \leq 5 \qquad \text{(constraint 1)}$$

$$6x_1 + x_2 \leq 22 \qquad \text{(constraint 2)}$$

$$x_1, x_2 \geq 0 \text{ and integer}$$

The dots in the left-hand graph of the figure correspond to all feasible integer solutions, while the area enclosed by the two constraints and the nonnegativity conditions on the variables is the feasible region for the initial corresponding relaxed LP. The optimal solution to the relaxed LP is at point A (3.21, 2.74) with a value of z_0. It is not integer, hence cut 1 of the form $x_2 \leq 2$ is added. It removes the shaded area in the right-hand graph. Point B is the new optimal LP solution with a value z_1. It is still not integer. Cut 2 of the form $x_1 + x_2 \leq 5$ removes the black area. Point C (3, 2) is

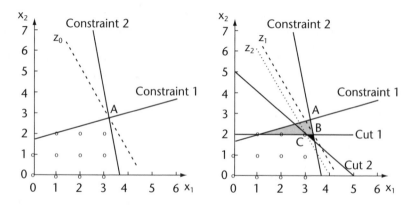

Figure: **Cutting planes**

the optimal LP solution to this further restricted LP with a value of z_2. It is integer and hence the optimal solution to the original IP.

Although the method is mathematically elegant, it is seldom used in commercial IP software because of the difficulty in determining good cutting planes. In practice, **branch-and-bound methods** are used instead. *D.K. Smith*

Reference
Schrijver, A. (1986) *Theory of Linear and Integer Programming*, New York: Wiley, ch. 23.

Cutting stock problems

The cutting stock problem (CSP) aims to find the most profitable way of cutting some material (e.g. paper, textiles, glass, timber, steel, film, foil) from rolls or blocks or boards (often called *raws*), of given widths or sizes, into smaller pieces (often called *finals*) of different widths, sizes, dimensions, shapes or values.

The simplest form of the CSP has a limited, but sufficient, supply of raws of one given width; the set of orders for the finals provides the number of each specified width required; the raws must be cut via a cutting pattern in such a way that there is minimal raw wastage generated. For example, suppose that raws of width 80 cm must be cut according to the following set of orders: 25 finals of width 15 cm, 60 finals of width 20 cm, 110 finals of width 30 cm and 50 finals of width 40 cm. In this example, there are only nine 'sensible' cutting patterns, where a sensible pattern is one which yields no more than 14 cm of wastage per raw (e.g. cutting patterns {2 of 15 cm, 1 of 20 cm, 1 of 30 cm, 0 of 40 cm}, {5,0,0,0} or {0,2,0,1}).

However, for larger problems involving raws of many different sizes and a wide range of final sizes, the number of possible cutting patterns quickly becomes unmanageable. Most solution techniques for the CSP therefore generate cutting patterns 'on the fly', as required, rather than tabulating them all in advance. But this idea does not make the CSP any easier to solve optimally, as the further practical requirement that each cutting pattern be used an integral number of times means that realistically sized problems cannot be solved via **linear programming**. Fortunately, a common feature of many simple CSPs is that their optimal integer-valued solutions have total costs that are frequently close to those of their corresponding optimal fractional solutions, provided that the number of finals ordered is not too small.

The most common techniques for solving CSPs are based upon *column generation*, **branch-and-bound** and **dynamic programming**. These techniques are able to cope with extensions to the simple model, such as the raws having different widths and costs, also called the *assortment problem*; raws of different widths being cut on several different machines of limited availability; there being a limit on the maximum amount of wastage per non-sensible cutting pattern used; and particular cutting patterns being admissible only on some machines.

CSPs which have a small number of raw sizes, but a large number of final sizes, are also known as **bin-packing problems**. These problems occur frequently in applications involving vehicle loading, warehousing and publishing. Many efficient **heuristic solution methods** have been devised for these problems, both in situations where the final order sizes are known beforehand and where they arrive gradually over time.   *J.W. Giffin*

References
Chvatal, V. (1983), *Linear Programming*, New York: W. H. Freeman.
Winston, W.L. (1994), *Operations Research: Applications and Algorithms*, 3rd edn, Belmont, CA: Duxbury, pp. 562–568.

Cybernetics

The field of cybernetics was founded by Norbert Wiener in the 1940s with the purpose of establishing basic principles of automatic control or response mechanisms used by living systems and autonomous operations of complex electromechanical systems. The concepts of *communication, feedback, goals* and *controls* are fundamental to cybernetics. Any action towards a goal must be controlled to realize that goal. Progress of the action towards the goal at any moment cannot be known without some form of communication. Communication within the system and with the environment enables the system to control its action if a deviation from the designated goal has been observed. This is the concept of **feedback**.

Although cybernetics began as an engineering discipline, it quickly found strong appeal for the study of the behaviour and control of all sorts of organizations. Any organization can be viewed as a system, with management providing the input or control needed to steer the organization towards desired goals, such as return on investment, market share, employee satisfaction, etc. Feedback may come from different sources: market factors, such as level of sales, employee factors, such as rate of turnover, customer factors, such as rate of complaints, use of funds, such as inventory levels, and so on. Management uses this feedback information to adjust current actions and/or initiate new actions that promise to steer the organization closer to its goals. Thus, the whole organization can be viewed as a system that regulates itself and adapts to changes in its performance and environment, as depicted in the diagram below.

System dynamics – a technique to study the dynamic behaviour of continuous systems through **simulation** – is strongly based on the principles of cybernetics. Cybernetic principles are also fundamental to the **viable systems model**, developed by Stafford Beer.   *A.D. Tsoularis*

References
Beer, S. (1970) *Cybernetics and Management*, London: The English University Press.
Strank, R.H.D. (1983) *Management Principles and Practice: A Cybernetic Approach*, New York: Gordon and Breach Science Publishers. Management principles examined from a cybernetic viewpoint.
Wiener, N. (1948) *Cybernetics: or Control and Communication in the Animal and the Machine*, Cambridge, MA: MIT Press.

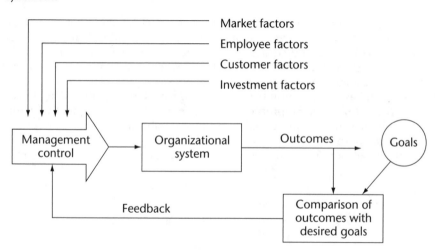

Figure: **Feedback system for organizational control**

D

Dantzig-Wolfe decomposition, see **decomposition**

Data envelopment analysis

Data envelopment analysis (DEA), developed by A. Charnes and W.W. Cooper in 1978, is a method for assessing how a given operating unit compares to similar units, in terms of the **efficiency** with which it uses resources (termed inputs) to provide goods or services (termed outputs). Examples of units whose efficiencies have been assessed by DEA are the branches of a bank, supermarkets, restaurants, university departments, hotels, schools, hospitals and tax collecting offices. We can assess a set of units such as bank branches by DEA because all branches perform the same function in terms of the kinds of inputs they use and outputs they produce. For example, their inputs may be expenditure on staff and on capital equipment, such as information technology. Their outputs may be the numbers of deposit and loan accounts served and the number of new accounts opened.

Measures of comparative efficiency reflect either the extent to which a unit can lower the levels of its inputs to produce given levels of outputs (*input orientation*), or alternatively, the extent to which a unit could increase its outputs for given levels of inputs (*output orientation*). These comparisons are based on observed inputs and outputs of the operating units rather than on some theoretical proposition of maximum output levels attainable from given input levels. Hence the efficiency measures obtained are relative to the comparative set of units used rather than absolute.

One of the basic principles underpinning DEA is that we can interpolate between two production units so that if we take 50 per cent of the input–output levels of one unit and add them to 50 per cent of the input–output levels of a second unit we will create a third *virtual* unit which could exist in principle. Using this and other assumptions we can identify efficient benchmarks, some observed, some virtual, relative to which we can measure the scope for savings on the inputs or the scope for raising the outputs. These are expressed as *efficiency measures* so that an efficiency of, say, 80 per cent in the input orientation means the unit concerned can lower its inputs to 80 per cent of their observed levels while still producing at least the output levels observed.

In using DEA in practice, we typically go beyond the computation of efficiency measures. We can identify best operating practices to be disseminated to all units, most productive scale of operation for each selected unit, role models some inefficient unit may emulate to improve its performance, and so on. If we have data over time we can also measure productivity changes, both at operating unit level and at sector or industry level.

The mathematical principles underpinning DEA are implemented using **linear programming** models which are formulated and solved automatically on specialist computer software. All the user has to do to assess a set of units is merely to identify their inputs and outputs and import the corresponding data for each unit. Two such software packages are DEA Software from the UK and IDEA from the USA.   *E. Thanassoulis*

References

Cooper, W.W., Seiford, L.M. and Tone, K. (1999) *Data Envelopment Analysis: A Comprehensive Text with Models, Applications, References and DEA-solver Software*, New York: Kluwer.

Thanassoulis, E. (2001) *Introduction to the Theory and Application of Data Envelopment Analysis: A Foundation Text with Integrated Software*, New York: Kluwer.

Data mining

Advances in computers, and especially reductions in the cost of storage of data (files measured in 'terabytes' are now common), have allowed organizations to keep huge files of client or customer information. Every purchase with a credit card, every item scanned as part of your supermarket purchases can be filed away for future analysis. Existing databases, such as insurance company or bank records, can be scanned for evidence of fraud, or to determine characteristics of bad credit risks. Data mining (also sometimes referred to as *knowledge discovery in databases*, or KDD) seeks to discover patterns or relationships in such databases, e.g. how many customers who have bought one particular item will buy another item at the same time, or within the next six months – this is an example of *association rule mining*. Other techniques include the generation of **decision trees**, based on a sequence of 'if-then' classifications, and methods, frequently based on **neural networks**, which search for previously unknown patterns in the data or attempt to predict future behaviour. The rise of data mining could not have occurred without improvements in database technology and search tools. *Data warehousing* refers to collecting and cleaning transactional data to make them available for analysis and decision support. *OLAP* (online analytical processing) tools are programs which can quickly provide pre-specified reports from databases.

With varying degrees of accuracy, most major **statistical analysis** packages, such as SAS or SPSS, have also claimed recently that they have data-mining capability. This may range from pointing out that their existing programs, such as discriminant analysis or **clustering techniques**, can be used to classify and look for relationships in data, to specific programs or suites incorporating the tree and pattern-search methods described above. Increasingly these techniques will be an important selling point for existing statistical packages.

A bewildering array of techniques, often highly mathematical, which are applicable to data mining have been published in a variety of journals on computers, database engineering, and recently specific data-mining journals.   *D.C. McNickle*

References

Cios, J., Pedrycz, W. and Swiniarski, W. (1998) *Data Mining Methods for Knowledge Discovery*, Boston: Kluwer Academic. Survey the range of techniques.

Han, J. and Kamber, M. (2001) *Data Mining: Concepts and Techniques*, San Francisco: Morgan Kaufmann.

Decision analysis, decision theory

Decision analysis – also referred to as *decision theory* – deals with the principles of choosing from among a set of mutually exclusive alternative courses of action the one that is optimal or best for a given decision **criterion**, when the decision has to be made in the face of uncertainty about the possible outcomes. Different decision criteria are suitable for different degrees of uncertainty. In decision analysis, **uncertainty** is expressed as an exhaustive list of mutually exclusive possible world-states, often called states of nature or *future states*. The decision problem may involve choosing a strategy, i.e. a sequence of decisions, where later decisions may depend on intermediate outcomes of earlier choices.

Consider the following example. A software firm is threatened with a law suit for copyright infringement for a piece of software used. The three alternative actions are an out- of-court settlement with (1) either a licensing agreement for continued use of the software or (2) abandoning the use of the software immediately and accelerating development of their own, or (3) going to court. The future states are the four combinations of 'win court case' or 'lose court case', and 'new software ready in 4 months' or 'new software ready in 8 months'. For each action and each possible future state, there is a *payoff* – a monetary outcome, ranging from heavy court costs and loss of sales, cost of out-of-court settlement, and so on, to continued high profits. A **decision tree** is often a convenient and instructive way to represent such a problem.

If the probability of each state of nature is known, we talk about *decision making under risk*, which is the domain of statistical decision analysis. The probabilities may be **objective** in the sense that they are derived from past data (e.g. past sales) or **subjective**, reflecting the strength of belief of the decision maker in the occurrence of the various future states. For the risk case, the best decision criterion is to maximize the expected benefits (e.g. profits) or minimize the expected cost.

If no information about the likelihood of future states is available, such as may be the case for a novel investment venture, other decision criteria must be used that do not need probabilistic information. The **minimax criterion** selects the action that has the best 'worst outcome' over all future states, i.e. plays safe rather than sorry. The *minimax regret criterion* selects the action with the lowest opportunity loss, defined as the difference between the best outcome for a given state and the outcome for each action. The *threshold criterion* eliminates all those actions that could lead to unacceptable outcomes, such as bankruptcy or loss of life, and then uses some other criterion for selecting among those remaining.

A variant of statistical decision analysis is **Bayesian decision analysis**. It explores whether the best expected monetary outcome can be improved by obtaining better probabilistic information obtained through sampling of data sources of known but imperfect reliability.

Decision analysis has been and is being used to assess risky investment projects, new product developments, product launch strategies, oil exploration, military and political conflict situations, medical diagnoses, etc. Several software packages are commercially available for decision analysis, Bayesian decision analysis, and decision trees.   *A.D. Tsoularis*

References

Behn, R.D. and Vaupel, J.W. (1982) *Quick Analysis for Busy Decision Makers*, New York: Basic Books. Entertaining text.

Daellenbach, H.G. and McNickle, D.C. (2001) *Systems Thinking and OR/MS Methods*, Christchurch: REA, ch. 8. Short basic introduction.

Samson, D. (1988) *Managerial Decision Analysis*, Homewood, IL: Irwin. Case studies and use of Arborist software.

Decision conferencing

Decision conferencing denotes a meeting in which people work on a decision. Although commonly conceptualized as a three-stage process (*intelligence – design – choice*), real-life decision making is a complex, nonlinear process with many steps and iterations. For simple problems, a decision conference might encompass this entire process, but more likely the participants will work on a few steps only.

Decision conferencing takes place in a special kind of environment – one that provides information and communication technology to support a variety of activities commonly performed by people who work on a decision. Today, there exists a range of commercially available electronic meeting systems that can be used for 'same time, same place' (people meet in one room) as well as 'different time, different place' (people interact from where and when it is convenient for them) decision conferences. These systems comprise a set of interlinked computers, together with specific electronic meeting support software, which can be used to harness the collective thoughts and opinions of a group. Such systems allow brainstorming, categorizing, commenting, voting and review to take place via the computer network. In a 'same time, same place' environment, the computer system allows a second level of expression to take place simultaneously with any group activity – be it a presentation, discussion, selection or planning session.

Such systems facilitate the exchange of ideas amongst participants in several ways:

- Ideas and comments can be input from all participants simultaneously, which improves the efficiency of activities that require *divergent thinking*, such as making an inventory of system failures, or *brainstorming* on potential solutions. Participants can build on each other's ideas, with many such developments taking place simultaneously.
- Input is anonymous, which reduces the inhibition that participants may otherwise feel to present their ideas or voice their criticism, especially in groups with high 'power distance'. This leads to a more full and frank contribution.
- Everyone is equal. No individual dominates, as is often the case in traditional meetings.
- Information is not filtered. Everyone is 'heard' on the basis of merit.
- All input is captured in the words of the participant who generated it. Transcription and retyping are avoided.
- Participants can contribute far more ideas in the time available than would be possible in a conventional meeting.
- Participants are able to vote on ideas.
- All contributions of all participants are recorded, providing a collective memory.

To be successful, decision conferences require even more careful planning and preparation than regular meetings. It must be clear in advance not only which particular step(s) in the decision-making process are to be worked on, but also which electronic tools are most suited for the task. Electronic meeting systems typically provide generic tools for generating ideas, organizing them in **hierarchical structures**, and evaluating them using different ranking techniques. Other MS/OR tools, both 'hard' and 'soft', can be, and have been, made available to the participants to explore aspects, such as **stakeholder analysis** and **cognitive mapping** to support the identification of relevant actors and factors in a decision-making situation, **multicriteria decision analysis** and **negotiation theory** to evaluate choices or maximize stakeholder utility and so on.

Practical experience shows that there is no single recipe for success. Each conference will require deliberate design, motivated participants, and, most of all, professional facilitation. *V. Mabin*

References

Frey, L.R. (ed.) (1995) *Innovations in Group Facilitation: Applications in natural settings*. Cresskill, NJ: Hampton Press. Discusses role of the facilitator.

Nunamaker, J.F. *et al.* (1997) 'Lessons from a dozen years of group support systems research: a discussion of lab and field findings', *Journal of Management Information Systems*, 13(3):163–207. A good overview of the history of decision conferencing.

Special issue on 'Decision support in the public sector' (2000) *Journal of Multi-Criteria Decision Analysis*, Fall. Review of recent applications.

Web site
http://www.groupsystems.com.

Decision criteria, see criteria

Decision support systems

A decision support system (DSS) is an interactive **information system** that incorporates MS/OR models and solution techniques to help end users learn about semi-structured or ill-defined problems. A DSS is designed to help managers make decisions by allowing them to access and use internal and external data and analytical models. A DSS is interactive, computer-based, menu-driven, and emphasizes flexibility, effectiveness and adaptability. Most DSSs are either large-scale software systems, constructed to facilitate well-defined and repetitive decision-related tasks, or small PC-based software systems for quick and economic routines that support one-time decision making.

The distinctive aspect of the DSS approach as an MS/OR tool is that, by definition, a DSS is controlled by its potential end users from conception, to implementation, to use. It is designed to give these end users the appropriate data and MS tools so that they can make useful decisions. Effective DSS design relies heavily upon the realization that a DSS is created to deliver problem analysis capabilities, not merely to respond to informational needs. Also, a genuine understanding of the decision-making environment of the end users and the type of support (in terms of data, models and solution techniques) is often crucial to good design. To this end, Keen (1988) has stated that there is a need for balance between each of the three DSS elements: *decision, support* and *systems*. As system technology is not a bottleneck, it is often the decision component that requires the most attention to achieve effective DSS design. To achieve the mission of helping people to make better decisions, Keen stressed the need for the DSS to play an active supporting role for making decisions that really matter.

This emphasis on the end user in DSS design and implementation is in contrast with many other MS/OR approaches which are dominated by MS/OR consultants building systems which are narrowly focused on a single type of decision.

Application areas of DSSs in MS/OR include: **facilities layout**, **vehicle routing**, **timetabling**, political decision making, public spending, commercial aircraft construction, economic policy planning, manufacturing, financial investment, airline route and price selection, corporate planning and **forecasting**, investment evaluation, pricing and advertising, price evaluation, production optimization, train scheduling, drilling site evaluation, flight scheduling, and defence contract analysis.

All of the above-mentioned systems contain the four essential ingredients of any DSS. These are: *information representation* (often in the form of computer graphics), *data manipulation* (often using MS/OR models and solution techniques), *memory aids* (often involving computer checking and alerting), and *control aids* (such as user-friendly computer languages). These features make the DSS an attractive complement to more traditional MS/OR approaches, through enhanced flexibility and end-user empowerment.   *L.R. Foulds*

References

Keen, P.G.W. (1988) 'Decision support systems: The next decade', *Decision Support Systems*, B3: 253–265.

Laudon, K.C. and Laudon, J.P. (1996) *Management Information Systems*, 4th edn, Englewood Cliffs, NJ: Prentice Hall.

Decision trees

Decision trees are diagrammatic representations, used in **decision analysis**, of the chronological sequence of events in multistage decision processes when some or all outcomes are uncertain.

Suppose a television licence costs £100 per year. If someone is caught using a TV without a licence, the fine is £1000. According to a recent report, it appears that only about 15 per cent of people who do not have a licence get caught and have to pay the fine. Should you buy a licence or not? The situation is shown in the decision tree below.

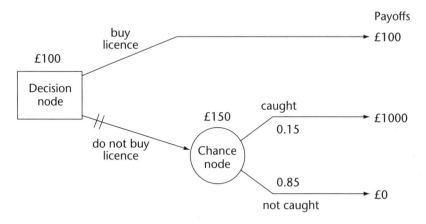

Figure: **Decision tree**

Rectangles denotes *decision nodes*. The branches coming off it indicate the alternative decisions at that point. Circles denote *chance nodes*. Each branch corresponds to one of the possible outcomes which may occur and has the associated probability attached to it. The numbers shown at the endpoints of each right-most branch are the *payoffs* associated with the path coming from the initial node. In this example, the payoffs are all monetary, but they could equally well be point scales, utilities, or any other numeric outcome.

The tree is analysed using a process called *backward induction*. We start at the end-points and work backwards from right to left. At each chance node we compute the expected value of the payoffs over its branches. At each decision node we select the action that offers the best value, either highest benefit or lowest cost, whichever is relevant. In the figure, at the chance node, the expected outcome is £1000(0.15) + £0(0.85) = £150, shown above the circle. At the decision node, the top branch has a cost of £100, the bottom branch one of £150. Buying a licence has the lowest cost. The bottom branch is blocked off, indicating that it is not the decision to take.

For situations that are repetitive – the TV licence is due every year – and where the chances and payoffs remain the same, the expected payoffs associated with all nodes can be interpreted as the average outcome over many repetitions. Hence the optimal payoff has the same interpretation. If we are dealing with unique one-off situations the interpretation changes. The expected payoffs cannot be seen as aver-ages; they are only expectations prior to the events. The actual payoff will be one of the payoffs of an endpoint that has an unblocked path from the initial node. All we can say is that, if we consistently use this approach for all decisions involving uncertainty then, in the long run, using the best expected value criterion will per-form better than any other criterion. Naturally, if some of the potential outcomes are completely outside our normal everyday experience or even spell disaster, such as bankruptcy or a threat to life, other approaches may be more appropriate, such as expressing all outcomes in terms of their *intrinsic values* or **utilities**, or using different criteria, such as a **minimax** approach, i.e. the best of the worst possibilities from any decision node.

Decision trees are a useful device for structuring and getting initial insights into a multistage decision problem, albeit for a usually simplified reality. *S. Stray*

References

Daellenbach, H.G. and McNickle, D.C. (2001) *Systems Thinking and OR/MS Methods*, Christchurch: REA, ch. 7.

Jennings, D. and Wattam, S. (1998) *Decision Making*, London: Financial Times Prentice Hall.

Decomposition

As the name suggests, the basic idea of decomposition is to break up the overall problem into its smaller and more manageable subproblems, suitably connected to each other. This concept has wide applicability in a multitude of mathematical disciplines, e.g. **mathematical programming**, probability, **graphs**, spectral analysis theory. Although the basic purpose is essentially the same, the context, mathematical treatment, and outcome are quite distinct. The discussion here, therefore, is confined to decomposition of mathematical programming problems.

The original motivation of decomposition came from the observation that many real-life optimization problems can be viewed as being composed of loosely connected simpler and smaller (sub)problems. There are different ways of breaking a problem into simpler subproblems and hence there are different decomposition schemes. They fall into two broad categories: decomposition by grouping the constraints into separate sets (the original *Dantzig–Wolfe decomposition* scheme, named in honour of the inventors) or decomposition by variables into separate sets (*Bender's decomposition*). The latter is particularly suitable if one subproblem becomes an integer problem, while the other involves real variables. Pioneered in the early 1960s, both have been extended, generalized and applied to a large number of problems over the past four decades. Both schemes are based on surprisingly simple ideas that are best understood in the form of practical examples.

Consider, for example, a cost minimization problem involving a dozen factories that are producing a common set of products to meet a pre-specified total demand requirement. This problem may be 'decomposed' into 12 individual factory-level optimizations, each dealing with its own production decisions, and a *master problem* that coordinates their activities. Each factory is offered prices for the products and then finds its best production mix, given its own production constraints and raw material costs and availabilities. The master problem then takes these offers for outputs and tries to satisfy the total demand requirements. If a product is in oversupply, it lowers the price offered; if not enough is produced, it increases the price offered. These new prices are fed back to the factories, which in turn find new optimal output levels. This iterative process continues until all total demand constraints (of the master problem) are met. The scheme resembles the physical process of a central planner trying to coordinate the activities of individual producers via a *price mechanism*.

This, in fact, is the basic idea behind Dantzig–Wolfe decomposition. These methods and their generalization are applicable to **linear programming**, as well as special **nonlinear programming** problems, that have a constraint structure involving a (large) set of independent blocks connected by a few coupling constraints. Many practical linear programming applications possess this type of constraint structure, formally known as the block angular structure, e.g. **transportation**, **cutting stock**, multi-commodity **network flows**, **vehicle routing**, **crew scheduling** problems. *D. Chattopadhyay*

References

Dantzig, G.B. and Thapa, M.N. (1997) *Linear Programming*, New York: Springer Series in Operations Research.

Geoffrion, A.M. (1972) 'Generalized Bender's decomposition', *Journal of Optimization Theory and Application*, 10:237–260.

Lasdon, L.S. (1970) *Optimization Theory for Large Systems*, London: MacMillan.

Decreasing or diminishing returns to scale, see economies of scale

Delivery person problem; Chinese postman problem

The delivery person problem (DPP) is also known as the *travelling repairperson* problem. It is a variant of the **travelling salesperson problem** (TSP), with the objective of finding a service route which minimizes the sum of the times all customers have to wait before being serviced. The simplest version of the DPP has one delivery person, and the service route begins at a specified location (the time taken to return to the starting location may or may not be included in the objective).

A common application is in the delivery of perishable commodities such as pizzas. It may also be a dynamic situation, in which customers are gradually assigned to a route as the calls for service occur, subject to service quality guarantees. Other applications include the routing of automated guided vehicles in **flexible manufacturing systems** and finding a job processing sequence on a single machine which minimizes the mean flow-time. The DPP is **NP-hard** (i.e. the computational effort to solve it increases exponentially with the number of nodes). Hence, only small problems may be solved optimally, usually by **dynamic programming** or *Lagrangian relaxation* techniques. Several **heuristic** approaches have also been developed for larger (and the dynamic) problems.

The Chinese postman problem (CPP) is the simplest example of an arc-routing problem. In the DPP and the TSP, it is the nodes that are being serviced. Arc-routing problems service the arcs themselves, in applications such as street sweeping, snow ploughing, postal delivery, patrol cars, electric meter readers, and automated guided vehicles in factories. The CPP in particular seeks to traverse all the arcs of a given network that require service, starting at and returning to the initial position, in a sequence which minimizes the total distance travelled (or the total time taken). If every node in the network has an even number of arcs meeting at it, if the service 'demand' of each arc is proportional to its length, and if every arc in the network allows two-way travel, then it is easy to find a CPP tour; in this case the tour is called an *Euler tour*. When some arcs are one-way, the CPP may still be solved efficiently, but the optimal tour will usually include deadheading, i.e. the traversal of at least one arc more than once. However, if the arcs have arbitrary (or dynamic) service demands, or if more than one vehicle of finite capacity must be routed, the problem becomes again NP-hard, and heuristic techniques must be applied to obtain good solutions within reasonable computing time.

Finally, why is it referred to as the 'Chinese' postman problem? Simply because the first person to describe the problem was a Chinese mathematician!   *J.W. Giffin*

Reference

Evans J.R. and Minieka, E. (1992) *Optimization Algorithms for Networks and Graphs*, 2nd edn, New York: Marcel Dekker.

Delphi method of prediction

Delphi is an iterative method, using judgements by experts, to predict the likely outcome of highly uncertain future events, such as the price of volatile world commodities, advances in technology, or social phenomena. Consider the example of oil price predictions in the late 1970s, needed to evaluate the economic desirability of the expansion of the only oil refinery operated in New Zealand. A group of recognized economic and energy experts from all over the world was asked to participate in a survey about the likely levels of oil prices over a 30–year span. Those who agreed were given a questionnaire with a set of precise questions. Their responses were collated and expressed in statistical form by a researcher. The results of the first questionnaire were then communicated in summarized form (averages, medium responses, distribution of responses, etc.) to the experts. They were asked if, in the light of the first-round results, they wanted to change their original responses and how. Their new answers were again processed in the same manner. The Delphi method usually repeats this procedure through two or three iterations, with the responses of the last iteration being used as the final predictions. Note that complete anonymity of any responses received is preserved throughout the procedure.

The Delphi method has seen many successful applications. It is not a cheap method and, unless the experts are all locally present, it takes considerable time to reach a conclusion. Each iteration can easily take weeks. It is therefore only suitable for relatively important projects. Before embarking on such an exercise, the analyst should perform considerable sensitivity analysis to determine how crucial it is to get a reliable estimate. In the oil price exercise reported above, the price for a barrel of the type of crude oil processed at the refinery in early 1978 was around US$35. The final predictions for the price by the end of the 1980s covered a range of US$60 to US$95. The analysis for the expansion option chosen established that it would remain economically viable as long as the price remained above about US$29. The actual price in 1989 was well below US$20, reaching lows of US$15 at times. So we see that even judgements by experts may be far off the mark.   *H.G. Daellenbach*

Reference

Linstone, H.A. (1975) *The Delphi Method: Techniques and Applications*, Reading MA: Addison-Wesley.

Democratic corporation, the

The more educated a workforce and the more technical its work, the more a democratic form of organization is appropriate to obtain the quality of work required and to provide those employed with a satisfying work environment and challenging work.

A democratic political system, including corporations, has three properties: (1) everyone in or out of that system who can be directly affected by a decision (i.e. the **stakeholders**) can participate either directly or indirectly (through elected representatives) in making that decision; (2) anyone who has authority over others taken separately, is subject to their collective authority; therefore, there is no ultimate authority; (3) everyone can do whatever they want providing doing so does not adversely affect anyone else; if it does, agreement of those affected is required.

A corporation can be democratized by the use of boards. Each board consists of the manager whose board it is, the immediate superior, and all immediate subordinates, as depicted in the figure below. It follows that every manager, except those at the very top and bottom of a hierarchy, participates in his or her own board, that of the immediate superior, and in the boards of each of the immediate subordinates. This implies that every manager – except those at the two top and bottom levels of a multilevel organization – interacts with members from four other levels of management: the two above and the two below. The members of every board are free to select others within or outside the organization to participate in their board provided that the subordinates on any board constitute the plurality of voting members.

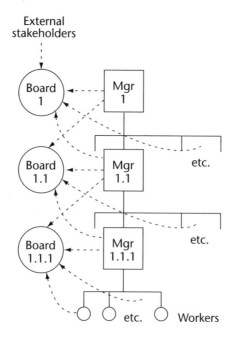

External
stakeholders

Figure: A circular organization
(*Source*: adapted from R.L. Ackoff, 1999, p. 336).

Each board has seven functions: (1) planning for the unit whose board it is; (2) making policy that applies to all subordinate units; (3) coordinating the plans and policies of subordinate units; (4) integrating the plans and policies of their own and lower level units with those of higher level units; (5) establishing quality of work life conditions for its members; (6) suggesting improvements in the behaviour and practices of board members and, in particular, the manager whose board it is; and (7) for the subordinates on every board acting separately to evaluate the performance of the manager whose board it is and relieve him or her from his or her position if that performance does not satisfy them.

Decisions should be made by consensus on an acceptable alternative, not necessarily on the best one. Lack of agreement usually stems from a different interpretation of facts. These can often be tested by experiments. Board members must agree to abide by the results obtained. If no experiment can be run, boards should have procedures for overcoming unresolvable differences, such as that the manager's declared choice is implemented if agreement is not reached. Then the manager agrees to accept whatever the board members agree to, whether or not it agrees with his or her own position.

Participation in these boards normally reduces or eliminates the need for any other meetings, hence it increases the discretionary time available to their members for other work. It also tends to convert antagonistic relationships between superiors and subordinates into collaborative ones. It also makes participants aware of how their behaviours affect the whole of which they are a part. It shifts focus from the behaviour of parts considered independently to the behaviour of the wholes of which they are part and to the interdependencies of these parts.   *R.L. Ackoff*

Reference

Ackoff, R.L. (1999) *Re-creating the Corporation: A Design of Organizations for the 21st Century*, Oxford: Oxford University Press.

Deterministic, stochastic

These two concepts are partial opposites of each other. In MS/OR, deterministic means that everything is known with certainty about an event, a sequence of events, or the behaviour of something, e.g. a **system**. (It does, however, not have the philosophical connotation of lack of free will.) For example, given the detailed production schedule for how many vacuum cleaners a firm is going to produce each week over the next three months, it is possible to determine exactly how many motors, how many plastic cases, how many electric cord retraction devices, etc. are needed each week. The requirements are deterministic. Similarly, barring any accidents, train movements of the Swiss railroads are known exactly each day. They are deterministic.

Stochastic means that there is uncertainty or randomness about certain aspects of an event, a sequence of events, or the behaviour of something. The uncertainty is not complete, in the sense that we know nothing at all, but stochastic is usually interpreted as implying that we have some information about the probability law underlying the event(s) or behaviour. For example, the number of vacuum cleaners sold by all outlets may be stochastic, following a normal distribution with a known mean and a known standard deviation. Although train movements are deterministic, the number of passengers taking a given train is stochastic, e.g. following a Poisson distribution with a mean of 135.

A deterministic system always follows exactly the same behaviour for a given starting position or initial **state of the system**, while a system whose behaviour is affected by stochastic events will have a completely different behaviour pattern each time, even if its initial state is the same. Interestingly, however, the long-run behaviour of a stochastic system may tend towards a state of equilibrium or **steady-state** that is independent of the initial state.

In real life, few things are really deterministic. The assumption that something is deterministic may be made as a convenient simplification, since it is often easier to find the best set of decisions for a deterministic system. As long as the stochastic influence is minor, i.e., the random fluctuations are small, this may be perfectly adequate for many types of decision situations. (See also **random variable**, **uncertainty**.)

Dialogue

Dialogue can be treated as having different functions in human existence. How dialogue is conceived depends on the orientation of those proposing its value. Some people believe that reality is capable of being represented in a more or less accurate fashion through human language. For such people, dialogue is a way of people checking their views of reality against those of others, with the intention of minimizing the different biases that each party may have in their encounters with the world. In other words, dialogue is a way of organizing 'reality checks' so that biases can be progressively ironed out in the process of developing knowledge about reality. The idea is that through the process of dialogue a better appreciation of reality will (eventually) result.

Other people, however, believe that it is not possible for humans ever to experience the world without the mediation of their (interested) consciousness. They argue that any view of the world is mediated through human consciousness: humans cannot access reality other than through their way of seeing it. For such people, dialogue is a way of people coming to appreciate the differing perceptions that may be forwarded by others. This appreciation may in turn enable them to enrich their own repertoire of responses as they act in the (experienced) world. Seen in this light, dialogue can be a way of extending one's appreciation of options for both seeing and acting.

Sometimes dialogue is presented as ideally directed towards achieving a consensus of opinion on matters of concern. That is, agreements are sought through the process of dialogue. If consensus is not sought, then sometimes it is suggested that at least some accommodation should be striven for – which the various participants in the dialogue can work with as a basis for coordinating their activities. It is suggested that the parties can then 'live with' the results of the dialogue. However, the notion that dialogue must be oriented in principle towards the quest to reach consensus, or (alternatively) some form of accommodation, has been challenged by some people (sometimes labelled as postmodernist in orientation). The latter argue that if conversation is guided by these quests, mutual toleration of difference (with the aim of preserving diversity of thought and experience) might indeed become threatened. To prevent domination in the form of people imposing a view of reality as they engage in discourse with others, it is suggested that dialogue be understood primarily as an opportunity for people to excavate a multitude of ways of experiencing the world.

N.R.A. Romm

References

Checkland, P.B. (1981) *Systems Thinking, Systems Practice*, Chichester: Wiley.

Freire, P. (1985) *The Politics of Education: Culture, Power, and Liberation*, Westpoint, CN: Bergin and Garvey.

Jackson, M.C. (1982) 'The nature of "soft" systems thinking: the work of Churchman, Ackoff and Checkland', *JORS*, 9:17–28.

Lyotard, J.F. (1984) *The Postmodern Condition: A Report on Knowledge*, Manchester: Manchester University Press.

Romm, NRA (1996) 'A dialogical intervention strategy for development', in J.K. Coetzee and J. Graaff (eds), *Reconstruction, Development and People*, Johannesburg: International Thomson Publishing.

Young I.M. (1990) *Justice and the Politics of Difference*, Princeton, NJ: Princeton University Press.

Diet problem, feed-mix problem

The diet problem deals with preparing a balanced mixture of foods, for human or animal consumption, that provides a desirable composition of various nutrients, such as proteins, fats, calcium, vitamins, minerals and water, as prescribed for good health and growth by human and animal nutritional experts. It has found widespread application in feed mix industries all over the world and in New Zealand has even been used to come up with a new type of meat sausage for human consumption that met government specifications as well as human taste buds. The diet problem was one of the first problems formulated as a **linear program**, as early as 1945.

In its most general form, it has the following structure: There are n different foodstuffs, called ingredients, available at a cost of c_j each. In addition there are m nutritional requirements, each specifying a minimum or a maximum diet content of b_i units of a given nutrient per day. Each ingredient j contains a_{ij} units of the nutrient for requirement i. While the b_i's are determined by nutritional experts through extensive experiments, the a_{ij}'s are obtained from laboratory analysis of the various ingredients and may need to be reassessed for each batch of the ingredient used. If we denote by x_j the number of units of ingredient j in the diet, then the problem is to find x_j's so as to minimize the total cost of the daily diet, i.e.

$$\text{minimize } \sum_{j=1}^{n} c_j x_j$$

subject to m upper and/or lower limits on various nutritional requirements, i.e.

$$\sum_{j=1}^{n} a_{ij} x_j \leq \text{ or } \geq b_i, \; i=1, \ldots , m,$$

$$\text{and } x_j \geq 0 \text{ for } j=1, \ldots , n.$$

In feed-mix applications, there may be limits on the available ingredients during certain periods of the year, resulting in additional constraints. Furthermore, several feed-mix formulations, e.g. feed mixes for egg-laying hens, birds raised for chicken meat, pigs, etc., may all compete for the same limited ingredients, resulting in the need to incorporate all feed mixes into one single linear program. Attempts to use the diet problem for human diet formulation, e.g. in hospitals, retirement homes or other residential institutions, have met with mixed success, largely because humans not only need a balanced diet but also daily variety. The latter aspect leads to a scheduling problem over time which is difficult to incorporate into the diet problem.

B. Chen

References
Luenberger, D.G. (1984) *Linear and Nonlinear Programming*, Reading: Addison-Wesley, ch. 2.
Stigler, G. (1945) 'The cost of subsistence', *Journal of Farm Economics*, 27.
Winston, W.L. (1994) *Operations Reseqrch Applications and Algorithms*, 3rd edn, Belmont, CA: Duxbury, pp. 70–73.

Difference and differential equations

Difference and differential equations are the standard mathematical tools for analysing processes that evolve over time. Planetary orbits, rocket trajectories, speed and acceleration of cars, flows of fluids, population dynamics and economic growth are all typical examples.

Difference equations involve differences between consecutive values of an entity or phenomenon as a function of a discrete variable, often time counted in fixed intervals, such as months or years. The annual growth of an initial deposit of F_0 dollars in a savings account that is compounded annually at a rate r is a difference equation. The difference between two consecutive years is the interest added at the end of each year, e.g. the increase at the end of the first year is $\Delta F_1 = F_1 - F_0 = rF_0$, for the second year $\Delta F_2 = F_2 - F_1 = rF_1$, ... , and for year n $\Delta F_n = F_n - F_{n-1} = rF_{n-1}$. The reason for formulating something in terms of difference equations is to find a solution for the function F_n itself in terms of the initial F_0 and some constants – r in our example. For this simple case the solution for F_n can easily be derived by rearranging the difference equations above as follows: $F_1 = F_0(1+r)$, $F_2 = F_1(1+r) = F_0(1+r)(1+r) = F_0(1+r)^2$ or in general $F_n = F_0(1+r)^n$. For example, for $F_0 = \$1000$ and $r = 0.1$, $F_1 = 1000(1+0.1) = 1100$; $F_2 = 1000(1+0.1)^2 = 1210$, and so on.

System dynamics, a technique to study the behaviour of dynamic systems over time, usually involving **feedback loops**, expresses the changes in the **state of the system** in terms of difference equations.

Differential equations express the behaviour of something in terms of rates of change as a function of a continuous variable, again usually time. Differential equations involve derivatives. Finding solutions needs advanced methods of differential and integral calculus. If we allow time to become a continuous variable in the savings growth example, with compounding also occurring at a constant rate, we get a differential equation (i.e. $dF/dt = (e^{rt} - 1)F$). Its solution is $F = e^{rt} F_0$.

Differential equations are mainly used in the highly mathematical fields of **control theory** and **cybernetics**. Most differential equations can only be solved numerically on a digital computer by approximating them by difference equations.

A.D. Tsoularis

References
Blanchard, P., Devaney, R.L. and Hall, G. (1998) *Differential Equations*, Pacific Grove Brooks/Cole Publishing Company. An elementary treatise with many graphical examples solved using state of the art software.
Goldberg, S. (1986) *Introduction to Difference Equations*, New York: Dover. An outstanding primer on difference equations with lots of examples.

Discounted cash flows, net present values, internal rate of return, and equivalent annuities

These concepts are closely related to the concepts of present value, future value, discounting and *compounding*. If the interest or return, r, that can be earned in a year is 10 per cent (or 0.1), then £100 now will grow in one year's time to be £100 + 100(.1) = £110. In general the future value of a sum P will grow in one year to $F = P + Pr = P(1+r)$. This is called compounding. The reverse operation is called discounting. £110 received one year from now is worth now only £100. So the present value of a sum F_1 that is received one year from now is $P = F_1/(1+r)$. In this case r is referred to as the *discount rate*, and $\alpha=1/(1+r)$ is called the *discount factor*. Since $r \geq 0$, α is between 0 and 1, where $\alpha=1$ implies no discounting.

These operations easily generalize to several years. P compounds in two years to $F_2 = P(1+r)(1+r) = P(1+r)^2$, and F_2 received two years from now is worth $P = F_2/(1+r)^2 = F_2 \alpha^2$ now. In general, F_n received n years from now has a present value of $P = F_n \alpha^n$.

The net present value (NPV) of a future stream of positive and/or negative monetary values is the sum of their present values. Consider a project that requires an investment of £10,000 at the beginning of year 1, produces losses of £2500 and £1365 at the end of year 1 and year 2, respectively, and then generates revenues of £8000 and £12,000 at the end of years 3 and 4, respectively. The initial funds could be invested in another venture that earns 12 per cent each year. Can the proposed project match that same return? The answer is 'yes', if the proposed project has a positive NPV at a discount rate of $r = 0.12$, and 'no' otherwise. The diagram below demonstrates the discounted cash flow calculations.

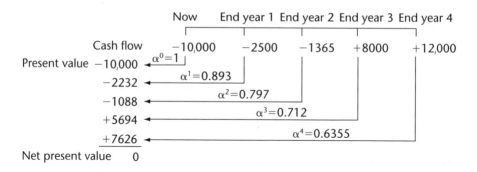

Figure: **Discounted cash flows**

The arrows show how the NPV is calculated. Obviously, the initial cash outlay is not discounted, hence its discount factor of $\alpha = 1$. It turns out that the NPV is equal to zero. The project offers exactly the same return as the alternative venture.

In general, if C_n is the cash flow occurring at the end of year n, then the NPV over N years, including the initial investment C_0 is

$$\text{NPV} = \sum_{n=0}^{n=N} \alpha^n C_n$$

This exercise demonstrates another concept related to discounting. The discount rate for which a cash flow has a NPV of zero is called the *internal rate of return* (IRR), a concept frequently used in economics. The IRR of this project is, therefore, $r = 0.12$ (or 12%).

If the cash flow in all N periods is identical, i.e. $C_n = C$, then this is an *annuity*. Often we are interested in what annuity would give the same NPV as a given irregular cash flow (known as the *equivalent annuity*, A). This is given by multiplying the NPV by the annuity factor, i.e.

$$A = \text{NPV}\frac{r}{1-(1 + r)^{-N}}$$

In our example, the NPV of the cash flow of C_1, C_2, C_3 and C_4 is equal to £10,000 if the discount rate is 0.12, i.e. the same as C_0. Hence, the equivalent annuity is £10,000(0.12)/[1−(1 + 0.12)$^{-4}$] = £3292. Obviously, the NPV of a cash flow of this amount (received at the end of each of four years) is £10,000.

All spreadsheet software provide a whole range of financial functions to compute future and present values, NPVs, IRRs, and equivalent annuities.

Discounted cash flows are widely used in economic evaluations of potential investment projects, both in the private and public sector. A major problem is the determination of an appropriate discount rate. Economic theory tells us that the correct rate to use is the one based on **opportunity cost** considerations. This means using the rate that can be earned on the best alternative use of the funds. This may appear to be a relatively easy concept, but is far more difficult to establish in practice.

S. Stray

References

Brealey, R. and Myers, S. (1999) *Principles of Corporate Finance*, New York: McGraw-Hill.
Daellenbach, H.G. (2001) *Systems Thinking and Decision Making*, Christchurch: REA, ch. 10.

Discrete event systems

Differential and **difference equations** have been cherished tools for scientists and engineers for a long time. Indeed, whenever a process varies continuously over time, differential equations are well posed to mirror the continuous evolution of a system over time, whereas difference equations are ideal for modelling systems that undergo discrete transitions at regular time intervals. The industrial, commercial, communications and computer industries of the 20th century gave rise to a new breed of systems that can be analysed neither by differential nor difference equations.

For example, the service operation of a teller system at a bank or supermarket can be characterized by three key features: the time of arrival of a customer, the time to serve a customer, and the number of customers waiting in the queue to be served. Although the number of customers in the system, i.e. waiting in the queue or being served, is discrete, arrivals do not occur at regular intervals and service times do not start or end at these same regular intervals. In the jargon of systems theory, the **state of the system**, referred to as the *state space*, consists of two discrete sets, like {0,1,2,...}, one recording the number of customers waiting in the queue, the other recording the number of servers being busy. The state changes when a customer arrives, a service begins, or a service ends and the customer departs. These occurrences are called *events*. They can occur at any time, not just at regular set intervals. It is immediately apparent that such systems are not amenable to be modelled by sets of difference equations.

In terms of a formal definition, a discrete event system is a system with a discrete state space, where changes in the state space are the result of discrete events, and which evolves according to asynchronous occurrences of events. **Queueing theory** describes the behaviour of discrete event systems in the form of waiting lines. **Markov processes** deal with such systems, and most **simulations** of industrial and commercial operations model discrete event systems.

A.D. Tsoularis

Reference

Cassandras, C.G. (1993) *Discrete Event Systems: Modeling and Performance Analysis*, Homewood, IL: Richard D. Irwin, Inc. and Aksen Associates, Inc.

Diseconomies of scale, see economies of scale

Diversity management

The diversity in culture, knowledge, skill, personality, value, expectation, etc. within an organization has always been a challenge for management. Traditionally, management has aimed to achieve an alignment of the values and expectations or even a homogeneous culture, e.g. through selective recruitment. However, due to an increase in the educational level of the workforce, as well as globalization of business, it is becoming difficult to achieve such alignment. Consequently, there is a need for managing the diversity.

Clearly a number of people skills are necessary in order to ensure that the broader objectives of an organization are achieved despite a lack of alignment among the organizational members. Besides, there is also a need to design appropriate coordination mechanisms, organizational structures and management systems that would help the members contribute to an organization's functioning, despite differences of opinion on values and culture, on goals and strategies to achieve them, on the organization's responsibility towards its owners, the community and the environment. This implies a need for appropriate methods for managing the debate and open **negotiations**. Anything that inhibits such debate, such as the absence of a free flow of information, becomes an obstacle to overcome.

Although it is important to develop the requisite skills for doing the above, an organization has also to be concerned with how to develop and deploy such skills. The process should be such that the organization progressively has the capacity to learn from its experience and mistakes of managing the diversity to confront similar and new issues of diversity. Flood and Romm have presented a framework for diversity management that seeks to accomplish multiple types of learning. This framework identifies four major tasks in diversity management:

- designing organizational structures;
- designing organizational processes;
- managing debates within organizations; and
- redressing the imbalance of power which tends to restrict debate in organizations and society.

The performance of these four tasks can produce organizational learning if the manager is conscious of the cycles or loops through which learning is produced. This suggests that the broad managerial tasks should be viewed as too complex for any-one to master fully, recognizing, however, that managers, as well as organizations, can continuously progress in their capacity to accomplish the tasks despite an ever-uncertain organizational environment.   *D.P. Dash*

References

Carr-Ruffino, N. (1998) *Managing Diversity: People Skills for a Multicultural Workplace*, 2nd edn, Needham Heights, MA: Simon and Schuster.

Flood, R.L. and Romm, N.R.A. (1996) *Diversity Management: Triple Loop Learning*, Chichester: Wiley.

Dominance, efficient solutions, efficient frontier, Pareto optimality

These four terms relate to **multicriteria decision making** and **multiobjective mathematical programming**, where the decision maker wants to pursue several, usually conflicting **objectives** and the most preferred solution must be a compromise between these objectives.

Some decision choices can be ruled out as potential candidates for the most pre-ferred solution. If a decision choice A performs no better than another decision choice C with respect to all objectives and worse for at least one, then A is *dominated* by C. A can be eliminated from further consideration. A rational decision maker will never consider A as a potential candidate for the most preferred solution (provided the objectives used capture everything that counts about the decision choices).

The graph below depicts dominance for the case of two conflicting objectives. Each axis measures the achievement level on one objective. The higher the level, the more desirable it becomes. The joint achievement levels on both objectives for each deci-sion choice are points in the positive quadrant.

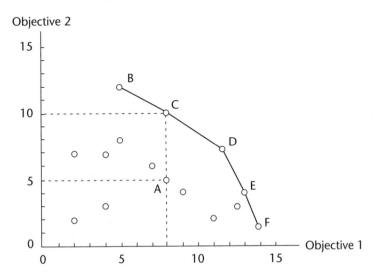

Figure: **Dominance and efficient solutions**

Note that alternative A achieves an outcome of 8 on objective 1 and 5 on objective 2, while alternative C achieves 8 on 1 and 10 on 2. A is no better than C for objec-tive 1, but worse for objective 2. Hence, A is dominated by C. On the other hand, alternative B with outcomes of 4 on objective 1 and 12 on objective 2 does not dominate C and is not dominated by it.

A solution that is not dominated by any other solution is called *efficient*. The solid line in the figure from B to F connects all those alternatives that are not dom-inated by any other alternative. They form the efficient solution set and should be the only candidates for the most preferred solution. The solid line is referred to as the *efficient frontier*. In economics, an efficient solution is also referred to as *Pareto optimal*. The decision maker should only consider *trade-offs* between efficient solu-tions, i.e. giving up some of the achievement towards one objective for a gain in another.

One of the principles of multiobjective mathematical programming approaches is to make sure only efficient solutions are ever considered as candidates for the most preferred solution.   *H.G. Daellenbach*

Reference
Keeney, R.L. and Raiffa, H. (1976) *Decisions with Multiple Objectives*, New York: Wiley, pp. 69–77.

Double-loop learning, see single-loop learning

Drama theory

Drama theory uses the metaphor of drama to make sense of human interactions in situations where the outcome is the result of action taken by several 'actors' or *players* who have different aspirations, motives, opportunities and alternatives. It does so by focusing on the way in which people read the situation, their intentions and how they anticipate interacting with others, before the action is actually 'played' out in real life. It is a radical development from **game theory** in that it brings the paradoxes of the situation and *dilemmas* faced by individuals on stage by playing out dramatic episodes, with all the emotions and irrationality that such situations include.

Howard (1998) provides the underlying mathematical theory of what he also calls *soft game theory*. This examines the dilemmas that can occur when each player has chosen an intended final position – an offered position that the player hopes will be accepted as a joint strategy by all players – and a fall-back position that the player says he or she will pursue if not convinced that the joint strategy will be implemented. Each dilemma arouses specific emotions in a player. Drama theory assumes strong associations between expressed emotions and preference change. In particular, it posits that positive emotion is expressed to render credible a promise that otherwise would be viewed as questionable by the other players, while negative emotion is used to support an otherwise doubtful threat. By having to rationalize such emotion, players tends to redefine the game in a way that eliminates the dilemma in question. Howard's paper also proves that, when all dilemmas have been eliminated, all players take the same position and can trust each other to implement it. In this way, the analysis of the dilemmas leads to an understanding of the situation and a resolution of the drama.

The potential range of application is very wide. The process used has been found relevant in fields such as international relations, human–computer communication, management training, and counselling. The most recent applications considered have been to situations arising in military interposition and humanitarian relief operations.

Game theory has often been criticized for its name. Drama theory might similarly be regarded as a trivial label for something very serious to those involved. Alternative labels include *confrontation analysis*, *interaction analysis* and *conflict resolution*. The second of these seems relatively safe, since both 'confrontation' and 'conflict' are pejorative labels to some.

K.C. Bowen

References

Bennett, P. (1998) 'Confrontation analysis as a diagnostic tool', *EJOR*, 109:465–482. Published version of reference 6 in Howard (1998).

Bryant, J. (1997) 'All the world's a stage – using drama theory to resolve confrontations', *OR Insight*, 10(4):14–21. Explains key concepts and the core technique of the analysis, with examples.

Howard, N. (1998) 'N-person soft games', *JORS*, 49:144–150.

Howard, N., Bennett, P., Bryant, J. and Bradley, M. (1993) 'Manifesto for a theory of drama and irrational choice', *JORS*, 44:99–103. Gives detailed overviews of aims.

Drum–Buffer–Rope, OPT, synchronized manufacturing

The Drum–Buffer–Rope or OPT system is a production planning and control method, popularized in the novel *The Goal*, and subsequently generalized by its inventor, Eliyahu Goldratt, into the 'theory of constraints', a systems-based problem solving/**problem structuring methodology**. Its principles are based on an analogy of how fast a troop of marching soldiers progresses, namely only as fast as its slowest walker. That soldier will hold up all soldiers behind. If the ones in front walk faster, the troop simply spreads out. Therefore, the slowest soldier is given the drum to beat the pace and a rope is attached to the front soldier to prevent the troop from spreading out. To account for the ups and downs in the path the rope is given some slack, i.e., a *buffer* is built in.

Goldratt starts from the premise that the output rate (= the walking speed of the troop) of all manufacturing facilities is restricted by a single constraint (= the slowest soldier), called the *capacity-constrained resource* or CCR. Usually, the CCR is a piece of equipment or an operation that forms a *bottleneck*. The first task is to identify the CCR, for instance by analysing overtime records. The CCR work schedule becomes the drum and all other activities must work to its beat. This means that all upstream operations that feed the CCR (i.e. the soldiers ahead of the drummer) must synchronize their schedule such that the CCR gets the requirement inputs when needed, not late, but also not early. This is achieved by scheduling the release of raw materials to the first operation (i.e. the soldier at the very front), offset by the total processing time of all upstream operations (= the length of the rope). Finally, to protect the CCR from upstream production disruptions and safeguard its own production schedule, some slack is added to the rope, i.e. a *time buffer* is provided in the form of having material arrive at the CCR some time (e.g. an hour) prior to the scheduled usage. This is depicted in the diagram below.

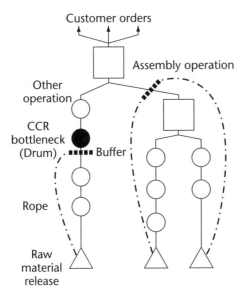

Figure: **Drum–Buffer–Rope production arrangement**

These are the basic principles of OPT. Its implementation will lead to looking for means to increase the capacity of the CCR, e.g. by reducing setup times through processing families of parts with similar setups, by engineering changes that save setup and production times, and training of operators. Any defective parts processed by the CCR will waste precious capacity. Hence, OPT puts great emphasis on zero-defects. Appropriate buffers are found through experience. Downstream operations are scheduled to follow the CCR schedule, avoiding lengthening of the total production lead time and increase in work in process. Whenever possible, sublots go to the next operation before the whole batch is completed, allowing operations to overlap, hence shortening the length of the rope needed. Priority is given to products that yield the highest profit contribution per hour of CCR time.

If these activities remove a CCR as the constraint, another operation will become the new CCR and the whole process of synchronization and improvements starts again. Ultimately, the customer demand may become the CCR, in which case efforts are made to raise demand or use spare capacity for new product lines.   *Jeff Foote*

References

Goldratt, E. and Cox, J. (1990) *The Goal: a process of ongoing improvement*, Croton-on-Hudson, NY: North River Press.

Umble, M. and Srikanth, M. (1990) *Synchronous Manufacturing: principles for manufacturing*, Cincinnati: South-Western Publ. Co.

Dual prices, see shadow prices

Dual simplex method

The **simplex method** and the dual simplex method are **algorithms** used in **linear programming** (LP), and they are very closely related to each other through **duality** theory. Consider the following two LPs:

(P): maximize $z = 6x_1 + x_2$ (D): minimize $w = 5y_1 + 6y_2$

subject to $x_1 + x_2 \leq 5$ subject to $y_1 + 2y_2 \geq 6$

$2x_1 + x_2 \leq 6$ $y_1 + y_2 \geq 1$

$x_1, x_2 \geq 0$ $y_1, y_2 \geq 0$

The original problem (P) is called the *primal LP*, and (D) is called the *dual LP*. Variables y_1 and y_2 can be interpreted as the **shadow or dual prices** on the first and second constraints of (P). If the simplex method is applied to (P), it starts with the initial **basic solution** $x_1 = 0$ and $x_2 = 0$; the corresponding solution in (D) turns out to be $y_1 = 0$ and $y_2 = 0$. Note that the given solution for (P) satisfies the constraints of (P) (i.e. it is **feasible**), whereas the corresponding solution for (D) does not satisfy the constraints of (D) (i.e. it is infeasible). At the optimal solution of (P), $x_1 = 3$ and $x_2 = 0$; the corresponding solution of (D) has $y_1 = 0$ and $y_2 = 3$, and is feasible. Both have an objective function value $z = w = 18$.

These observations now lead to an interpretation of the (primal) simplex method (PSM) as follows: It starts with a feasible primal solution (and a corresponding infeasible dual solution). It iterates through a sequence of basic primal feasible solutions until it finds the optimal primal solution, whose corresponding dual solution will also be feasible. The optimal objective function values of both problems are equal. The dual simplex method (DSM) essentially does this in reverse: it starts with a dual feasible solution (and a primal infeasible solution); it iterates through a sequence of basic dual feasible solutions until it finds the optimal dual solution (and the corresponding optimal primal solution).

The PSM is used to solve linear programs, but the major role of the DSM is in *post-optimal* or **sensitivity analysis**, particularly when adding a new constraint, or modifying the values of constraint resources, after (P) has been solved. For example, in (P) above, suppose the right-hand side of the second constraint is doubled to 12. Applying the PSM to (P) now yields $x_1 = -1$ and $x_2 = 0$; but the corresponding solution to (D) is unchanged. Hence (P) is infeasible and (D) is feasible, which are the necessary conditions for initializing the DSM and using it to find the optimal solution for (P).

As long as any right-hand-side changes are not too great, the DSM will start with a near-optimal dual feasible solution and a close-to-feasible primal solution. Optimality will therefore be restored quickly – certainly much more quickly than resolving (P) from scratch.   *J.W. Giffin*

Reference

Winston, W.L. (1994) *Operations Research: Applications and Algorithms*, 3rd edn, Belmont, CA: Duxbury.

Duality

Duality is the study of the relationships between two very closely related **mathematical programming** problems. Every **linear program** (LP) has an 'opposite' LP. The original is referred to as the *primal LP*, its opposite as the *dual LP*. The primal and dual are like two sides of the same coin – they both solve the same underlying problem, but look at it from different points of view.

Consider a decision maker who has access to a fixed level of resources (such as time and materials). These resources can be used to produce some good(s) that return a profit. Given that the production of any one good uses up a certain amount of resources, the best solution to this primal problem is to find the quantity of each of the goods which provides the maximum total profit, using up no more than the available resources.

Consider now another decision maker who wishes to value the resources, say as a potential buyer. In order to do this, she must decide on unit 'prices' for each of the resources that represent the value or worth of that resource. The purchaser wishes to minimize the total amount paid for the resources, but is constrained by the fact that the owner would be unwilling to sell the resources at a price any lower than the profit they contribute via the production of goods in the above primal problem. Stated another way, the value of all the resources used in the production of a certain good must at least be equal to the profit earned from that good. The minimization of the total amount paid for the resources and the restrictions on the prices of each resource combine to become the dual formulation.

The following simple production problem demonstrates this. A firm produces three products that require processing on two machines. The table below shows the net profit and machine usage per unit and daily capacity:

Primal	A: x_1 / day	B: x_2 / day	C: x_3 / day	Daily capacity	Dual variables
Machine 1	1.5 hrs/unit	3.6 hrs/unit	0.9 hr/unit	21 hrs	y_1 / hr
Machine 2	2.1 hrs/unit	2.8 hrs/unit	1.2 hrs/unit	23 hrs	y_2 / hr
Profit/unit	$180	$320	$150		

The primal LP is to find values for the x_i variables that maximize their daily profit contribution. The dual LP is to value the two machines so as to minimize the daily cost. Hence

Primal LP:

maximize $180x_1 + 320x_2 + 155x_3$

subject to $1.5x_1 + 3.6x_2 + 0.9x_3 \leq 21$

$2.1x_1 + 2.8x_2 + 1.2x_3 \leq 23$

$x_1, x_2, x_3 \geq 0$

Dual LP:

minimize $21y_1 + 23y_2$

$1.5y_1 + 2.1y_2 \geq 180$

$3.6y_1 + 2.8y_2 \geq 320$

$0.9y_1 + 1.2y_2 \geq 150$

$y_1, y_2 \geq 0$

For each constraint in the primal, the dual has a variable, and for each variable in the primal the dual has a constraint. These linkages between the primal and the dual are known as *primal–dual relationships*. From the solution of one, the solution of the other can be inferred.

The dual variables are often called **shadow prices**, since they represent the marginal productive value of a unit of a resource. If the firm were given the opportunity to purchase more resources, and thus produce more goods, the shadow price or dual variable associated with that resource represents its unit profit contribution, or the maximum price the firm should be willing to pay to obtain it. If the firm paid any more, it would not recoup the cost in the extra sales it would make.   S. Batstone

Reference

Hillier, F.S. and Lieberman, G.J. (2001) *Introduction to Operations Research*, 7th edn, New York: McGraw-Hill, ch. 6.

Dynamic economic order quantity model

In contrast to the ordinary **economic order quantity model**, this **inventory control model** assumes that the demand in each period is known, but may differ from period to period. It then determines the sequence of production batches to cover that demand so as to minimize the sum of the setup costs and holding costs over a given planning horizon of N periods. Holding costs are assessed on end-of-period stock only, and production in a period is available to meet demand in that period. Although the **algorithm** used is derived from **dynamic programming**, it is deceptively simple and can easily be done by hand.

It can be shown that it is never optimal to carry stock into a period, unless it covers the entire demand for that period. As a result, in each period n, the only candidate batches that need to be considered are those that cover the demand in period n, d_n, or the demand in periods n and $n+1$, $d_n + d_{n+1}$, or periods n, $n+1$, and $n+2$, and so on.

The following algorithm finds the optimal batch sequence:

Step 1: Set $Q_1 = d_1$, $T_0 = 0$, $T_1 = A$, $n = 1$, and $m = 2$, where A is the fixed setup cost.
Step 2: For all k from n to m compute

$$T_k = T_n + \sum_{k=n+1}^{k=m} d_k h(k-n)$$

where h is the unit holding cost per period.

Step 3: Find k^* which has the minimum value of T_k.
Step 4: If $k^* > n$, set $n = k^*$ and $T_n = T_{k^*}$; otherwise leave n unchanged.
Step 5: If $m = N$, stop and trace the optimal batch sequence backwards, starting in period N; otherwise increase m to $m+1$, and return to step 2.

Consider the following small example for a setup cost $c = \$250$ and a holding $h = \$1/\text{unit}$. The algorithm performs the computations column by column. To find the optimal sequence, we find k^* for the last period. It is 6. So a batch of $Q_6 = 210$ is made in period 6. In column 5, $k^* = 3$. This implies a batch is produced in period 3 to cover the demand in 3, 4 and 5, so $Q_3 = 260$. In column 2, $k^* = 1$, hence $Q_1 = 250$ covers periods 1 and 2. Note that the periods have to be sufficiently short, otherwise the truly optimal sequence might have more than one replenishment per period.

H.G. Daellenbach

Reference

Silver, E.A., Pyke, D. and Peterson, R. (1998) *Inventory Management and Production Planning and Scheduling*, New York: Wiley, ch. 6.

n	k	$m = 1$	2	3	4	5	6
		$d_m = 170$	80	120	40	100	210
1	1	250*	330*	570*	690		
1	2		500	620	700		
3	3			580	620*	820*	1450
3	4				820	920	1340
3	5					870	1080
6	6						1070*
Q_m		250	0	260	0	0	200
Covering periods		1, 2		3, 4, 5			6

Dynamic programming

Dynamic Programming (DP), invented by R. Bellman in the 1950s, is a computational method for optimizing multistage decision problems.

Consider scheduling production of a hand-built car over a four-month period to meet known customer orders of 1, 1, 0, and 3 cars, respectively. In each period a decision is made whether to produce 0 cars at a cost of 0, 1 car at a cost of 50, or 2 at a cost of 90 thousand dollars. Any cars carried into the next period incur a storage cost of 8. The network below shows the possible production combinations. The number on top of each arrow is the amount produced, the number below the combined production and storage cost. The number in each circle is the number of cars carried forward from the preceding period. For example, if two cars are produced at a cost of 90 in the first period, one of which is sold, one car will be carried into the next period. If no car is produced in the second period, the car left over from period 1 is sold, and the cost is only the holding cost of 8.

The basic idea of DP is to break the problem into a sequence of *stages* and solve a large number of smaller and simpler problems. In our case, the stages are the periods. The smaller problems are finding the optimal number of cars to produce for each

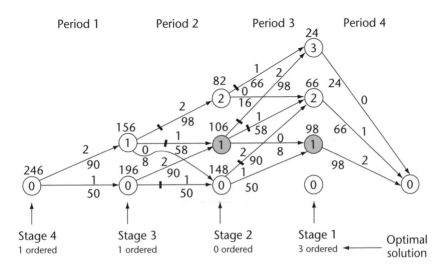

Figure: **Network diagram for dynamic programming formulation**

possible number of cars carried into the period – referred to as the *states* at that stage. The computational method works recursively backwards, starting with the last period labelled stage 1. At stage 1, the optimal decision for each state is usually trivial since there is only one choice. For example, referring to the shaded circle, if 1 car is carried into the last period (state = 1), 2 cars have to be produced in that period to meet the demand of 3. The total cost is 8 plus 90, or 98 – the number shown on top of the circle.

Knowing the optimal decision for each state at stage 1, we can now evaluate the optimal decision for each state at stage 2. For example, for state 1, the choices are: (1) produce 0 at a cost of 8, plus the cost from the state at stage 1 resulting from this decision (i.e., state 1), which is 98, giving a total of 106, (2) produce 1 car at a cost of 58 plus the cost of 66 from state 2 at stage 1, totalling 124, or (3) produce 2 at a cost of 98 plus the cost of 24 from state 3 at stage 1, totalling 122. The optimal choice is 0, hence the number 106 shown on top of the shaded circle at stage 2. The other two options are blocked off.

These computations are performed recursively for all states at each stage. The overall optimal sequence of decisions – called a *policy* – can easily be evaluated by following the unblocked arrows from the starting node. In our example, there are two optimal policies, both with a minimum cost of 246, i.e. produce 1, 2, 0, 2 (solution shown) or 2, 0, 1, 2 in periods 1, 2, 3 and 4, respectively.

Note that the optimal policy is built only from optimal *subpolicies*. This is known as the *principle of optimality*. In this formulation, we worked backwards from the last period – a so-called *backward formulation*. The problem can also be reformulated as a *forward formulation*.

More complex problems may require more than one variable to describe a state. For example, if the production process consists of two phases and partially completed cars can be stored from one period to the next, the state is described by two state variables: the number of partially completed cars and the number of finished cars brought forward. The computational effort may become impractically huge if the number of state variables increases beyond 2. This is known as the *curse of dimensionality*.

In the above example, the demand is known with certainty. If the demand in each period is **stochastic**, i.e. we know only its probability distribution, then each decision will lead to one of several new states at the next stage, each with a given probability. This implies that we compute an expected value over the costs associated with these states. This is added to the immediate cost of the decision itself. The decision that has the lowest total expected cost is the optimal one. The basic computational procedure therefore remains the same, except that the minimum cost associated with each state at each stage is now an expected value.

This is known as *stochastic or probabilistic dynamic programming*. There are two further consequences. The first is that we can only use a backward formulation to solve the problem. The second is that, except for the first decision in the planning horizon, it is not possible to indicate which decision we will take at each stage. All we can give is a policy or *strategy of conditional decisions* that depend on which state eventuates at a given stage. *S. Dye*

Reference

Hillier, F.S. and Lieberman, G.J. (2001) *Introduction to Operations Research*, 7th edn, New York: McGraw-Hill, ch. 11.

E

Economic order quantity (EOQ) models

The most basic deterministic inventory replenishment or batch size model is the EOQ model, formulated independently by Harris in 1915 and Wilson a few years later. The EOQ is also known as the Wilson lot size and the *square root formula*. It minimizes the sum $T(Q)$ of annual inventory holding costs, annual setup or ordering costs, and annual product costs, i.e.

$$T(Q) = 0.5Q\ h\ v(Q) + A\ (D/Q) + D\ v(Q),$$

where h = annual holding cost fraction per dollar invested, $v(Q)$ = average unit product cost in stock, A = fixed setup or ordering cost per replenishment, and D = annual rate of demand or usage. If $v(Q)=V$ for all Q (i.e. there are no *quantity discounts* for large Q), then the last term of $T(Q)$ is constant and the optimal Q^* = EOQ that minimizes $T(Q)$ is given by the well-known square root formula:

$$EOQ = \sqrt{\frac{2DA}{hV}}$$

Since D is assumed constant, $t = Q/D$ years, and the time stocks are depleted can be predicted exactly. The graph below shows the shape of the first two terms of $T(Q)$, i.e. the annual holding cost, the annual setup cost, and their sum. The optimal Q^* always occur where these two costs are equal.

If there are quantity discounts, the last term of $T(Q)$ depends on Q. The EOQ and its total associated cost $T(EOQ)$ has to be computed separately for each price range of Q. If the EOQ falls into its corresponding price range it becomes a candidate. $T(Q)$ is also computed for all Q's equal to a price break (i.e. the order size where the unit price decreases), which also become candidates. The optimal Q^* its the one that has the lowest cost of all these candidates. If the quantity discounts are substantial, the optimal Q^* is usually equal to one of the higher price breaks, since the annual saving of a large price discount far outweighs the increase in the other costs.

If demand is not constant, but changes in a known pattern each period, the **dynamic economic order quantity model** can be used to find the optimal sequence of replenishments.

The EOQ model is very **robust**. It can be shown that even if a pseudo-optimal EOQ is used which has been computed for a demand or a fixed setup cost that has been

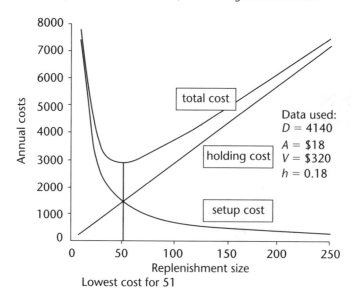

Figure: **Annual holding and setup costs for economic order quantity model**

erroneously overestimated by 50 per cent, the increase in actual costs, based on the correct input data, is only about 2.1 per cent. (See also **sensitivity analysis**.)

H.G. Daellenbach

Reference

Silver, E.A., Pyke, D. and Peterson, R. (1998) *Inventory Management and Production Planning and Scheduling*, New York: Wiley.

Economies of scale, diseconomies of scale, diminishing returns to scale

The concepts of economies of scale, synergy and 'biggest is best' are all widely misunderstood. Synergy – the ability to exploit economies of scale effectively – is more difficult to achieve in practice than portrayed in theory. 'Biggest is best', as the dinosaurs found to their cost, invariably makes adaptation to change more problematic. Large size often makes organizations more vulnerable to competition from those able to adapt to change more quickly.

Economies of scale or *increasing returns to scale* means that a given increase in inputs increases outputs more than proportionately. *Constant returns to scale* implies that outputs increase by the same proportion as inputs. Finally, *decreasing returns to scale* or *diseconomies of scale* means that outputs increase by a smaller proportion than inputs. It is generally true that as inputs increase, most productive activities initially enjoy increasing returns to scale, followed by constant returns to scale, but ultimately will experience decreasing returns to scale. This is depicted in the figure below.

The latter phenomenon is due to the *law of diminishing returns*. That law states that if the amount of one or more factors used in a particular form of production is fixed, and increasing amounts of other factors are combined with the fixed factor (or factors), then both the average output in relation to the variable factors and successive additions to the total output will eventually diminish. The conclusion from this is that there is always an optimal output level for any given productive activity.

Economies and diseconomies of scale arise from the interaction of specific, but very different causes. Economies of scale are associated with more efficient use of

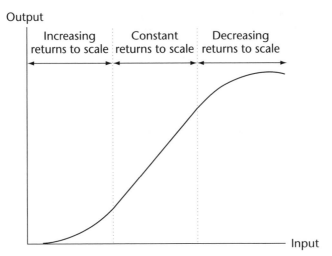

Figure: **Increasing, constant, decreasing returns to scale**

resources and with technical developments. They also arise from the effect of the law of multiples. This law states that the smallest rate of output at which a number of machines can be used together with maximum **efficiency** is the lowest common multiple of their separate capacities. Other examples include the use of reserves and the spreading of risks, bulk buying and the benefits of specialization. It should, however, be noted that specialization also has disadvantages; for example, it can easily make organizations vulnerable if a change takes place that undermines the value of the specialism (e.g. large sailing ships were completely replaced over a relatively short time period by engine-driven vehicles, using very different technology).

Diseconomies of scale are associated with mutual interference of activities and the increasing complexity of managing change and innovation successfully as a function of the size of an organization. Radical change involves renegotiation of formal and informal relationships and this increases exponentially with the size of the organization.

The more change there is, the harder it is to exploit economies of scale. At the same time, those organizations that can use economies of scale to their benefit invariably achieve a competitive advantage. All too frequently mergers and acquisitions fail to achieve the expected benefits because the potential for economies of scale has been overestimated and the potential diseconomies of scale have been ignored.   *B. Lloyd*

References

Baldock, R. (2000) *The Last Days of the Giants? A Route Map for Big Business Survival*, Chichester: Wiley.

Campbell, A. and Goold, M. (1998) *Synergy: Why Links Between Business Units Often Fail and How to Make Them Work*, Oxford: Capstone.

Cassimatis, P. (1996) *Introduction to Managerial Economies*, London: Routledge.

Lloyd, B. (1976) 'Economies of scale', *Mergers and Acquisitions*, Winter: 4–14.

Effectiveness, efficiency, efficacy

Systems or organizations use resources (staff, equipment, funds, etc.) to engage in activities or operations in order to achieve certain goals or **objectives**. *Effectiveness* is concerned with the degree to which these activities or operations achieve the desired goals or objectives. *Efficiency* looks at how well the system's resources are used in a given activity or operation. For example, one of the objectives of the fire service is fire prevention through education, fire drills, and inspection. The service's effectiveness can be measured by the fraction of private houses with operational fire alarms, the fraction of compulsory fire drills actually undertaken, the fraction of fire drills that meet performance standards, and the fraction of buildings that meet fire regulations. Efficiency compares specific performance measures with resource use, mainly staff time and funds. The efficiency of staff time use is measured by the fraction of staff time available for fire prevention that has actually been used for that purpose, or the number of each type of activity executed per, say, 100 staff hours.

Efficacy is the ability of an activity to produce the desired effect, e.g. increasing the capacity of a bottleneck operation will increase output, but adding another machine when the first already has excess capacity will not produce more output – its efficacy is nil.

A narrow concern with efficiency may be detrimental to effectiveness. For example, the fewer checkout operators that are on duty at a supermarket, the busier they are, the lower is their idle time, and hence the more efficient is their time usage. However, the resulting long customer-waiting queues may reduce customer patronage, thereby reducing the operation's effectiveness in achieving high sales and profits, the primary objectives of the supermarket. Similarly, trying to achieve the highest effectiveness of a system by maximizing the efficiency of each operation or each subsystem is a fallacy, unless each subsystem is separable, i.e. is not a proper subsystem, but an independent system in its own right. Efficiency gains in some subsystem may result in loss of efficiency or effectiveness in other parts of the system or the system as a whole (as well as have adverse effects on the wider system). True concerns for efficiency take the overall goals of the system into account by allowing the resources to go further without detriment to overall effectiveness – in fact, enhancing it. Effectiveness and efficiency are thus complementary. Effectiveness deals with 'doing the right things', efficiency with 'doing things right'.   *H.G. Daellenbach*

Reference

Daellenbach, H.G. (2001) *Systems Thinking and Decision Making*, Christchurch: REA, ch. 2.

Efficient frontier, efficient solutions, see dominance

ELECTRE methods

There is a family of **multicriteria decision methods** (MCDM) or aids known collectively as *outranking methods*. The most prominent of these are the ELECTRE methods, developed by Bernard Roy, the 'father' of outranking methods. The outranking methods are distinguished from **multiattribute value function methods** (MAVF) in that they can work with poorer information, but also yield a weaker result, allowing two alternatives to be declared incomparable. In common with other MCDM approaches, outranking methods begin with the definition of a set of alternatives to be evaluated and a set of **objectives** to form the basis of the evaluation. The performance of each alternative against each objective, $g_j(A)$, is specified and usually weights, W_j's, are allocated to objectives. These weights do not describe acceptable trade-offs in the same manner as the weights in MAVF. They are more akin to a 'voting power'. The first and simplest outranking approach, ELECTRE I, is a good basis for explaining the underlying principles. For each ordered pair of alternatives, say, A and B, a *concordance index*, c(A,B), and a *discordance index*, d(A,B) is calculated. The concordance index measures the strength of evidence that A outranks B, i.e. is at least as good as B; the discordance index is the strength of the evidence against. The concordance index is defined as:

$$c(A,B) = (\Sigma_{j:\ g(A) \geq g(B)}\ W_j) / \Sigma_j\ W_j$$

i.e. the proportion of the total weight allocated to those objectives for which A performs at least as well as B. The discordance index may also be defined by a formula, or by reference to a veto threshold v_j, where $d_j(A,B) = 1$ if $g_j(B) - g_j(A) > v_j$

Alternative A is said to outrank B if:

$$c(A,B) \geq c^* \text{ and } d_j(A,B) = 0 \text{ for all }_j$$

where the outranking threshold, c^*, is specified by the analyst.

The outranking relation can be visualized as a graph, as illustrated below. An arrow from A to B indicates that A outranks B. In ELECTRE I the aim is to identify the set of alternatives which are not outranked. In the figure this is the set {B, F}.

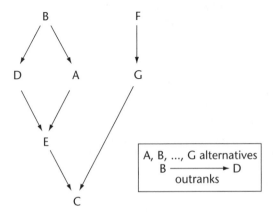

Figure: **Example of an outranking relation**

Later ELECTRE methods retain these basic principles, but incorporate more sophisticated modelling of preferences and to deal with other problem types (e.g. ranking or classification of alternatives). ELECTRE III incorporates **fuzzy** preferences, i.e. preferences that have a degree of vagueness. ELECTRE IV operates without weights and ELECTRE TRI is used to assign alternatives to categories.   *V. Belton*

Reference
Vincke, P. (1999) 'Outranking approach', in T. Gal, T.J. Stewart and T. Hanne (eds) *Multicriteria Decision Making: Advances in MCDM Models, Algorithms, Theory and Applications*, Dordrecht: Kluwer, pp. 11–1 to 11–29.

Electricity, see **applications of MS/OR in electricity sector**

Emergency services, see **applications of MS/OR to emergency services**

Emergent properties

An emergent property is a behaviour or output of a **system** that none of its components or groups of components individually exhibit. Such behaviours or properties are new or different from the behaviours or properties of the individual components.

They are a result of synergy between components and only emerge from their inter-action, hence the label 'emergent properties'. This phenomenon is often summarized by 'the whole is greater than the sum of its parts'.

Consider a telephone system. It consists of telephones, capable of sending and receiving electric impulses over telephone lines and wireless transmission devices, which are processed by telephone switching systems. Working together they become a communication system for people not simultaneously present in the same loca-tion. None of its parts or activities between parts has that property. It emerges from their interactions. In fact, the purpose of putting these components together was to create a such a communication system.

Similarly, a training system consists of classrooms, teachers, equipment, and a cur-riculum to transform students without certain skills into graduates with those skills. Again it is the synergy between the parts, not the parts alone, and their interactions that makes it a training system. This is the system's purpose.

In fact, **human activity systems** are generally 'created' in order to produce desired emergent properties. Unfortunately, such systems also produce emergent properties that are not desirable or even planned. For example, a training system also produces 'drop-outs' and psychological stress for both teachers and students. The classic exam-ple for the lack of a comprehensive systems approach was the building of the Aswan High Dam in Egypt in the 1960s. Hailed as the 'great' technical achievement, it not only increased the availability of irrigation water, but also had several unplanned undesirable emergent properties, with consequences that may outweigh the benefits, such as the loss of fertile silt to the farm areas, also causing the sea to begin to encroach on the land and the killing of the sardine fishery which depended on the nutrients in the silt, a massive increase in the use of fertilizers, which in turn caused salinization and consequent loss of arable land, and so on.

One of the compelling reasons for using a **systems thinking** approach to problem solving, rather than only **reductionism** and **cause-and-effect thinking**, is exactly to predict and control both planned desirable emergent properties and unplanned undesirable emergent properties more effectively and before they happen.

H.G. Daellenbach

Energy, see applications of MS/OR in energy modelling

Entity cycle diagrams, see activity cycle diagrams

Environment, see applications of MS/OR to environmental issues

Equifinality, equilibrium, see steady-state

Equivalent annuity, see discounted cash flows

Error analysis, see sensitivity analysis

Ethics and MS/OR

Ethics is the philosophical study of how moral decisions are made – the moral prin-ciples and values that govern the behaviour of a person or a group with respect to what is judged right or wrong by the society they live in. Different societies may abide by different ethical standards.

Ethics enters into decision making in two ways. First, ethical principles should form the basis for all decision making, or else it will become egotistical, opportunistic and harmful to other **stakeholders** – 'what you do not want done to yourself, do not do to others'. Other stakeholders includes future generations and nature. Second, the process of decision making should abide by both personal and professional ethical standards. On a personal level, managers and analysts should not promise more than they can reasonably expect to deliver. They also must take moral responsibility for the effect of their actions on all stakeholders and third parties that benefit from or are the voiceless victims of undesirable consequences of the recommended decisions. At a minimum, this means alerting the decision maker to these effects.

Many professions have a formal code of ethics, e.g. accountants, engineers, doctors and lawyers. Others, such as management scientists, have informal codes. Most would agree on the following rules:

- Disclose any vested interest in the project or its outcomes.
- Approach the problem situation from the **world view** of the project sponsor.
- Keep the project sponsor regularly informed about progress and immediately report the discovery of new aspects or undesirable consequences that calls for a reassessment of the project.
- Fully document the analysis done, such that it can be verified, i.e. record and justify all assumptions and simplifications made, data sources and data ignored or discarded. Disclose any weaknesses in the analysis (even if it is tempting not to do so). This also applies to weaknesses or mistakes discovered after the report has been submitted. Your own loss of credibility may be of minor importance in comparison with the possible risks for other stakeholders if you fail to come clean.
- Test, **verify** and **validate** the model. Perform **sensitivity analysis** and establish ranges of critical input parameters for which the recommendations remain valid.
- Scrupulously observe any ground rules about confidentiality for the disclosure of data and any reports produced, as laid out at the inception of the project. This should also cover what material may be removed from the premises of the sponsor and by whom.
- Do not undertake a project that requires you to rubber stamp a conclusion or decision already reached or do it only with the clear (written) understanding that you are in no way bound by prior decisions or conclusions.
- The report and its analysis should be written in such a manner that neither can be easily misrepresented or used to imply more than it should.   *H.G. Daellenbach*

References

Daellenbach, H.G. (1994) *Systems Thinking and Decision Making*, Christchurch: REA, pp. 329–333.

Gass, S.I. (1991) 'The many faces of OR', *JORS*, January: 2–16.

Guidelines for Practice of OR, *Operations Research*, September 1971. Entire issue devoted to professional guidelines, with emphasis on ethical standards.

Wallace, W.A. (ed.) (1994) *Ethics in Modeling*, New York: Pergamon.

Expert systems

An expert system is a computer-based system using a knowledge base to help solve problems or give advice in narrowly defined fields of expertise. The knowledge base consists of rules and data dealing with the specific domain of expertise and is typically derived from human experts. The expert system is designed by a knowledge engineer, a programmer/analyst expert in designing knowledge-based systems. The term *knowledge-based system* is often used as a synonym for expert system.

Expert systems have been created in a wide range of fields, including medicine, engineering, **logistics** management, **production control**, system configuration, investment analysis and credit management. These systems are typically designed to capture the expert knowledge of the few top people within the organization, so as to make this scarce expertise more widely available. Thus expert advice is provided to the average worker, improving the organization's overall performance. In practice, such systems sometimes match or exceed human experts, and sometimes perform well below such expectations.

The components that typically make up an expert system are the *knowledge acquisition subsystem*, the *knowledge base*, the *inference engine* and the *explanatory interface*.

The knowledge acquisition subsystem provides the tools for the knowledge engineer to capture the human expert knowledge and structure it in the most effective way, often as rules that will be applied to specific diagnosis or problem situations. This subsystem may also include a machine learning component capable of inferring new rules through pattern recognition, from the growing body of data on situations to which the system has been applied.

The knowledge base contains the expert rules, data from past problems and the data of the current situation on which the system is being asked to provide advice.

The inference engine takes the rules from the knowledge base and applies them to infer a solution to a current problem. The inference engine may be designed to use forward chaining (inductive reasoning) or backward chaining (deductive reasoning) to reach its conclusion. The inference engine will often be designed to deal with uncertain situations using such methods as certainty factors or **fuzzy logic** to permit a decision to be reached.

The explanatory interface provides the means by which the advice or potential solution from the system is communicated to the user of the system as well as providing the reasoning behind the advice. In some applications the system may automatically execute a decision, for example rejecting a bad credit risk. However, typically the system provides advice to a human decision maker, may link to other decision support tools to further assist the decision maker and leaves the final decision to the human.   *J. Vargo*

References

Cawsey, A. (1998) *The Essence of Artificial Intelligence*, Upper Saddle River, NJ: Prentice Hall.

Durkin, J. and Durkin, J. (1998) *Expert Systems: Design and Development*, Englewood Cliffs, NJ: Prentice-Hall.

Jackson, P. (1999) *Introduction to Expert Systems*, 3rd edn, Harlow: Addison-Wesley.

Exponential smoothing, Holt's method, Winter's method

Some of the most successful **forecasting** methods are based on the idea of combining past history with the most recent observation. Exponential smoothing methods, first suggested by R.G. Brown in 1956, do this by expressing all past history by one or two numbers which are regularly updated. If A_{t-1} represents past history up to period t − 1, and X_t is the observation for period t, then history is updated as

$$A_t = \alpha X_t + (1 - \alpha)A_{t-1}$$

A_t is called the *smoothed average* and α the *smoothing constant* – a fraction usually set between 0.1 and 0.3. The larger α, the more weight is given to more recent observations. The impact of past observations on A_t decreases geometrically with age. A_t is used as the forecast for the next period, $t+1$. Note that A_t is the only piece of information carried forward from period to period. This method is called *simple exponential smoothing*. It is suitable for stable phenomena without a trend or seasonality.

Holt's method is an extension of exponential smoothing to cater for a trend in the time series. It applies exponential smoothing to both the average and the trend. Let L_t denote the *smoothed base level*, T_t the *smoothed one-period trend*, and β the smoothing constant for the trend. Then

$$L_t = \alpha X_t + (1 - \alpha)(L_{t-1} + T_{t-1})$$

$$T_t = \beta(L_t - L_{t-1}) + (1 - \beta)T_{t-1}.$$

β is usually smaller than α, e.g. $\alpha = 0.3$, $\beta = 0.1$. A forecast for period $t + k$ is obtained as

$$F_{t+k} = L_t + k\,T_t$$

Two pieces of data now need to be carried forward. The table below shows example computations, using $\alpha = 0.3$ and $\beta = 0.1$. It also shows the forecast error.

Period	Observation	Base level	Trend	Forecast in $t+1$	Forecast error
t	X_t	L_t	T_t	F_{t+1}	$X_t - F_t$
Carried forward		68	5.46	73.46	
1	80	75.42	5.66	81.08	80 − 73.46 = 6.54
2	94	84.95	6.04	90.99	12.92
3	100	93.70	6.31	100.01	9.01
4	98	99.41	6.25	105.66	−3.01
5	112	107.56	6.44	114.00	6.34
6	106	111.61	6.20	117.81	−8

Exponential smoothing can also be applied to the forecast error, which can then be used to determine an interval estimate for the forecast.

Winter's method extends this analysis to account for seasonality. For example, if demand in July is 40 per cent above the average, then it has a *seasonality factor* of 1.4. The equations of Holt's method are still used, but on the *deseasonalized* data. So rather than use X_t, it is first deseasonalized by dividing it by its seasonality factor. The deseasonalized forecast is then corrected for seasonality by multiplying it by its corresponding factor. The seasonality factors in turn are updated using exponential smoothing. *H.G. Daellenbach*

Reference

Makridakis, S., Wheelwright, S.C. and Hyndman, R.J. (1998) *Forecasting: Methods and Applications*, New York: Wiley.

Facilities layout

Facilities layout (FL) is concerned with the design of physical entities, such as factories or public buildings. One of the main aims of FL is to produce a scale plan (often called a *block plan*) of the entity to be designed. The plan depicts the physical components or facilities of the entity, laid out in relative position to each other. Each facility in the plan has a given area and, in some instances, shapes as well are pre-specified. The construction of an effective plan usually depends, in part, upon the relationships between its facilities, which may be either qualitative or quantitative, and are based on the relative location of pairs of facilities. Qualitative relationships are often reflected in what is known as an *REL chart*, which is a table whose numerical entries represent the desirability of locating pairs of facilities adjacent to each other. Quantitative relationships, reflecting penalties for travel distances between facilities or other manufacturing or handling costs, are represented in *from-to tables*.

FL solution techniques attempt to design a layout that optimizes a given objective function which measures the overall qualitative desirability or the sum of quantitative penalties resulting from the inter-facility relationships. Most FL models are **NP-hard**, which means that there are no **algorithms** that guarantee to find an **optimal solution** in a reasonable amount of computer time. This has reinforced the search for effective **heuristic solution methods**.

The first major FL heuristic methods to become widely used in industry were CRAFT, CORELAP and ALDEP, which were all developed in the 1960s for implementation on mainframe computers. Since that time, two main FL models have emerged, one based on the quadratic **assignment problem** (QAP) (Hillier and Connors, 1966) and the other on **graph theory** (GT) (Foulds, 1983). QAP models involve minimizing the total transportation cost between all pairs of facilities. The calculation of this cost is based on the work flow and the distance between each pair of facilities (Chiang, 1998). This objective makes these models more appropriate for manufacturing operations. GT models usually involve maximizing the sum of the REL chart scores corresponding to the pairs of facilities that are adjacent in the block plan (Merkler and Waescher, 1997). This objective makes these models more appropriate for service operations. There are also variations to these classical models which involve multiple floors (Johnson, 1982) and **multiple objectives** (Malakooti and Tsirisjo, 1989). The standard solution techniques for these models are often not very effective for some of the large-scale FL problems encountered in practice. This has given rise to approaches based on random search procedures, such as **simulated annealing** and **genetic algorithms** (Chiang, 1998; Suresh and Sahu, 1993; Suresh *et al.*, 1995).

Among the software packages that implement the solution techniques already mentioned there are **decision support** and **expert systems** (Malakooti and Tsirisjo, 1989; Foulds, 1997). FL has become increasingly important as organizations have struggled to control operating costs. This is especially true for many manufacturing plants, where global competition has spurred efforts to reduce production costs.

L.R. Foulds

References

Chiang, W.-C. (1998) 'Intelligent local search strategies for solving facility layout problems with the quadratic assignment problem formulation', *EJOR*, 106:457–488.

Foulds, L.R. (1983) 'Techniques for facilities layout: deciding which pairs of facilities should be adjacent', *Management Science*, 29:1414–1426.

Foulds, L.R. (1997) 'LayoutManager: a microcomputer-based decision support system for facilities layout', *Decision Support Systems*, 20:199–213.

Hillier, F.S. and Connors, MM (1966) 'Quadratic assignment algorithm and the location of indivisible facilities', *Management Science*, 13:33–40.

Johnson, R.V. (1982) 'SPACECRAFT for multi-floor layout planning', *Management Science*, 28:407–417.

Malakooti, B. and Tsirisjo, A. (1989) 'An expert system using priorities for solving multiple criteria facility layout problems', *International Journal of Production Research*, 27:793–808.

Merkler, J. and Waescher, G. (1997) 'A comparative evaluation of heuristics for the adjacency problem in facility layout planning', *International Journal of Production Research*, 33:447–466.

Suresh, G. and Sahu, S. (1993) 'Multi-objective facility layout using simulated annealing', *International Journal of Production Research*, 32:239–254.

Suresh, G., Vinod, V.V. and Sahu, S. (1995) 'A genetic algorithm for facility layout', *International Journal of Production Research*, 33:3411–3423.

Facility location

Studies indicate that physical distribution accounts for 10 to 30 per cent of the total cost of products on store shelves. The judicious choice of sites for manufacturing facilities and wholesale distribution warehouses may keep these costs to a minimum. Similarly, suitable locations for service facilities which have face-to-face client contacts will have a major effect on patronage and revenue.

Location choice is affected by proximity to customers, proximity to raw materials and suppliers, availability of land, availability of a suitable labour force, various amenities (water, housing, recreational facilities, etc.), access (by rail, road, or water) and business climate (local government by-laws, presence or absence of competing companies), as well as political risks and cultural barriers. Some of these factors become constraints, while others have the character of goals or objectives to minimize or maximize. Facility location is therefore a **multicriteria decision problem**, with total cost minimization the major focus for manufacturing facilities and revenue maximization the major focus of service industries.

A number of quantitative approaches have been developed:

1. *Factor rating methods*, such as the aggregate value function method, are multicriteria approaches. Both qualitative and quantitative factors are assigned an importance weight. Each site is then given a numeric score for each factor (e.g. between 0 and 100), that measures how well the site meets a particular factor. An overall score for each site is computed as the weighted average of the factor scores. The preferred site is the one with the highest overall score.
2. *Centre of gravity* or load-distance models determine the geographical coordinates for a single facility that represents the centre of gravity of the sum of load-distance products.

3. *Finite set location* models minimize total costs over a predefined set of potential feasible facility locations. The costs considered include some or all of annualized facility setup cost, operating costs, raw material supply costs and finished product distribution costs, the latter two based on actual road distances between suppliers and plants or between plants and customers. They are usually formulated as large **integer programming** models.
4. *Infinite set location* models also minimize some or all of the above costs, but assume not a potential feasible set of locations, but a region. The facilities can be located anywhere in this region. Distances between known supplier and customer locations are expressed as *Euclidian* distances (i.e. straight-line distances). Starting with an initial guess for a set of potential locations, the final set of locations is found using an **algorithm**. Since the cost function is not well behaved, the solution derived may only be a **local optimum**, nor may all locations be feasible (i.e. there may be no land available, etc.). Better solutions may be found by systematically altering the number of locations and using different starting locations.   H.G. Daellenbach

References

Francis, R.L. and White, J.A. (1992) *Facilities Layout and Location: An Analytical Approach*, Englewood Cliffs, NJ: Prentice Hall. Covers finite set approaches.

MacCormack, A.D., Newman III, L.J. and Rosenfield, D.B. (1994) The new dynamics of global manufacturing site location, *Sloan Management Review*, Summer: 69–77.

Xion, S.C. and Watson-Gandy, C.D.T. (1971) *Distribution Management: Mathematical Modelling and Practical Analysis*, London: Griffen. Extensive coverage of infinite set approach.

Feasible solution, solution space

The solution space is the set of all combinations of values that the decision variable can assume. For example, if funds can be allocated to two activities, X and Y, and x and y are the amounts allocated, then the solution space is given by all combination of values for x and y. If we assume that negative investments are not admissible, the solution space is the positive quadrant, including zero allocation for either or both, shown in the left-hand panel of the figure below.

If the amount of funds is limited to F, then only combinations of x and y that do not use more than F are admissible, i.e. x and y must satisfy the inequality $x + y \leq F$. Admissible solutions are called feasible solutions. The right-hand panel shows the boundary of the inequality. A feasible solution has to lie on or below that boundary, i.e. in the shaded area. The shaded area is also known as the *feasible region*.

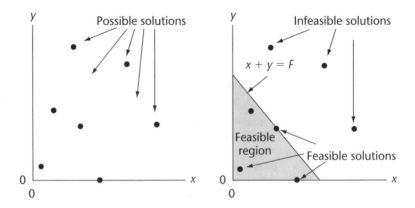

Figure: **Solution space, feasible solutions, feasible region**

The solution space may also be made up of discrete points, rather than a continuous space as in the above example. For instance, the **knapsack problem**, i.e. which combination of items on a wish list a hiker may take along on a wilderness trip, has a solution space that can be expressed as a list or vector of zeroes and ones. Element $x_i = 1$ if item i is taken along, and 0 if it is left out. The list [0, 1, 0, 0, 1] implies that items 2 and 5 are included, items 1, 3 and 4 are excluded. If there are N items, the solution space consists of 2^N discrete solutions. If the knapsack has a given volume, then only those solutions that do not fill more than the available volume are feasible solutions.

Feedback loops, feedback control, self-regulation

You may have experienced the feedback occurring in an audio system, when the microphone picks up the sound of the speakers and gradually amplifies the sound to a high pitch level. In feedback, signals or outputs from a process or **system** are fed back as inputs, directly or indirectly, into the process or system, affecting the behaviour of the system. Many natural biological and ecological systems depend on *self-regulation* in the form of feedback loops for their continued survival. For instance, warm-blooded animals maintain their temperature through feedback loops, as shown in the bottom panel of the figure below.

Engineering systems and **human activity systems** use feedback loops for control of the system, so-called *feedback control* (or *closed-loop control*, because the feedback loop is closed). A thermostat controls the heat output of a heating system. It monitors the discrepancy between its temperature setting and the room temperature. If the discrepancy exceeds a set amount, the thermostat initiates an action, signalling the heater to start or stop producing heat. The way we drive a car, e.g. steering or pressure on the accelerator, is in part based on feedback control. Many decision processes use feedback. For instance, in a production-inventory system, sales reduce stock levels. When stocks for a given product fall below a given critical minimum, a production run is initiated, which in time will bring stock levels back up again, and a new cycle begins. The top panel in the figure depicts feedback control.

The feedback may be negative or positive. The feedback loop in the heating system is negative, i.e. it decreases the discrepancy between the target temperature and the room temperature. So is the feedback loop in the production-inventory system. A positive feedback loop reinforces or accelerates the process that is occurring. The **state of the system** deviates more and more from its initial state. For instance, the nuclear melt-down of a reactor is based on positive feedback. Positive feedback leads to instability – the system explodes or implodes. Natural systems are mainly based on negative feedback. It returns the system back to a natural state of equilibrium or **steady-state**. Negative feedback is also the basis of control for most human systems.

Feedback may not be instantaneous, but lagged. It takes time for the results of the feedback to occur and to affect the system. For instance, it may take several days or weeks to produce the goods and add them to inventory. This is called a *transport lag*. The feedback may begin to show effects immediately, but the intensity of the effects changes over time, i.e. increasing gradually (as e.g. the room temperature) or decreasing gradually (as e.g. the rate of filling the toilet tank decreasing as the ballcock is raised).

Feedback loops are at the heart of **cybernetics** and **system dynamics**, a simulation tool to study the dynamic behaviour of systems. *H.G. Daellenbach*

Reference
Daellenbach, H.G. (2001) *Systems Thinking and Decision Making*, Christchurch: REA, pp. 47–53.

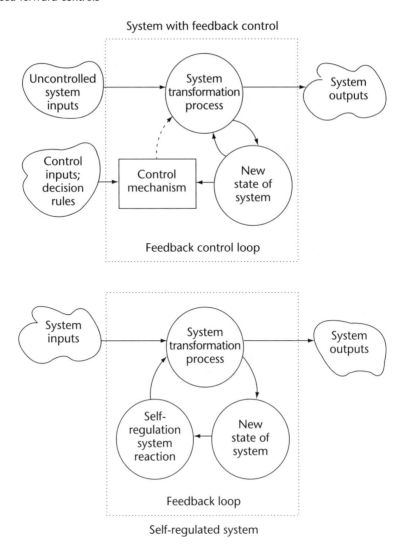

Figure: **Self-regulation and feedback control**
(*Source*: H.G. Daellenbach, 2001, p. 49)

Feed-forward controls

In contrast to **feedback controls**, which react to **system** outputs only, feed-forward controls predict or anticipate changes or disturbances in the behaviour, performance or output of a system and send control signals that will maintain system behaviour on its desired course or target or steer it back to that, thereby counteracting the effects of disturbances. This is the type of control used by an experienced car driver. It is also extensively used for the control of chemical processes. Planning in view of expected or predicted future events is feed-forward control. For instance, economic planners attempt to forecast future economic trends so as to steer the economy on the desired course of prosperity; business people predict demand trends to schedule production and plan expansion of facilities.

Feed-mix problem, see diet problem

Fishbone diagrams, see seven tools of quality management

Fixed charge problem

In a fixed charge problem, there is a fixed cost associated with undertaking an activity at a positive level that does not depend on the level of the activity. **Facility location** and production **scheduling** problems involve fixed charges. If all other costs are proportional to the level of activity and all constraints are linear, such problems can be formulated as **integer programs** with zero-one variables, where a 1 means that the activity level is positive and a 0 that it is not undertaken.

Consider a firm that wants to set up a distribution network for its products which are produced in several geographically dispersed plants. Products will be shipped from the plants to distribution warehouses and from warehouses to the final customers. A number of suitable potential warehouse sites have been identified. If a site is used, there is the fixed cost of building a warehouse of a given capacity. This cost is not incurred if the site remains unused. Each warehouse that is built at a site incurs a variable operating cost and with each site there are associated transportation costs from the plants and to the customers. The problem is to find the number of warehouses and their locations so as to minimize the sum of all total fixed and variable costs while meeting all customer demands without violating the capacity constraints of any warehouse operated. The problem can be extended to cover several time periods, leading to a dynamic facility location model where the costs are usually **discounted** over the **planning horizon**.

Another instance of a fixed charge problem is the selection of concentrator locations in a computer communication network that involves connecting a large number of given remote terminal sites via low-speed lines to one of several concentrators, which in turn are connected to the central site via high-speed lines. A fixed cost is incurred for each concentrator site set up.

In production scheduling, several machines with different fixed setup and variable production costs are capable of producing the parts required. The problem is to determine which machines to set up and which products to process on each machine set up so as to minimize the sum of fixed setup and variable production costs. The formulation can be extended to cover requirements over several periods, where requirements can be met from stock produced by a setup in an earlier period and carried forward, and stock holdings are penalized by a holding cost. In the special case of scheduling a single product over several periods, the latter problem can be easily solved via the **dynamic economic order quantity model**.

R. Sridharan

References

Hillier, F.S. and Lieberman, G.J. (2001) *Introduction to Operations Research*, New York: McGraw-Hill, ch. 12.

Winston, W.L. (1994) *Operations Research Applications and Algorithms*, 3rd edn, Belmont CA: Duxbury, ch. 9.

Fixed costs, see costs, types of

Flexible manufacturing system

Flexibility measures the ability to adapt to a wide range of possible environments. In today's dynamic, competitive and uncertain world, flexibility becomes an important requirement for long-term survival. Microprocessor technology has been a prime factor behind the development of equipment and information technologies that enable flexibility in manufacturing.

A flexible manufacturing system (FMS) is a set of computer-numerically-controlled (CNC) machine tools and workstations, connected by an automated material handling system, such as an automatic guided vehicle, where everything is controlled by a central computer. FMS technology represents an evolutionary step beyond the assembly lines, offering manufacturing a means to address the growing customer demand for quick delivery of customized products, particularly when the volumes are not large. Small batches of finished goods are expensive to make in traditional manufacturing facilities; production systems designed for producing high-volume, low-variety products are simply uneconomical, unresponsive, and noncompetitive in the small batch, higher variety marketplace. Increasingly, customers are demanding greater variety of products and shorter delivery times.

An FMS is generally appropriate when: (1) all products are variations of a stable design, (2) all products utilize the same family of components, (3) the number of components is moderate, say 10 to 50, and (4) the volume of each component is moderate, say 1000 to 30 000 units annually, but in lot sizes as small as one unit.

The key elements of a typical FMS are:

- automatically reprogrammable machines,
- automated tool delivery changing,
- automated material handling both for transferring parts between machines and for loading/unloading parts at machines, and
- coordinated control.

Many parts can be simultaneously loaded onto the system because the machines have the tooling and processing information to work on any part. Therefore, parts can arrive at a given machine in any sequence. By reading a code on the part, or following supervisory instructions, the part type can be identified, the proper processing sequence retrieved from the machine's computer and the part appropriately processed.

FMSs are expensive to implement but yield significant savings. In conventional systems, equipment utilization normally runs at most at 30 per cent, but may be at 85 per cent or higher in an FMS. It is more flexible than conventional high-volume production systems (millions of units annually of one or a few products). It is less flexible than a job shop that specializes in one-of-a-kind products. An FMS is a 'mid-range' system, appropriate for moderate variety and moderate volume markets.

R. Sridharan

References

Adam, E.E. and Ebert, R.J. (1996) *Production and Operations Management*, 5th edn, Englewood Cliffs: Prentice-Hall.

Askin, R.G. and Standridge, C.R. (1993) *Modeling and Analysis of Manufacturing Systems*, New York: John Wiley.

Ford–Fulkerson algorithm, see maximum flow problem

Forecasting and prediction

Much of the **uncertainty** faced by decision makers deals with future events, such as future demand, technological advances, moves by competitors. How can we predict the future? There are several approaches:

- *Persistence prediction* extrapolates stable, persistent characteristics about the phenomenon, such as stable past sales, and is good for phenomena that show small

random fluctuations around some averages, such as sales of a product that is in its mature phase of its life cycle.

- *Trend prediction* is without doubt the most successful and most used forecasting approach, often outperforming more sophisticated econometric approaches. It assumes that the change itself is stable. If wine consumption in each of the past few years has increased by 10 per cent, it is fairly safe to assume that this trend will continue, at least for another few years. The favoured methods are based on **exponential smoothing**. Trend prediction provides usually good short-run forecasts, but may lead to absurd long-range predictions for phenomena that are temporarily in an exponential-type growth phase which is bound to come to an end sooner or later.
- *Cyclic prediction* is based on the principle that history repeats itself. Cyclic predictions for seasonal phenomena – the weather, beverages and certain types of foods, outdoor activities – are highly successful. They may be superimposed onto a trend prediction and hence give considerably better short-run forecasts. But cyclic predictions are equally fickle when they try to forecast long-range phenomena, such as *business cycles*.
- *Associative prediction* uses past data from one type of process (the independent factor(s)) to predict another type of process (the dependent factor). It uses **regression analysis** and econometrics. Associative prediction greatly enlarges the area that is searched for clues. Much of regression analysis deals with establishing which clues are helpful. Sometimes the association is causal. For instance, an increase in building permits issued causes a delayed increase in construction activity. However, often there is no causal relationship, but both variables are affected by a third for which no data are available.

All these approaches use past data to predict future events. The crucial **systemic** assumption is that the past is a valid basis for predicting the future. If there are indications of structural or behavioural changes in the phenomenon observed, the validity of these methods becomes highly questionable. Structural changes could be due to changes in legislation, technology, or economic relationships governing the phenomenon in question.

Predictions should not be restricted to single estimates. The only way to assess the reliability of predictions is to obtain also a measure of the variability inherent in the phenomenon predicted. If the predictions are part of an ongoing repetitive process, then it will be possible to compute the prediction error, such as an estimate of the standard deviation of the differences between corresponding pairs of predicted and actual values.   *H.G. Daellenbach*

References

Armstrong, J.S. (2001) *Principles of Forecasting*, Norwell, MA: Kluwer Academic. Particularly ch. 12.

Makridakis, S., Wheelwright, S.C. and Hyndman, R.J. (1998) *Forecasting: methods and applications*, New York: Wiley.

Forestry, see applications of MS/OR in forestry

Fuzzy programming

Many decision-making situations involve a set of decision variables, a set of constraints on the decision variables, an **objective** to be maximized or minimized, and can be formulated as **mathematical programming** problems.

In conventional mathematical programming, the objective function and the constraints are unambiguously specified. However, it is not hard to imagine situations where such crisp specification is untenable due to inadequate data accuracy, genuine data uncertainty, or vagueness in restrictions on decisions or in the aims, and the specification has the character of a **fuzzy set**. A typical example is a company whose board of directors is trying to determine the 'optimal' dividend to be paid to the shareholders. They must meet shareholders' minimal expectations in terms of what is viewed as a satisfactory return on investment, while at the same time retain sufficient funds for further expansion of the operations and for new projects. Assume that the fuzzy set {satisfactory dividend} contains all possible dividends from the highest, say $10, to the lowest, say $5 per share. It in turn breaks into two subsets, one for dividends between $7 and $10, judged 'very satisfactory', and the other between $5 and $7.5, judged 'satisfactory'. Dividends below $5 are viewed as unacceptable by the shareholders and do not belong to the fuzzy set. The returns offered by expansion opportunities and new projects involve varying degrees of uncertainty and risk, forming also a fuzzy set. The board of directors must strike a compromise between a 'satisfactory' and a 'very satisfactory' dividend, taking into account the fuzziness of the new investment set.

Fuzzy set theory is equipped to deal with such imprecision. In 1970, Bellman, the inventor of **dynamic programming**, and Zadeh, the father of fuzzy logic, formulated the first mathematical programming problem with a fuzzy objective function and fuzzy constraints, but crisp decision variables. Some years later, Zimmermann formulated the first fuzzy **linear program**, which allows fuzzy constraints to be met to a specified degree, similar to meeting a constraint with a given probability.

Another obvious candidate for the use of fuzzy logic is **stochastic programming** where the effects of decision variable choices on the constraints or the right-hand sides of the constraints, such as the amounts of available resources, are **random variables**. Given the philosophical battle between probabilists and fuzzy theorists, fuzzy versions of stochastic programming received special attention. Advocates of the fuzzy doctrine point out that decision makers are often unable to provide a precise probabilistic model of their problem, and allowing for a degree of vagueness in the model would give them more flexibility. The probabilists' defence is to indicate that degrees of belonging have to be determined for all the fuzzy constraints and objective functions, so the same problem reappears in a different guise.   *A.D. Tsoularis*

References

Bellman, R.E. and Zadeh, L.A. (1970) 'Decision-making in a fuzzy environment,' *Management Science*, 17(4):141–164. Remarkable for its foresight.

Zimmermann, H.-J. (1996) *Fuzzy Set Theory and its Applications*, Boston: Kluwer, ch. 13.

Fuzzy sets and fuzzy logic

Sets are normally defined as a collection of elements or objects. A given object can either belong to or not belong to a specific set. For example, the collection of employees of a company can be viewed as a set. Any given individual either works for the company or not.

In contrast, fuzzy sets are collections of elements or objects where set membership is not absolute or unqualified, but vague, measured by a degree of belonging to the set. For example, different people may ascribe the fuzzy set {all young males} to different age ranges. Males in the group 10–25 years will be assigned a much higher degree of belonging to the set than the age group 40–50. A degree of 1 might be assigned to the first, implying that anyone in that group can be undoubtedly classified as young, whereas the second group gets a low degree of 0.2, based on the common perception of what constitutes 'young' and 'old'.

To deal with and manipulate fuzzy sets, L.A. Zadeh, a control system theorist in the US, introduced *fuzzy set theory* in 1965, together with a new form of logic, called *fuzzy logic*. Fuzzy logic is loosely based on the ways humans make decisions and take actions in the real world. For example, a quality inspector classifies each item coming off the production line as 'good' or as 'defective'. If classified as defective, the item is scrapped. The decision of whether an item is defective involves a certain degree of judgement. Therefore, {defective} is a fuzzy set that leads to the corrective action 'scrap product', where due to the fuzzy membership, some good items end up being scrapped. The 'if-then' rule used in the above example is unambiguous, but the phenomenon it relates to or information it is based on is fuzzy.

Fuzzy logic made significant inroads in business and industry in the 1990s across a wide range of areas. It is used in **expert systems**. In Japan washing machines are designed with fuzzy logic chips which adjust the washing cycle according to the dirtiness of the clothes. Other applications occurred for auto-focus and video cameras, anti-lock brakes and lift systems. In the US a division of General Motors patented a fuzzy automatic transmission system based on how humans tend to change gears rather than the conventional automatic systems, and the home appliance company Whirlpool designed an environmentally friendly refrigerator that also operates on a fuzzy logic chip.   *A.D. Tsoularis*

References

Grint, K. (1997) *Fuzzy Management*, New York: Oxford University Press.

Kosko, B. (1993) *Fuzzy Thinking*, New York: Hyperion. Popular science book by fuzzy logic theorist with the occasional flights of fancy.

Zimmermann, H.J. (1996) *Fuzzy Set Theory and its Applications*, Boston: Kluwer.

Game theory

Game theory deals with situations of conflict between two or more people. Each person can choose between several actions and the final outcome depends on the actions taken by all. This terse description obviously fits the situation of a parlour game. The choice of name is rather unfortunate, for it suggests that the theory deals with trivial human pursuits, when in fact its domain of applications in social, economic and international conflict situations is far more important. Game theory uses mathematics and axioms of economic rationality to study such conflict situations.

John von Neumann, a German mathematician, is generally credited with founding game theory in 1928, by proving the **minimax theorem** – the cornerstone of game theory. Surprisingly, it caused little interest. Only his joint publication with the economist O. Morgenstern in 1944, entitled *Theory of Games and Economic Behavior*, really launched game theory on its path.

Game theory makes certain strong informational assumptions. Each player not only knows all his own possible actions or *strategies*, but also those of all other players, including the rewards (or punishments), called *payoffs*, that result from all combinations of strategies of all players. Note that a strategy is not necessarily a simple decision, but can be a sequence of conditional decisions. For example, for a game of chess, a particular strategy consists of the first move if white, followed by a second move to each possible first move of black, and so on to the conclusion of the game. The number of possible chess strategies is larger than estimates of the number of atomic particles in the entire universe.

The simplest game is the two-person **zero-sum game** – zero-sum because the payoffs of player 2 are exactly the negative of the ones for player 1. It is for this class of games that von Neumann proved the minimax theorem, which states that there always exists a strategy for each player such that neither has an incentive to depart from it. The strategy may be *pure*, i.e. it is a unique action, or *mixed*, i.e. the strategy is chosen randomly from a subset of strategies, so as to keep the opponent guessing.

This existence of a minimax strategy may not hold for non-zero-sum games. In such situations the players could do better if they cooperated to achieve higher payoffs for both, or to agree on side-payments to entice the opponent to change to another strategy – so-called **cooperative games**. Most legal and diplomatic conflicts usually will do better if the players are willing to compromise and come to an agreement.

A further extension is the case where there are more than two players, so-called *N-person games*. Here there is not only an opportunity to cooperate, but also to form coalitions which again may involve side-payments to certain players.

A.D. Tsoularis

References

Casti, J. (1996) *Five Golden Rules*, New York: Wiley.
Owen, G. (1992) *Game Theory*, Orlando FL: Academic Press.
Poundstone, W. (1993) *Prisoner's Dilemma*, New York: Anchor Books, Doubleday.
Von Neumann, J. and Morgestern, O. (1944) *Theory of Games and Economic Behaviour*, Princeton: Princeton University Press.
Williams, J.D. (1986) *The Compleat Strategyst*, Mineola, NY: Dover.

Gaming, operational gaming, interactive simulation

Gaming is an interactive **simulation** or role-playing episode of two or more decision makers (which may be groups of individuals) who are faced with a decision situation where the final outcome or *payoff* is the joint result of the decisions of all decision makers. Gaming is used when the situation cannot be examined safely or adequately in the real world, or in preparation for real-world action. Certain aspects of the game are controlled by an umpire or controller, such as a carefully prepared starting situation, the information made available to the participants at the beginning and during the game, how and when action may be taken, the payoffs. Depending on the purpose, the rules may specify the kind of decisions that can be taken or the game may be free-form, allowing the participants to invent rules as the game proceeds. Gaming consists of five phases:

- *Game plan:* Transforming the situation to be studied into a game; setting the game rules, the number of participants, and the starting point.
- *Pre-game phase:* Preparing the physical arrangements, including computer support; selecting the participants.
- *Actual play of game:* Initial briefing of participants on their roles and the game rules; playing the game, usually through a number of rounds, where at each round the participants make one or more decisions, the game controller determines the joint outcome and feeds that information back to them; introducing rule changes (if that is part of the game plan).
- *Post-game activity:* debriefing of participants, collecting feedback from them on the game.
- *Analysis* of the game to draw conclusions.

The purpose of gaming can be varied. Operational gaming has a long tradition in the military. Educational games have teaching, learning, and attitude-changing aims. Job training games are geared to train for specific tasks or to improve performance, e.g. civil emergency training. Research and experimental games are intended to obtain empirical findings to test hypotheses or theories or make forecasts. They are usually repeated a number of times. Operational games are played to improve decision making, aid in planning and implementation in situations of immediate or imminent concern, such as preparation for upcoming **negotiations**, policy formation and **scenario planning**. The clients for operational gaming are international organizations (e.g. United Nations Environment Program, World Health Organization), governmental agencies (e.g. US Department of Energy, national health services), regional and local government, and more recently also private enterprise. Areas of application cover agriculture, disaster planning, energy, environment, education, health, management information systems, marketing strategy, tourism, and so on.

Gaming is considered a fairly costly tool.

K.C. Bowen

References

Bowen, K.C. (1978) *Research Games: an Approach to the Study of Decision Processes*, London: Taylor and Francis.

Ståhl, I. (1988) 'Using operational gaming', in H.J. Miser and E.S. Quade (eds) (1988) *Handbook of Systems Analysis: Craft Issues*, Amsterdam: Elsevier Science, pp. 121–171. See also pp. 78–80.

GAMS, see **algebraic modelling languages**

Gantt charts, see **critical path method**

General equilibrium analysis

The theory of general equilibrium deals with finding a set of prices for all commodity markets in the economy that ensures supply and demand are in balance for all the markets simultaneously. This is in sharp contrast with the more traditional partial equilibrium framework, wherein the objective is to locate the equilibrium price and quantity for that specific (and somewhat artificially isolated) market alone. Although finding the equilibrium for a single market is remarkably simple, a partial equilibrium analysis effectively implies that the market under question is completely immune to what is happening in the rest of the economy! It proved to be restrictive not merely in terms of the economic theory, but also one that has major practical consequences. A good example would be the US automobile market in 1974 immediately after the oil price shock in 1973. Automobile sales projections that ignored the impact of oil price grossly overestimated the sales figures when in fact US manufactures in that year lost out a significant market share to imported fuel-efficient Japanese cars. The concept of general equilibrium relaxes this restrictive assumption by capturing the connectivity of all the markets in an economy. This elegant theory, originally proposed by Leon Walras back in 1874, has proved to be a key development in modern economics.

Mathematically speaking, the demand and supply of a commodity is a function not merely of its own price, but also that of several (or all) other commodities. The overall economy is in general equilibrium if there exists a set of (equilibrium) prices for all the commodities which simultaneously satisfies all the demand–supply balances. Although it appears to be a relatively easy task to locate a set of prices that satisfies all the demand–supply balance conditions, it is in fact far from being so, primarily due to the nonlinear nature of the demand and supply/production functions. To start with, whether a solution exists at all needs to be examined, following which the question of uniqueness arises. There is a related important issue of stability, i.e. whether a particular price mechanism will drive the economy towards, or away from, this cherished equilibrium state. Finally, there are **efficiency** implications of the general equilibrium outcome that allows economists to study the ideal way of deploying resources in an economy.

General equilibrium theory and its application in the form of computable general equilibrium (CGE) models for several countries have employed the ideas and techniques in real analysis. The developments in OR/MS theory and practice have significantly contributed to the development of practical CGE models. Some of the very early models used **linear programming**. The advancement in large-scale **nonlinear programming**, and more importantly linear/nonlinear **complementarity theory** together with the development of **algebraic modelling languages** and solver engines, saw development of advanced CGE models for a variety of policy analyses, including sector-specific analysis, tax/subsidy issues, employment, energy and environmental policies.   *D. Chattopadhyay*

Reference
Starr, R.M. (1997) *General Equilibrium Theory: An Introduction*, Cambridge: Cambridge University Press.

General system theory

System is used to denote an arrangement of interrelated parts, and general system theory (GST) aims at formulating and deriving principles which are valid for systems in general.

Until well into the 20th century, research into organisms was based largely on **reductionist**, mechanistic **cause-and-effect thinking**: understanding the behavioural structure of an organism was attempted by analysing its parts and the causal relationships between them. This implies that organisms are independent, **closed systems**, i.e. systems that have no connection with an environment.

The acknowledged founder of GST, Ludwig von Bertalanffy, motivated by his disagreement with the reductionist approach, developed his theory of *open systems* in the mid-1920s. This theory is based on the manner in which organisms relate to a wider environmental context, assimilating inputs from the environment and excreting outputs to it. The principle of *equifinality* arising from this theory states that, for open systems, a similar **steady-state** or equilibrium of behaviour or organization can be reached which is independent of the initial conditions which triggered the path toward the steady-state. For example, living things tend to grow to a more or less standard (i.e. steady-state) weight, irrespective of reasonably diverse environmental conditions. Only severe disruptions, such as prolonged deprivation, may result in a different weight. This is in contrast to closed systems which follow a deterministic path to a final state that depends on the starting state.

GST arose from this theory of open systems, believing that there could exist an effective methodological means of controlling and instigating the transfer of principles from one field to another, thus minimizing duplication of similar principles in isolated fields of thought. The means of doing this was based on the view that similar forms and structures exist between otherwise radically different systems, different fields of science as well as completely different fields of knowledge – what is known as *isomorphy*, the condition of being identical or similar in form or shape. A simple example is found in physics where the same equations apply to the flow of liquids, of heat and of electric currents in a wire. Similarly, the statistical exponential equation – which states that given a complex of a number of entities, a constant percentage of these elements decay or multiply per unit time – applies to phenomena as diverse as the money in a bank account, radium atoms, and individuals in a population. In addition, GST stresses the **emergent properties** of systems, that is, properties which none of a system's parts, or subset thereof, exhibit by themselves. For example, telling the time emerges as a property of a clock viewed as a whole system.

The vision of von Bertalanffy was that GST, being a general theory, would become the unifying theory of the natural and social sciences. Though this did not materialize, the theory's generality has allowed for the use of its key concepts in particular problem areas.
I. Georgiou

References
Laszlo, E. (1972) *Introduction to Systems Philosophy: Toward a New Paradigm of Contemporary Thought*, New York: Gordon and Breach.
Laszlo, E. (1972) *The Systems View of the World*, Oxford: Blackwell.
Skyttner, L. (1996) *General Systems Theory: an introduction*, Hampshire: Macmillan.
von Bertalanffy, L. (1968) *General System Theory: Foundations, Development, Applications*, New York: George Braziller.

Generalizability/transferability

Sometimes a distinction is drawn in the literature between statistical and analytic generalization. *Statistical generalization* is based on considering the likelihood that results achieved from a sample of a population studied do indeed apply to the population. For example, one can assess statistically the likelihood that if one takes a sample of, say, 100 people out of some population (say, university graduates from a particular university 1996–2001) any results obtained from the sample will hold in the population. So, if we find in the sample that, say, 70 per cent of the male graduates earn higher salaries than the female ones, we can assess (using statistical reasoning) the likelihood that this holds for the population (in this case, all the graduates of the university from the relevant years). (Statistical validity of inferences assumes random sampling of observations.)

In the case of *analytic generalization*, empirical material gathered during the study of some unit(s) of analysis is examined with a view to considering its (wider) theoretical relevance. The results are addressed in terms of their importance for developing more general insights (that extend beyond the study). For example, one can study a company and examine the interrelationships between people in terms of the way they use gender categories in their conversations. One might then try to relate any insights gained from this case to wider discussion in the literature about gender in the workplace. Some authors argue that while statistical generalization is the logic used in survey research to achieve generalization of results, other forms of research (such as experiments, case studies, and **action research**) rely more on the skill of analytic generalization.

However, critique can be (and has been) levelled against attempts to 'generalize' in the social sciences, especially if this implies efforts to discover general regularities (or law-like patterns) holding in social life. So, for instance, we can question whether the aim of social science is to uncover the way in which certain factors isolated for attention might be generally associated with having effects on certain other factors ('gender' and 'income' being examples of factors considered). We can then try to leave open the extent to which results regarding chosen factors from any given study may be regarded as transferable to other situations. This means leaving it open for audiences (anyone seeing the results) to decide in what respects they think they can justifiably make comparisons across different contexts, and to what extent insights generated in some setting might help them to extend their appreciation of other ones. The term *transferability* suggests that audiences can play a more active part in interpreting the significance of any findings generated for making sense of different contexts.

N.R.A. Romm

References

Bryman, A. (1989) *Research Methods and Organisation Studies*, London: Allen and Unwin.

Pawson, R. and Tilley, N. (1997) *Realistic Evaluation*, London: Sage.

Remenyi, D., Williams, B., Money, A. and Swartz, E. (1998) *Doing Research in Business and Management*, London: Sage.

Romm, N.R.A. (2001) *Accountability in Social Research: Issues and Debates*, New York: Kluwer/Plenum.

Schofield, J.W. (1993) 'Increasing the generalisability of qualitative research', in M. Hammersley (ed), *Social Research*, London: Sage.

Yin, R.K. (1994) *Case Study Research: Design and Methods*, London: Sage.

Genetic algorithms

Genetic algorithms (GAs) are a class of **search heuristics** that were developed by John Holland in 1975. Their inspiration came from evolutionary theory where poor performing species die out while better performing species flourish and adapt. In a biological cell, chromosomes contain all information about the cell. The chromosome genes represent different characteristics of the organism. Holland modelled these ideas mathematically. Chromosomes become vectors or lists of numbers, with each element in the vectors analogous to a gene. Each chromosome represents a point in the search space. For example, in the **knapsack** problem, i.e. the problem of deciding which items to pack into a knapsack from a set of n possible items, a chromosome is a vector of zeros and ones, with a 1 in the ith position meaning the ith item is included and a 0 meaning it is not. The search space is the set of all possible combinations of items included and excluded.

One difference between GAs and some of the other search heuristics is that it uses a set of points in the search space, called the population, to generate improved solutions rather than a single point as in the case of **simulated annealing** and **tabu search**. Being able to maintain a diverse set of chromosomes stops the GA from becoming irretrievably trapped in a **local optimum**.

Three *genetic operators* are used to produce a new improved population of chromosomes: *reproduction*, *crossover* and *mutation*. The reproduction operator implements a 'survival of the fittest' philosophy, whereby the stronger members of the population have a higher probability of making it through to the next generation than the weaker members. The strength of a member is measured by the value of the objective function – the higher the value the fitter the member. The crossover and/or mutation operators subsequently use population members assessed as fit for further manipulation.

The crossover operator tries to combine the good parts of different chromosomes to produce even better chromosomes, by taking two or more members, called *parents*, and recombining them into two new chromosomes, called *children*, which go through to the next generation. For example, in the knapsack problem a child may consist of the first k elements of the chromosome from the first parent and last $(n-k)$ elements of the second parent, where k is a randomly chosen number between 1 and $(n-1)$.

The mutation operator takes a single member of the population and changes it in some way. For example, in the knapsack problem this may involve switching a 0 in a chromosome to a 1 or vice versa. Mutation is important in ensuring that the population remains diverse and therefore the search does not get trapped in a local optimum.

There are additional schemes that can be used to control which members will go on to the next generation. Depending on the problem, different schemes perform better or worse. Once a new generation is completed, the process starts again until the stopping criterion for the search is met. *R. James*

References
Goldberg, D.E. (1989) *Genetic Algorithms in Search, Optimization, and Machine Learning*, Reading, MA: Addison-Wesley.

Holland, J.H. (1992) *Adaptation in Natural and Artificial Systems*, Boston: MIT Press.

Reeves, C.R. (ed.) (1995) *Modern Heuristic Techniques for Combinatorial Problems*, London: McGraw-Hill.

Global optimum, see optimization

Goal programming

Goal programming is an adaptation of **linear programming** (LP) to deal with multiple objectives. The brain-child of Charnes and Cooper, published in 1957, it is the first such formal **multicriteria decision-making** technique.

Consider the following example. An investor has 2 million dollars to invest in a combination of shares X, offering an average return of 15 per cent, and shares

Y, offering 6 per cent. She wants to maximize the overall return, while at the same time minimizing the risk, measured as the fluctuation in returns of each investment, expressed as the variance of returns, i.e. 9 for X and 1 for Y. Letting x and y be the fraction of millions invested in X and Y, respectively, this problem becomes:

$$\text{maximize} \quad 15x + 6y$$

$$\text{minimize} \quad 9x + y$$

$$\text{subject to} \quad x + y \leq 2$$

$$\text{and} \quad x, y \geq 0$$

There are two objective functions. The solution technique for LP can only handle one. Charnes' and Cooper's approach is to replace each original objective by a *target* and create a new objective function which minimizes a weighted sum of the deviations from the targets, where each weight reflects the importance of its objective. Assume that our investor aims for an average return of 12 per cent and a risk of 5 per cent, with return being twice as important as risk. The new problem is now:

$$\text{minimize} \quad 2s_1^+ + 2s_1^- + s_2^+ + s_2^- \qquad \text{(weighted sum of deviations)}$$

$$\text{subject to} \quad 15x + 6y - s_1^+ + s_1^- = 12 \qquad \text{(return target)}$$

$$9x + y - s_2^+ + s_2^- = 5 \qquad \text{(risk target)}$$

$$x + y \leq 2 \qquad \text{(budget constraint)}$$

$$\text{and} \qquad \text{all variables nonnegative}$$

where s_1^+, s_2^+ represent overachievement and s_1^-, s_2^- underachievements.

This is an ordinary LP that can be solved by the **simplex method**. Care must be taken to fix targets that do not force the LP to a so-called **dominated solution**, i.e. there are other solutions that are **feasible** for the original constraints and which offer a better performance on at least one objective and are no worse on all others. Note also, that the decision maker may be mainly concerned with deviations in one direction, rather than both, or assign them different weights.

This formulation has become known as the *weighted-sum formulation*. It allows *trade-offs* between objectives. The decision maker may not allow trade-offs, but rank the objectives in order of priority. Say, return has priority 1, risk priority 2. This is expressed by the following objective function:

$$\text{minimize} \quad P_1 (s_1^+ + s_1^-) + P_2 (s_2^+ + s_2^-) \qquad \text{(priority ordering)}$$

where P_1 is infinitely larger than P_2. This is known as *preemptive goal programming*. The problem is first solved for the highest priority objective, ignoring all lower priority objective (i.e. for minimize $s_1^+ + s_1^-$ in our example). If **alternative optimal solutions** exist, they become the feasible set of solutions for the second highest priority objective. This process continues until either a unique solution is found or all priority levels have been optimized. Experience indicates that the process rarely goes beyond two or three levels of priority. *H.G. Daellenbach*

References

Hillier, F.S. and Lieberman, G. (2001) *Introduction to Operations Research*, 7th edn, McGraw-Hill, ch. 8.

Ignizio, J. (1976) *Goal Programming and Extensions*, Lexington, MA: Lexington Books.

Goals, see criteria

Gradient methods, steepest ascent methods

Gradient methods are nonlinear optimization **algorithms** using differential calculus. The easiest way to explain their general principle is to take a geographical analogy. You are at the bottom of a hill shrouded in fog and you want to climb to its highest point. There is no track and visibility is limited to a few feet or metres. But you reason that as long as you move upwards, you will sooner or later reach the summit, provided the shape of the hill is such that there is only one peak, i.e. there is always a direction going up no matter where you are, except when you have reached the peak.

You decide to use the following iterative procedure. You reason that the best direction to take from the point you are is the one with the steepest slope. Owing to the dense fog, you only take one step up that direction. At that new point you look again for the steepest slope, take another step up, and repeat this process until you reach a point where all directions go downhill. You have reached the summit (and can now wait for the sun to burn away the fog).

This is what steepest ascent methods (also appropriately named *hill-climbing methods*) do. The starting point is the initial solution. Its elevation corresponds to the objective function value. The slopes away from the point are measured by the gradients (also a mathematical concept). You find the maximum gradient. The step size you take depends on how fast that gradient changes (the curvature of the terrain) – small for a fast change, larger for a slow change. The new elevation reached is the new value of the objective function. You have completed one iteration of the algorithm. The new solution is the next starting point.

The figure below demonstrates that process for a problem in two variables. The circles represent *contour lines* or the combinations of (x_1, x_2) values that produce the same value of the objective function z. Each circle farther in is a higher value. Their curvature indicates how fast the gradient changes. The first step is from point P_0 to P_1, increasing z_0 to z_1, the second from P_1 to P_2. As the curvature increases, the steps become smaller.

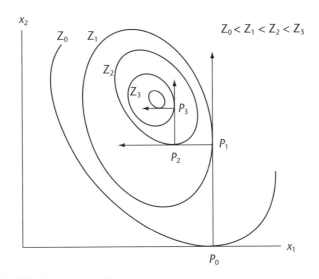

Figure: **Steepest ascent method for a two-variables problem**

If the objective function is well behaved, i.e. **concave**, gradient methods will find the **global optimum**.

Gradient methods can also be used when the decision variables are subject to constraints. In terms of the hill-climbing example, cliffs and other obstacles represent constraints for moving beyond them. So the size of the step now also takes this into account. If the best size step (based on the curvature) would take you over a cliff, you reduce the step size to only reach the cliff edge, which now becomes the next starting point. *H.G. Daellenbach*

Reference
Winston, W.L. (1994) *Operations Research Applications and Algorithms*, Belmont, CA: Duxbury, pp. 674–683.

Graph theory

A graph is defined by two sets of symbols: points or *vertices*, and lines or *edges*. Each edge joins two vertices. The figure below shows five types of graphs.

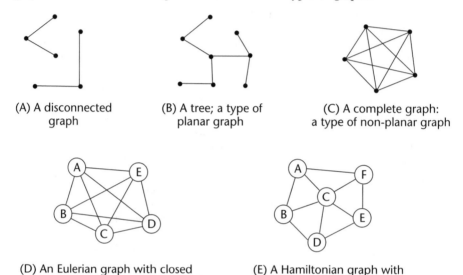

(A) A disconnected graph

(B) A tree; a type of planar graph

(C) A complete graph: a type of non-planar graph

(D) An Eulerian graph with closed edge tour A–C–E–B–D–A–B–C–D–E–A

(E) A Hamiltonian graph with closed vertex tour A–B–C–D–E–F–A

Figure: **Graphs**

Graph theory (GT) was founded by the Swiss mathematician, L. Euler, in 1736 when he solved the so-called Königsberg Bridge Problem (KBP) by modelling it in terms of a specific graph. The problem involves starting at any point in the inner city of Königsberg, crossing the bridges across the river Pregel, including those to Kneiphof Island in the river, exactly once, and returning to the starting point. Since no one in the city could solve the problem it came as no surprise that Euler proved it to be impossible.

The main concepts of GT that have applications in MS/OR are: *connectivity, trees, traversability, planarity, coverings, colourings, matchings,* and *matroids.* Let G denote a graph. G is connected if every pair of vertices in G is joined by a path of edges (see graphs B, C, D and E). G is a tree if it is connected and its number of vertices is one more than its number of edges (see graph B). There are two types of traversability that are of interest in MS/OR. G is *Eulerian* if it contains a closed tour containing each of its edges exactly once (see closed edge tour in graph D). This term is used in honour of Euler who used this notion in his model of the KBP mentioned earlier. A graph G is *Hamiltonian* if it contains a closed tour containing each of its vertices exactly once (see tour in graph E). G is planar if it can be drawn on a plane in such a way that no two edges intersect (see graphs A, B and E). A vertex (edge) of G is said to cover the edges (vertices) with which it is incident. A colouring of G is an assignment of colours to its vertices so that no two vertices that are joined directly by an edge have the same colour. A subset of the edges of G is called a matching if no two of the edges have a vertex of G in common. A matroid is an algebraic system with a certain precisely defined independence structure. Matroids are related to GT because certain collections of subsets of edges of graphs are matroids, e.g. the subsets of edges of G that do not contain any closed tours of edges.

Many MS/OR problems can be modelled in GT terms. These include: shortest path (connectivity) (Floyd), **minimum spanning tree** (trees) (Prim, 1957), the **Chinese postman** and **travelling salesperson** (traversability), the **facilities layout** (planarity), location (covering), **timetabling** (colouring), the **assignment problem** (matching), and the **greedy algorithm** (matroids). For more general applications of GT in MS/OR see (Boffey, 1982; Robinson and Foulds, 1980; Chachra *et al.*, 1979; Foulds, 1998).

There are two areas, closely related to GT, that also have significant application in MS/OR: *digraphs* and **networks**. A digraph is a graph in which each edge is directed, i.e. can be represented by an arrow, referred to as an arc. A network is a digraph with exactly one source vertex (with all arcs orientated away from it) and one sink vertex (with all arcs orientated towards it) and at least one commodity flowing along its arcs.
L.R. Foulds

References
Boffey, T.B. (1982) *Graph Theory and Operations Research*, London: Macmillan.
Chachra, V., Ghare, P.M. and Moore, J.M. (1979) *Applications of Graph Theory Algorithms*, Amsterdam: North Holland.
Floyd, R.W. (1962) 'Algorithm 97 – shortest path', *Comm. ACM*, 5:345.
Foulds, L.R. (1998) *Graph Theory Applications*, Berlin: Springer-Verlag.
Prim, R.C. (1957) 'Shortest connection networks and some generalizations', *Bell Syst. Tech. J.*, 36:1389.
Robinson, D.F. and Foulds, L.R. (1980) *Digraphs: Theory and Techniques*, London: Gordon and Breach.

Greedy algorithm, see algorithm

Group dynamics

The term 'group dynamics' has two different but related meanings. It is used to describe the scientific study of groups and the behaviour of people in groups. It is also the popular general term for the ongoing processes in a group. The interaction among group members is characterized by constant change, and Kurt Lewin (1951), who is regarded as the founder of group dynamics, chose the word 'dynamic' to capture these complex social processes. They include such things as

- the level of participation in a group (is it equal, or are some people more involved than others?);
- communication patterns (who is talking to whom?);
- how the group makes decisions (by consensus, majority, or some other means?),
- conflict (what is the level of conflict and how is it being handled?),
- group norms (what are the unwritten rules in the group?),
- group roles (what are the formal and informal roles which team members are filling?),
- leadership (what style of leadership is the leader displaying?), and
- stages of group development.

As a field of scientific inquiry, group dynamics emerged during a 50 year period between 1890 and 1940 when researchers from a number of different fields began to be interested in groups. But it was Lewin who more than anyone else made a major contribution to the study of groups. Lewin was a brilliant theorist who had a genius for designing experimental methods to research the dynamics of groups. He coined the term **action research** and was largely responsible for developing the experiential learning methods which are still widely used today to teach people about group theory and group skills. Lewin believed that 'there is no hope of creating a better world without a deeper scientific insight into the essentials of group life' (1943, p. 113). Today, that goal is being met by researchers in many fields. Business and management researchers study topics such as **team building**, the management of project teams, self-managing work teams and the functioning of boards of directors. Social workers study family counselling and team interventions. Psychologists study the effectiveness of counselling and therapy groups, criminologists the process of jury deliberations, and sociologists the impact of gender on group functioning. Their findings are published in a diverse range of journals, but in the field of business and management the best known are *Group Dynamics, Group and Organisation Behaviour, The Journal of Applied Behavioural Science*, and *Small Group Research*.

Group dynamics are also important in several **soft operational research methods**, including **gaming** and **drama theory**. *I. Brooks*

References

Forsyth, D. (1998) *Group Dynamics*, Belmont, CA: Wadsworth Publishing.

Johnson, D. and Johnson, F. (2000) *Joining Together: Group Theory and Group Skills*, Boston: Allyn and Bacon.

Lewin, K. (1943) 'Forces behind food habits and methods of change', *Bulletin of the National Research Council*, 108:35–65.

Lewin, K. (1951) *Field Theory in Social Science*, New York: Harper.

Health services, see applications of MS/OR to health services

Heuristics, heuristic problem solving

An **algorithm** is a set of logical steps and numerical procedures, applied iteratively, that is guaranteed to converge to a solution for a mathematical model in a finite number of steps. MS/OR optimizing techniques use algorithms that find the **optimal solution**, if one exists. However, many practical problems are so complex and/or require such a large number of computations that the use of these algorithms is computationally infeasible, i.e. they would take years to find the optimal solution even on the fastest computers. This includes the group known as **NP-hard** problems. There are a number of possible ways forward to overcome this difficulty. One can attempt to find an efficient algorithm for special cases of the problem that are easier to tackle than the general case. One can ignore the difficult aspects and only solve a simplified version of the problem, or one can abandon the quest to find the optimal solution and devise heuristics that find satisfactory, good, or close-to-optimal solutions, depending on the complexity of the problem.

This is not a recent idea. As early as 300 AD, Pappas, writing on Euclid, suggested the use of approximate methods to solve certain geometric problems. Through the work of Descartes and of Leibniz, the subject became known as *heuristics*. The name itself is derived from the Greek word *heuriskein* – to discover. Heuristic methods are based on characteristics of human problem solving, such as creativity, insight, intuition, experience, and learning. There are four basic heuristic strategies:

1. The *construction strategy* builds up a final solution one element at a time. Consider the so-called **travelling salesperson problem** (TSP) that involves visiting each of *n* cities once while keeping the total distance travelled as short as possible. A tour is constructed by adding one city at a time, e.g. by always choosing the one closest to the last city visited.
2. The *improvement strategy* starts with an initial solution and then iteratively tries to improve on it. In the TSP, after having found an initial itinerary to visit each of *n* cities, using a construction strategy, you could try to improve on it by interchanging each pair of consecutive cities. If an interchange decreases the total distance, it is retained, otherwise not.
3. The *component analysis or break-make strategy* breaks a large problem into smaller,

easier subproblems. For instance, rather than solve the *n* city visit problem as one, clusters of close-by cities are identified (i.e. break problem into subproblems). Each cluster is then solved separately, and then the clusters are linked together.

4. The *learning strategy* uses a tree-search approach to eliminate more and more of the solution space as learning occurs about the problem, until only a few branches are left for complete exploration. The way most players play 'battle-ships' or 'mastermind' uses a search-learning strategy. The information gained by each trial is used to eliminate portions of the solution space and helps in forming the next trial.

As pointed out, heuristics do not guarantee finding an optimal solution. In fact, they may at times terminate with a bad solution, because the procedure becomes trapped in a **local optimum**. The usefulness of a given heuristic is gauged by the computa-tional time required or the number of elementary steps needed to find a satisfactory solution, and the closeness of the final solution to a lower or upper bound derived by other means, as well as the simplicity of its use. In addition to the reduction in the computational effort, good heuristics may have other advantages over optimiz-ing approaches, such as lower training or knowledge needs for their use, and greater flexibility to cater for non-quantifiable aspects (Osman and Kelly, 1996; Foulds, 1983).

L.R. Foulds

References

Cook, T.M. and Russel, R.A. (1993) *Management Science*, 5th edn, Englewood Cliffs, NJ: Prentice-Hall, ch.17.

Foulds, L.R. (1983) 'The heuristic problem-solving approach', *JORS*, 10:927–934.

Osman, J.H. and Kelly, J.P. (eds) (1996) *Meta-Heuristics: Theory and Applications*, London: Kluwer.

Hierarchies, hierarchical trees or diagrams

When a concept, an entity, an organization or a **system** can be broken into a hierarchy of several levels where each lower level is completely contained in the next higher level, it may be useful to represent this structure as a hierarchical tree. The diagram below is

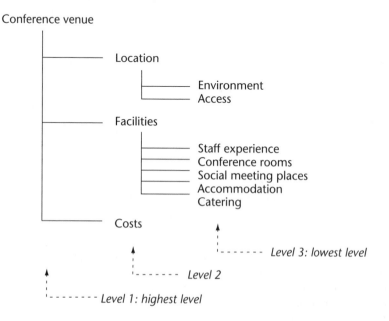

Figure: **Hierarchical tree**

an example, depicting which factors are important in the selection of a conference venue. Conference venue is the level 1 concept, the highest level. It can be broken down into lower-level concepts, i.e. location, facilities and costs, each of which in turn may be broken into even more basic concepts.

Hierarchical trees are useful to represent organization charts (shareholders, board of directors, managing director, vice-presidents, division managers, supervisors, workers), hierarchies of objectives (overall goal, individual objectives, attributes) as for the example in the figure, a **hierarchy of systems** (wider system, narrow system, subsystems, etc.), classification systems (categories, subcategories, sub-subcategories, etc.), and so on. (See also **hierarchy theory**.)

Hierarchy of systems, narrow and wider system of interest, subsystems

A **system** may be completely contained in another, larger system, which in turn may also be part of an even wider system. The 'workings' of a school class, viewed as a system, is contained within the school – a larger system, which in turn is part of a regional education system. The latter is one of the subsystems of the national education system, which is part of the national government system. They form a hierarchy of systems, one contained in the other like Russian dolls, as shown in the figure below.

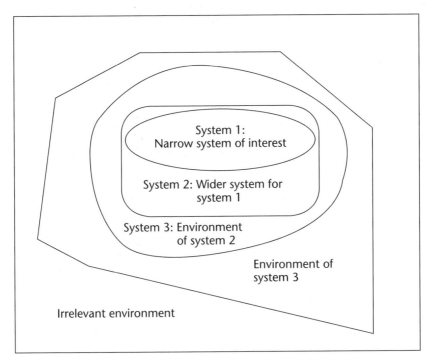

Figure: **Hierarchy of systems**

Control over the resources (particularly funds) that a given system needs for its activities or the setting of overall policy that shapes its functions (the core business the organization is in, e.g. fire fighting service), its purpose and objectives (e.g. minimize loss of life and property) is often located in the containing system which may also monitor its performance. The containing or controlling system is then referred to as the *wider system of interest*, while the contained system becomes the *narrow system of interest*. The wider system (including its own relevant environment) is the relevant environment of the narrow system.

The advantage of viewing two systems in a hierarchy of a narrow and a wider system is that their relationships are shown in their correct context. It may show that improvements in the performance of the narrow system may require action to be taken in the wider system, or that certain actions in the narrow system may be detrimental to the performance of the wider system. A narrow focus on **efficiency** may often cause such effects. Similarly, the relationships between various inputs into the narrow system are clarified. Two inputs first seen as independent of one another may turn out to be highly related or affected by the same factors when seen in their proper relationship within the wider system. For example, the allocation of funds for different projects in the narrow system has to meet the same critical minimum return, set by the wider system that provides the funds.

Hierarchy theory

The concepts of parts and wholes are basic in the study of **systems**. A whole system can be viewed as consisting of various parts or **subsystems** which in turn may be viewed in terms of even smaller subsystems. These different sets of subsystems can also be called the levels within the system. Hierarchy theory looks at these levels in a system and investigates how they relate to one another and how their interrelated behaviours come together to form the **emergent** behaviour of the whole system. Systems are hierarchical when they need more than one level to explain their complex behaviour (Allen and Starr, 1982).

Each of the levels in a system can be viewed as being at a different scale from the level above or below. When we look at an organization as a system, we choose a scale for the study of the system. Perhaps the level of interest is the purchasing division within a company. Perhaps it is at a lower level, looking at the work of only one individual within the purchasing department. Or perhaps it is at a higher level, looking at how the purchasing function operates within the company as a whole. No matter what scale is chosen, the goal is to understand how the levels fit together, and how the behaviour of the individual levels can be explained in terms of their role in the system as a whole.

Without hierarchy, complex systems would not be able to organize and maintain complex structures and processes. An example of how hierarchy facilitates the emergence of complex systems lies in the story of the two watchmakers. Both watchmakers have several hundred components to put together to form one watch. One of the watchmakers begins and puts the individual pieces together one at a time, one after another. The second watchmaker gathers several components together, perhaps ten or so, and assembles those pieces into a sub-part of the watch. Each watchmaker is interrupted several times during the day. Each time the first watchmaker is interrupted the work is stopped, and all of the watch pieces fall apart. The work must begin again with the first piece after each interruption. The second watchmaker is able to create several sub-assemblies for the watch between interruptions, and by the end of the day is able to then put together the sub-assemblies to form the whole watch.

Many authors have written about hierarchical systems. Some of these people have organized hierarchies from physical to chemical, to biological or man-made in order to explain how smaller units can be built to form extremely complex systems.

J. Wilby

References

Allen, T.F.H. and Starr, T.B. (1982) *Hierarchy: Perspectives for Ecological Complexity*, Chicago: The University of Chicago Press.
Boulding, K. (1956) 'General systems theory: the skeleton of science', *Management Science*, 2:197–208.

Checkland, P. (1981) *Systems Thinking, Systems Practice*, Chichester: Wiley.
Miller, J.G. (1978) *Living Systems*, New York: McGraw-Hill.

History of MS/OR and its precursors

Although there are earlier examples of the use of scientific methods to analyse **problematic situations** in organizations and inform managerial decision making, Taylor's scientific management being probably the best known, the term Operational Research (OR) was first used in the late 1930s. The development of radar technology led to the British military having to decide how best to take advantage of the valuable information it provided about the location of enemy aircraft. Research into the operation rather than the science of radar was carried out by groups of civilian scientists reporting to senior military staff. Such Operational Research Sections went on to play a vital role in the British, American and other allied forces throughout the Second World War and their organization and methods of working laid the foundations for the use of operational research in business organizations after 1945.

The contribution of OR to the British war effort was sufficient to guarantee its acceptance by major industrial sectors as regeneration of the post-war economy took place. The new National Coal Board made extensive use of OR in all areas of its activities and rapidly became home to one of the major OR groups in the world. The iron and steel sector, both in its Research Association and in individual companies, hosted various OR groups. By the 1960s managers in almost all sectors of British industry and commerce had access to and used some form of OR support. In the USA, the uptake of OR in industry was somewhat slower, largely because the majority of OR people remained attached to military organizations well into the 1950s. The aircraft, oil and chemical industries were the first to adopt OR. However, once it started to penetrate into industry, the uptake was rapid, and by the end of the 1950s a majority of large corporations and consulting firms had established internal OR groups.

The growth of OR on this scale required and encouraged a growth in the institutions necessary to support and develop the discipline. Societies were created as early as 1948 in Britain (the Operational Research Society) and 1952 (the Operations Research Society of America) and 1953 (the Institute of Management Sciences) in the USA. Management Science became synonymous with OR. Journals devoted to the dissemination and advancement of the discipline followed, *Operational Research Quarterly* (later *Journal of the OR Society*), *Operations Research*, and *Management Science* appearing in 1950, 1952 and 1954 respectively. Courses in universities to teach the methods and practice of OR began at around the same time, firstly in the USA and later in Britain. The first main textbook to offer a unified overview of the discipline was Churchman, Ackoff and Arnoff's *Introduction to Operations Research* published in 1957. Since the 1960s OR has been at the centre of business and management theory and practice. The late 1950s and early 1960s also saw a real explosion of ever more sophisticated and powerful **optimization** methods and techniques, largely made possible by the ever-increasing computational power of digital computers.

The list below gives the major MS/OR milestones, including some of the precursors, showing the person(s) mainly responsible. The dates are only approximate, indicating either the first published work on a topic or the first hearsay report about it.

1881	Scientific analysis to methods of production: F.W. Taylor
1900s	**Gantt charts**: H.L. Gantt (1861–1919)
1915	**Economic order quantity model**: X.X. Harris Westinghouse
1916	Aircraft warfare and N square law: F.W. Lanchester

1917	Avoidance strategies for merchant ships: T.A. Edison, US Naval Consulting Board
1917	First **queueing** formulas: A.K. Erlang, Danish Telephone Company
1920s	**Statistical quality control**: W.A. Shewhart (text published 1939)
1927	First table of **random numbers**: L.H.C. Tippett, Cambridge University
1928	**Game theory, minimax theorem**: J. von Neumann
1930s/50	**General systems theory**: L. von Bertalanffy
1930/2	**Steady-state** formulas for the M/G/1 queue: F. Pollaczek and A. Khinchine
1931/2	**Travelling salesman problem**: M.M. Flood
1939	First **linear programming** formulation: V. Kantorovich
1941	**Transportation problem**: F.L. Hitchcock
1944	von Neuman–Morgenstern **utility theory**: J. von Neumann and O. Morgenstern
1944/5	**Monte-Carlo methods**: J. von Neumann and S. Ulam, Manhattan Project
1947	**Simplex method** for linear programming: G.B. Dantzig, US Airforce
1948	**Cybernetics**: N. Wiener, MIT, C.E. Shannon, Bell Lab
1950s	**Gaming**, The Rand Corporation
1951	Analog **simulation**: US Airforce and University of Chicago
1951	First computer program for simplex method: A. Orden, Project SCOOP at SEAC
1951	Kuhn–Tucker conditions for **nonlinear programming**: H.W. Kuhn and A.W. Tucker
1951	**Efficient solutions**: T.C. Koopman, H.W. Kuhn and A.W. Tucker
1951	Beginning of stochastic **inventory control** models: K.J. Arrow, T. Harris and J. Marschak
1952/59/63	**Portfolio selection** models: H. Markowitz, W.F. Sharpe
1953	Revised simplex method: G.B. Dantzig
1954	42 city travelling salesman problem solved: G.B. Dantzig, D.R. Fulkerson and S.M. Johnson
1954	Stepping-stone algorithm for transportation problem: A. Charnes and W.W. Cooper
1954	**Dynamic programming**: R.E. Bellman
1954	**Dual simplex method** for linear programming: C.E. Lemke
1955	**Hungarian method** for **assignment problem**: H.W. Kuhn
1956	**Exponential smoothing**: R.G. Brown
1956	**Transshipment problem**: A. Orden

1956/9	**Quadratic programming** algorithm: M. Frank, P. Wolfe and E.M.L. Beale
1957	**Jackson networks**: J.R. Jackson
1957	**Goal programming**: A. Charnes and W. Cooper
1957/8	Simulation using digital computers
1958	**Decomposition** of linear programs: G.B. Dantzig and P. Wolfe
1958	**Stochastic programming**: A. Charnes, W.W. Cooper and G.H. Symonds
1958	**Cutting plane algorithm** for integer programming: R.E. Gomory
1958	**System dynamics**: Jay Forrester, MIT
1958/9	**Delphi method**: The RAND Corporation
1958/9	**PERT** method: US Navy special projects office & Booz, Allen & Hamilton
1958/9	**Critical path method**: duPont & Remington Rand
1959	**Risk analysis**: D.B. Hertz
1959	Reduced **gradient method**: P. Wolfe
1960	**Branch-and-bound algorithm**: A.G. Land and A. Doig
1960	**Markovian decision processes**: R.A. Howard
1961	GPSS simulation language: IBM Research
1961	**Little's queueing formulas**: J.D.C. Little
1962	Optimal **control theory**: L.S. Pontryagin
1962	**Max-flow/min-cut network flow**: L.R. Ford and D.R. Fulkerson
1962	Bender's decomposition, price decomposition: J.F. Bender
1963	**Chaos theory**: E. Lorenz
1964	**Multiattribute utility functions**: P.C. Fishburn
1965	**Fuzzy logic**: L.A. Zadeh
1968	Beginning of **multiobjective decision models**: E. Johnsen
1971	**ELECTRE** multicriteria method: B. Roy
1973	**Black–Scholes call option pricing model**: F. Black and M. Scholes
1975	**Genetic algorithms**: J. Holland
1978	**Data envelopment analysis**: A. Charnes and W.W. Cooper
1978	**Analytic hierarchy process**: T. Saaty
1978	Minos mathematical programming system: B. Murtagh, M. Saunders, Stanford OR Institute
1980s	**Kaizen**: Masaaki Imai, Japan
1980s	**Neural networks, simulated annealing**
1981	**Just-in-time** production systems introduced in the US; 1970s at Toyota by Ohno Taiichi

1984	**Interior-point methods**: N. Karmarkar, Bell Labs
1986	**Tabu search**: F. Glover, University of Colorado
1987	Algebraic modelling languages: AIMMS, AMPL, GAMS

H.G. Daellenbach and P. Keys

References

Churchman, C.W., Ackoff, R.L. and Arnoff, E.L. (1957) *Introduction to Operations Research*, New York: Wiley.

Fildes, R. and Ranyard, J.C. (eds) (1998) 'The foundation, development and current practice of OR', *JORS*, 49(4):303–443.

History of soft systems thinking, soft operational research, and problem structuring methods

The chronological list below of the first appearance of soft systems thinking, **soft operational research** and **problem structuring methods** is indicative of the three phases of development. The first phase going into the 1960s is a period of disquiet and unease over the ever-increasing mathematical emphasis of MS/OR, **general systems theory** and its several spin-offs, to the detriment of a fuller systems view. The main voices were those of C. West Churchman, R.L. Ackoff and I. Hoos in the US, and Stafford Beer, G. Vickers and F.E. Emery in the UK. They voiced their concern that this trend condemned MS/OR to suboptimization. The first step away from quantitative MS/OR was Beer's **viable systems model** which looks at the indispensable functions that any viable organization has to have and maintain.

The second phase from about 1969 to the end of the 1970s could be labelled 'the period of revolt'. It is driven by the recognition that quantitative approaches are unable to cope with the full complexity of organizational situations, characterized by a strong behavioural component. These approaches retain the traditional MS/OR reliance on rational analysis and the use of representations to capture the meaning of situations. They differ in that the data used within the representations are qualitative and consequently the tools of analysis are non-mathematical.

The third phase from about 1983 on takes a more critical view of the use of methodologies along several fronts, such as their neglect of their analysis on the voiceless groups of victims and beneficiaries of project outcomes, their inability to cope with power differences between stakeholders, the lack of guidance as to which method is most appropriate for a given situation, and the recognition that no single approach will address all the aspects of a situation, the latter giving rise to *methodological pluralism* and **multimethodology**. In the late 1980s this phase became known as **critical systems thinking**.

The references discuss significant developments of this side of MS/OR. The following list shows the approximate date of new developments, as well as the person(s) mainly responsible and, where appropriate, academic and research organizations associated them.

1950s	**Gaming**: The RAND Corporation
1955	**System analysis**: C. Hitch, E.S. Quade, H.L. Miser; RAND Corporation, International Institute for Applied System Analysis (IIASA)
1956	**System dynamics**: Jay Forrester, E.F. Wolstenholme, Massachusetts Institute of Technology (MIT)
1957	System engineering: H.H. Goode, R. Machol, A.D. Hall, Bell Telephone Laboratories
1959	**Viable systems model**: Stafford Beer

1960s	**Metagame analysis**: US Arms Control and Disarmament Agency
1969	**Strategic choice approach**: J. Friend, A. Hickling, IOR, Tavistock Institute
1969/81	**Strategic assumption surfacing and testing**: R.O. Mason, I.I. Mitroff
1971	Social systems design: C. West Churchman, University of California, Berkeley
1974	Social systems sciences (now known as **interactive planning**): R.L. Ackoff, University of Pennsylvania, Philadelphia
1975	**Soft systems methodology**: P. Checkland, University of Lancaster
1979	**Strategic option development and analysis**: C. Eden, University of Bath, now at University of Strathyclyde, Glasgow
1980	**Hypergame analysis**: P.G. Bennett
1983	**Critical systems heuristics**: W. Ulrich
1990/3	**Drama theory**: P.G. Bennett, M. Bradley, J. Bryant, N. Howard
1991	**Total systems interventions**: R.L. Flood, M.C. Jackson, University of Hull
1997	**Multimethodology**: J. Brocklesby, J. Mingers

P. Keys

References
Jackson, M.C. (2000) *Systems Approaches to Management*, New York: Kluwer/Plenum.
Keys, P. (1991) *Operational Research and Systems*, New York: Plenum.
Mingers, J. and Gill, A. (eds) (1997) *Multimethodology*, Chichester: John Wiley.
Rosenhead, J. (ed.) (1989) *Rational Analysis for a Problematic World*, Chichester: John Wiley.

Holt's method, see **exponential smoothing**

Human activity system

Human activity system is the term used to denote **systems** that are 'created' by and for humans, in contrast to natural systems. While most scientists would define a given natural system in very much the same terms, managers and decision makers would often define a human activity system in quite different terms. Note the quotation marks around created. Created does not imply that the system under study exists in the real world. The term is used in the 'inside-us-view' of systems language, i.e. that systems are **conceptualizations** – mental images, and this is the reason for differing views of the same entity or operation. This is not to deny that certain counterparts may exist in the real world or that it is not envisaged that they may exist at some future time.

At one extreme, a human activity system can represent the operation of a totally automated plant where all the activities are performed by robots and controlled by computers (a production system). So from this view, no humans are actually part of the system, but humans are involved in its creation. At the other extreme, it may only represent the activities and relationships between a group of people with no physical components as part of the system, such as a group of individuals playing a game of 'charades' (an entertainment system).

In contrast to physical or natural systems, human activity systems may be affected to a lesser or greater extent by human values and emotions. They often involve conflict and the additional uncertainty of human behaviour. So, while the methods of science are appropriate to study natural systems, they may badly fail when applied to human activity systems. In fact, one of the reproaches levelled at 'hard' MS/OR methods is that they try to transpose the presumably value-free methodology of the natural sciences onto studying human activity systems. Although there are many instances of a largely technical nature where the objective and the possible actions are clearly defined and 'hard' approaches prove successful, they stumble when faced with conflicting values and human emotions. Such situations need the approaches offered by **soft operational research** or **problem-structuring methods** that attempt to deal explicitly with values and different viewpoints. *H.G. Daellenbach*

Reference

Checkland, P. (1999) *Systems Thinking, Systems Practice: a 30–Year Retrospective*, Chichester: Wiley, pp. 115–121.

Hungarian method

The Hungarian method is a highly efficient solution algorithm for solving the standard **assignment problem** of allocating n candidates to n jobs, so as to minimize the total penalty $\sum_{\text{all } ij} c_{ij} x_{ij}$, where c_{ij} is the penalty for assigning candidate i to job j, and $x_{ij} = 1$ if candidate i is assigned to job j and 0 otherwise. The algorithm is a special version of a more general method, known as the *primal–dual method* for solving general **linear programming** problems. Published in 1955, H.W. Kuhn, its inventor, named it the Hungarian method in recognition of the work by J. Egerváry and D. König, two Hungarian mathematicians of the 19th century.

Let C denote the $n \times n$ table of penalties c_{ij}. The algorithm consists of the following steps:

Step 1: Subtract from each element in C the minimum c_{ij} in its row and then subtract from each (new) element in C the minimum c_{ij} in its column. (The resulting elements in C are called the **reduced costs**.)

Step 2: Draw the minimum number of lines (either horizontal or vertical) that are needed to cover all the zeros in the reduced cost table. If fewer than n lines are needed, proceed to Step 3. Otherwise, set the x_{ij}'s corresponding to positive reduced cost in C equal to 0 and those of zero reduced costs equal to 1. If there are more than one $x_{ij} = 1$ in any given column or row, eliminate all but one such that each row and each column only has one $x_{ij} = 1$ element. (If this happens, then there are **alternative optimal solutions**.)

Step 3: Find the smallest uncrossed reduced cost coefficient, d, in C. Subtract d from all uncrossed elements and add d to all elements that are crossed by two lines. Return to Step 2.

Steps 1 and 3 create a sequence of assignment problems with c_{ij}'s that are linear transformations of the original problem and hence have the same optimal solution. The optimality of the solution found at Step 2 is implied by the fact that any feasible solution with total cost of zero must be optimal.

The example below demonstrates the algorithm:

Step 1:

7	5	1	(-1)
8	6	3	(-3)
7	4	4	(-4)

6	4	0
5	3	0
3	2	0
(-3)	(-2)	(0)

Step 2:

3	2	0
2	1	0
0	0	0

3	2	0
2	1	0
0	0	0

Step 3:	2	1	0	Step 2:	2	1	0	Solution:	0	0	1	$x_{13} = 1$
	1	0	0		1	0	0		0	1	0	$x_{22} = 1$
	0	0	1		0	0	1		1	0	0	$x_{31} = 1$

An important development at the time of highly limited computational speed of computers, it is now rarely used and is more of historical interest. *B. Chen*

References

Kuhn, H.W. (1955) 'The Hungarian method for the assignment problem', *Naval Research Logistics Quarterly*, 2:83–97.

Winston, W.L. (1994) *Operations Research: Applications and Algorithms*, Belmont, CA: Duxbury, pp. 375–378.

Hypergame analysis

Traditional **game theory** assumes not only that relevant actors (*players*) are rational, but that all see the same 'game' – a common view of what the issue is, who the relevant parties are and what their aims are, what strategies they have and the payoffs for each combination of opposing strategies. In real life this is seldom so, suggesting that rationality can be a very subjective matter. Perhaps decision makers do conceptualize conflict situations in terms of actors, strategies and preferences, but see different games. Attempts to incorporate differing perceptions into game theory have a long history, notably through Harsanyi's (1968) model of games with incomplete information. However, most start from the idea that there is a well-defined game going on, but the players may be uncertain what it is.

Hypergame analysis adopts the simpler and more radical presumption that actors may construe the world in quite different terms. This is captured by each actor defining a *perceptual game* – her or his perception of the game, i.e. individual lists of the strategies and preferences perceived for all players (including self). A priori, these perceptual games may be completely dissimilar. Nevertheless, they are connected in that an action taken in one will impact on the others. Differences in perception may be localized to particular aspects of the situation, such as preferences for outcomes (i.e. each side really prefers to honour an agreement, but neither believes the other is sincere), or the feasibility of a particular strategy (e.g. strategic surprises in warfare). In other cases, there may be wholesale differences. For example, the Vietnam War may have been defined as 'the battle against communism' by some actors, but as 'national unification' (rather than 'battle for communism') by others. Hypergame analysis allows for the possibility of such wholesale differences. As part of the analysis, hypergame analysis explores all or a subset of (feasible) strategy combinations. One device used is a tree representation for each actor that shows all sequences of feasible actions and feasible counteractions. This analysis may reveal the relative strength and weaknesses of strategies, that a potential threat can be evaded or that seemingly irrational moves by others have a predictable logic of their own, and so on. By providing greater insight into the situation, this can lead to better decision making.

In practical applications, the analysis is usually done for a single client, i.e. only one party of the conflict situation is represented. Hence, possible or likely views of the perceptions of the other actors have to be inferred from the client, from research into the context of the situation, or from previously observed behaviour of the other actors.

More recently, insights from both hypergame and **metagame analysis** have been incorporated into a new approach known as **drama theory**, in which confrontations between players' positions are analysed dynamically (Bennett, 1998: Howard, 1999).

P.G. Bennett

References

Bennett, P.G. (1980) 'Hypergames: developing a model of conflict', *Futures*, 12:489–507, 1980.

Bennett, P.G. (1998) 'Confrontation analysis as a diagnostic tool', EJOR, 109(2):465–482.

Harsanyi, J.C. (1968) 'Games with incomplete information played by "Bayesian" players', *Management Science*, Theory 14:159–182, 320–334, & 486–502. Advanced mathematical treatment.

Howard, N. (1999) *Confrontation Analysis: How to Win Operations other than War*, Washington: CCRP Publications.

Implementation

The benefits of any MS/OR study can only be reaped if the results are implemented to a high degree. In fact, the whole effort and all activities needed to find a solution which offers positive net benefits must be focused on ensuring implementation. So although the physical implementation, i.e. putting the solution to work, is the last phase of a project, concerns for implementation start right from its inception.

Difficulties with implementation stem from three causes:

1. Those relating to the physical task of implementation, such as the complexity of the solution, its sensitivity to deviations and errors, and the extent it deviates from current practice. The greater any of these, the greater the difficulties.
2. Those relating to the **stakeholders** who could affect its proper working, such as their personalities, motivation (the less important the solution, the less attention it gets), educational background (does it place too high demands on them?), age (routine becomes more entrenched with age), and how they accommodate to the solution (does it take away their importance, their freedom?).
3. Those relating to the environmental context of the project, such as the support given by top management (the less visible the support, the more likely implementation fails), and the organizational implications (does the solution make the users more dependent on others? does it affect the activities of other departments in negative ways? does it pose a threat to job security?).

Generally, MS/OR analysts pay great attention to (1), which are questions of technology. They tend to overlook the human factors of (2) and (3) which then may become real constraints on implementation. Some can be partially overcome by good training of the users or simplifying the solution to fit their educational level. However, it is crucial that these human factors are dealt with right from the start by involving the stakeholders in the analysis such that they come to own the project and its results.

Planning for implementation therefore involves:

- identifying all stakeholders that could affect implementation;
- managing prior unrealistic expectations of the stakeholders and not promising more than can be delivered;
- involving the problem owners and solution users in the project;

- checking out and securing data sources so as not to delay implementation due to lack of data;
- planning execution of the actual process of physical implementation, such as preparing databases, software, new equipment, user documentation, training manuals, scheduling training, whether implementation can be done in one operation or should be done in stages (e.g. to alleviate stress on workloads or financial requirements);
- planning several follow-up sessions with problem users to monitor correct use of solution;
- setting up control procedures for maintaining and updating the solution.

H.G. Daellenbach

References
Daellenbach, H.G. (2001) *Systems Thinking and Decision Making*, Christchurch: REA, ch. 8.
Special Issue: Implementation (1987) *Interfaces*, May–June.

Implicit costs, see costs, types of

Increasing returns to scale, see economies of scale

Incremental analysis, see marginal and incremental analysis

Influence diagrams

An influence diagram (ID) is a network of nodes connected by arrows. An arrow from node i to node j implies that node i is relevant for node j. This relevance may be of the form that the **attribute** of node j is determined, influenced or affected by the attribute of node i; it could imply precedence in knowledge or events, or simply be qualitative in nature, in the sense that when considering node j, aspects of node i are known and relevant. The nodes are depicted by simple geometric shapes: rectangles, circles, ovals, hexagons and clouds, each having a distinct meaning.

R. Howard makes use of IDs for **decision analysis**, either in place of **decision trees** or in parallel to decision trees. While a change in sequence in which events occur or knowledge becomes available, or a reversal of the direction of influence, requires the entire decision tree to be redrawn, in an ID it may just mean a change in the direction of an arrow. Figure 1 below depicts the notational convention used for a pricing/quantity decision problem, where the demand and the production costs are **random variables**, but the sales revenue, once demand is known, is **deterministic**. Note that the pricing decision is made prior to the quantity decision. Reversing this arrow reverses the decision sequence.

Another use of IDs is to depict the transformation process in a **system**. It shows the uncontrollable system inputs (clouds), the controllable system inputs or decision choices (rectangles), the system variables (circles) that represent the details of the transformation, and the system outputs (ovals), as shown in Figure 2. An elongated 'tilde' signifies that the uncontrollable input is uncertain or random. Rectangles, clouds, and ovals are outside the system, circles are inside. As a consequence, a circle always has at least one arrow leading to it and at least one leaving it. Clouds and rectangles only have arrows leaving, and ovals only arrows entering. Feedback loops are entirely between circles, i.e. between system variables. Decisions or decision rules are 'imposed' on the system from outside.

The arrows only give the direction of influence, but no indication as to its form or strength. System variables and outputs are therefore seen as functions of the nodes

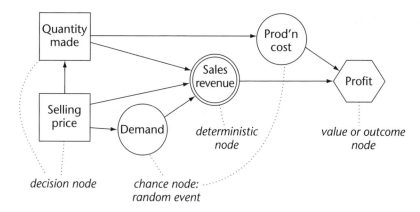

Figure 1: **Decision analysis influence diagram**

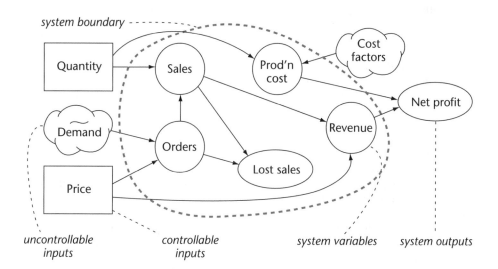

Figure 2: **Systems influence diagram**

where incoming arrows originate. For example, in Figure 2 the orders received are a function of the external demand and the price charged.

IDs help explore complexity. They are a highly effective means to define the relevant system for the problem analysed and can be used as input into computer programming and the actual **mathematical modelling** phase. They facilitate the communication process between the analyst and the clients or problem **stakeholders**.

H.G. Daellenbach

References

Daellenbach, H.G. (2001) *Systems Thinking and Decision Making*, Christchurch: REA, ch. 4.

Howard, R.A. (1990) 'From influence to relevance to knowledge', in R.M. Oliver and J.Q. Smith (eds) *Influence Diagrams, Belief Nets and Decision Analysis*, Chichester: Wiley.

Matheson, J.E. (1990) 'Using influence diagrams to value information and control', in R.M. Oliver and J.Q. Smith (eds) *Influence Diagrams, Belief Nets and Decision Analysis*, Chichester: Wiley.

Information systems

Information systems (IS) is a field focusing on business applications of computing systems. Terms that are often used as synonyms are *management information systems* (MIS), *information technology* (IT) and *information systems and technology* (IST). An information system for a business may include: *functional and enterprise systems*, **decision support systems** (DSS) and *e-commerce systems*.

Functional systems for an organization may include: transaction processing systems, accounting information systems, production and operations management systems, marketing and human resource systems. The category of enterprise resource planning systems is one that is capable of integrating some or all of these other functional systems into a cohesive whole working in conjunction with a central database.

Decision support systems provide a basis for improved decision making across the range of operational, managerial and strategic situations (problems and opportunities) faced by organizations. These situations may range from highly structured and routine problems, such as minimizing inventory holding costs, or maximizing the profits from an investment, to highly unstructured decisions, such as selecting research and development projects. Decision support systems often support the full process of decision modelling and decision-making, including: the intelligence phase (setting **objectives**, defining the problem, collecting data), the design phase (designing a model of the problem, determining alternatives and **criteria** for solution choice) and the choice phase (running the model, **sensitivity analysis**, selecting alternative to implement). This modelling and decision-making process may draw on internal data from the functional systems or data warehouses, as well as outside sources, including competitor information, industry trends and news services.

E-commerce systems permit the organization to conduct business online, buying, selling and paying over global networks, speeding up slow paper-based processes and permitting lower-cost processes and better service to customers. These systems typically use *Internet*-based technologies, enhanced with secure communication systems and online payment processes with customers and suppliers and other business partnerships across the organization's value chain.

Information systems are composed of six key components: hardware, software, data, people, procedures and networks. These components work together to achieve the common goal of the information system – to provide accurate, reliable and timely information for managing the organization. Hardware includes the computers, printers, servers, routers, firewalls and other physical components of the system. Software includes the operating systems (Windows and Linux, for example) and the applications such as databases, functional and decision support applications, spreadsheets and wordprocessors. Data consist of the records and documents the organization creates in the operation of the business. The people component includes designers of the systems, such as programmers and analysts, as well as users of the systems such as clerical staff and management. Procedures encompass the processes people must follow in effectively using the systems. Networks are the local and wide area linkages that permit the systems to communicate across space and time with branches of the business, customers and suppliers, as well as government and industry associations. *J. Vargo*

References

McLeod, R. (2000) *Management Information Systems*, 8th edn, Upper Saddle River, NJ: Prentice Hall.

Martin, E.W. *et al.* (1999) *Managing Information Technology*, 3rd edn, Basingstoke: Macmillan.

Stair, R.M. and Reynolds, G.W. (1999) *Principles of Information Systems*, 4th edn, Cambridge, MA: Course Technology.

Inputs, see **systems concepts**

Intangible costs, see **costs, types of**

Integer programming

Integer programming (IP) problems are a class of **linear programs** (LPs) with the additional requirement that some or all of the decision variables must be integer. In analogy to LPs:

- The objective function, to be maximized or minimized, is a linear function of the decision variables.
- The values of the decision variables are restricted by constraints in the form of linear equations or linear inequalities.
- All decision variables must be nonnegative, and some or all are restricted to assume integer values only.

If all variables are restricted to be integer, the problem is an *all-integer program*. If only some are required to be integer, we have a *mixed-integer program*. In many practical applications, the integer variables are permitted to assume only the values 0 or 1. In such cases, we have a *zero–one integer program*.

The use of integer variables, especially 0–1 integer variables, provides additional modelling flexibility, allowing a much wider range of important practical problems to be modelled, many of a strategic nature, such as

- **Fixed-charge problems**: A given activity incurs a positive setup or start-up cost at a nonzero level that does not depend on the level of the activity, but no cost if the activity level is zero. Examples of this occur when a machine has to be set up for a production run, a thermal power station unit has to go through an expensive warm-up period, or an activity at a positive level requires a piece of equipment to be purchased.
- *Set-covering problems:* Given a set of elements and various feasible groupings, we want to assign each element to a group so that all elements are 'covered'. An example is the **bin-packing problem**, i.e. loading and arranging items in several packing containers so as to minimize the number of containers needed.
- Problems with logical constraints, such as 'either-or' and 'if-then' constraints.
- **Facility location problems**: Siting of *n* identical facilities at *m* possible locations, where *n* is usually much smaller than *m*, so as to minimize the total **discounted** cost over a given **planning horizon**, including land, construction, operating, and transportation costs.
- **Knapsack problems**: Choosing which items to pack into a limited space, where each item packed contributes a given value or benefit, with the aim of maximizing the total value of benefits.
- *Investment problems*: Allocating limited sources of funds to various investment projects so as to maximize the net present value of their terminal wealth.

The cost of the added modelling flexibility is that IPs are much more difficult to solve than LPs. Favoured solution methods include **branch-and-bound** and **network** algorithms. **Cutting-plane algorithms**, developed in the late 1950s, are nowadays rarely used any longer. *B. Chen*

References
Nemhauser, G.L. and Wolsey, L.A. (1988) *Integer and Combinatorial Optimization*, New York: Wiley.

Winston, W.L. (1994) *Operations Research Application and Algorithms*, Belmont, CA: Duxbury, ch. 9.

Interactive multicriteria decision-making methods

Multicriteria decision making (MCDM) deals with problems where the decision maker has several conflicting goals or **objectives**. Two main approaches are used by MCDM methods: (1) aggregate value function methods which construct a single objective function as the sum of weighted individual objective functions, implying that *trade-offs* between objectives are possible, or (2) *outranking methods* which do not allow trade-offs.

In either case, the decision maker needs to provide information that reveals his or her preference structure. This information can be sought either (1) at the point a mathematical problem is formulated, (2) during the actual exploration of the **solution space** (the set of all possible **feasible solutions**), or (3) after all so-called **efficient solutions** (solutions where it is impossible to improve on one objective without a loss in another) have been identified. In case (2), the decision maker is usually asked to provide this information interactively.

Aggregate value function methods ask for trade-off information. A typical interactive session goes as follows: The constraints on the decision choice and the individual objective functions are first entered into the computer. The software then finds an initial feasible solution that (usually) is efficient. This solution or essential aspects of it are displayed on the screen. The decision maker is asked to answer one or more questions. One type of question has the form: 'How much of objective i are you willing to give up in order to gain an increase of z_j in (another) objective j?' Rather than ask for such numeric information that decision makers may find difficult to supply, an alternative form of question is: 'Do you prefer solution A to B, or B to A, or are you indifferent between them?' Solutions A and B differ only on two objectives – A offers more on objective i and less on j than B, but the same on all other objectives. Using the response given, the method then infers preliminary weights to assign to various objectives. It may solve the problem again with these new weights and submit a new set of questions. This process continues until the last answers given are consistent with the previous set of weights inferred. It is surprising that some of these methods need to ask only a few questions, typically between four and ten, to find the aggregate objective function that supposedly correctly reflects the decision maker's preferences. It is, though, questionable whether a decision maker can make valid comparisons between the contrived choices offered except for rather trivial problems. MCDM problems are usually of a strategic nature and hence far from trivial.

The *V.I.S.A.* software uses interactive visual display. Outranking methods also have interactive software for comparison of whole alternatives or aspects of alternatives.

H.G. Daellenbach

References

Belton, V. (1990) 'Multiple criteria decision analysis – Practically the only way to choose', in L.C. Hendry and R.V. Aglossa (eds) *Operational Research Tutorial Papers 1990*, Birmingham: Operational Research Society.
Miettinen, K. (1999) *Nonlinear Multiobjective Optimization*, Boston: Kluwer.
Steuer, R.E. (1986) *Multiple Criteria Optimization: Theory, Computation and Application*, New York: Wiley.

Interactive planning

Interactive planning is aimed at designing a desirable present and inventing ways of bringing it about as closely as possible. The process has six phases:

1. *Formulation of the mess* (Ackoff's term for a collection of interconnected problems): identifying the current policies and behaviour that may become the seeds of the **system's** self-destruction.

2. *Ends planning:* producing the best ideal seeking design currently conceivable that is technologically feasible, capable of surviving in the current environment, and adaptable to change as a result of learning. When the design is completed, the closest approximation that is currently achievable becomes the target of the remainder of the planning process.
3. *Means planning:* selecting or inventing the policies, programmes, practices and courses of action for realizing the approximation to the idealized design.
4. *Resource planning:* determining the resources (people, money, plant and equipment, supplies and services, and information) needed.
5. *Implementation:* assigning responsibility for who is to do what, by when and where, and what results are expected and on what assumptions these expectations are based.
6. *Controls:* designing and implementing processes for determining when these expectations and assumptions are in error, adjusting the plan accordingly and reducing the likelihood of similar errors in the future. Of particular importance are assumptions about possible futures which require contingency planning. This eliminates the need for forecasting, since forecasts are about probable events, while assumptions are about possible events.

The interactive planning process has three major characteristics:

1. It implies continuous revision in response to (a) imbalances between resource required and available, and (b) adjustments required by failure of expectations to be realized and assumptions to materialize.
2. It is participative: All **stakeholders** have an opportunity to participate in the planning process. Each level of an organization prepares a plan based on the plan prepared at higher levels, but it can suggest changes to the latter which it believes would improve their plans as well as its own.
3. It is holistic: The plans produced by each part must be compatible with those produced at every other level. Units are permitted to plan anything they want that does not affect any other unit, but if what they want affects another unit, that unit must agree or the difference between the units must be settled at a higher level of the system.

In interactive planning, process is the most important product because it yields improved understanding of the system planned for, its structure and how and why it functions as it does. Because planning is continuous, plans are not finished products, but snapshots selected from a moving picture.

Systems engaged in interactive planning are engaged at every moment of time in closing the gap between where they are and where they would most like to be at that moment of time. In this way they create their future because, in general, any system's future depends more on what it does than on what is done to it.

R.L. Ackoff

Reference
Ackoff, R.L. (1999) *Re-creating the Corporation: A Design of Organizations for the 21st Century,* Oxford: Oxford University Press.

Interactive simulation, see gaming

Interior point methods for linear and nonlinear programming

The well-known **simplex method** has been the dominant solution technique in **linear programming** since its development by Dantzig in 1947. The **algorithm** starts at a corner point or extreme point of the **feasible** region – the set of all solutions that satisfy all constraints which in three decision variables has the shape of an irregular polyhedron. It then moves along an edge to an adjacent extreme point, provided this improves the value of the objective function. The new extreme point becomes the starting point for the next iteration, and this process repeats itself until no edge can be found that offers a further improvement in the objective function. The optimal extreme point has then been found.

Algorithms are classified as *polynomial* or *exponential* in terms of their computational efficiency or computer time needed. Polynomial-time algorithms require a computer run time that is no more than a polynomial function of the 'size' of the problem, where size may be defined by the number of variables and constraints. In contrast, the computation time for exponential-time algorithms increases exponentially as a function of 'size'. The former are referred to as efficient and are more desirable than the latter which can only find optimal solutions for relatively small problems.

In the 1970s it was demonstrated via pathological examples that the simplex algorithm is not a polynomial-time algorithm. This result was more of a theoretical interest, since in practice the simplex method has proven itself as highly efficient, but nevertheless sparked an interest in alternative methods. In 1979, the Russian mathematician Khachian published his *ellipsoid algorithm*, which exhibited polynomial-time performance, but in use proved to be less efficient than the simplex method owing to the high precision calculations required for its implementation.

A paper by Karmarkar in 1984 initiated new research into what has become known as *interior point methods* (IPMs). IPMs differ from the simplex method in that they start from the interior of the feasible region and attempt to find directions of successive improvements of the objective function while remaining strictly within the feasible region. Karmarkar's algorithm is polynomial-time and has also found efficient implementations in practice.

IPMs can also handle **nonlinear programming problems**. Their modus operandi is identical to the linear case. They operate by introducing a *barrier function* (see **barrier methods**) close to the boundary of the feasible region that prevents the search procedure from leaving the feasible region.

Since Karmarkar's seminal paper in 1984, more than 3000 papers have been published related to interior point methods. It remains to be seen whether IPMs will eventually displace the simplex method which has reigned supreme for the past 50 years.

A.D. Tsoularis

References

Hooker, J.N. (1986) 'Karmarkar's linear programming algorithm', *Interfaces*, 16 (July–August): 75–90. A well-written and informative paper.

Karmarkar, N. (1984) 'A new polynomial-time algorithm for linear programming', *Combinatorica*, 4(4):373–395.

Roos, C., Vial, J.-Ph. and Terlaky, T. (1997) *Theory and Algorithms for Linear Optimization: An Interior Point Approach*, New York: Wiley. Excellent introductory text.

Terlaky, T. (ed.) (1996) *Interior Point Methods of Mathematical Programming*, Boston: Kluwer. Aimed at the mathematically sophisticated reader.

Web-site

Interior Point Methods Online: http://www-unix.mcs.anl.gov/otc/InteriorPoint/

Internal rate of return, see discounted cash flows

Interpretive systems approach, soft systems thinking

Interpretive systems thinking (or soft systems thinking) is based on a subjectivist approach. Systems are seen as convenient human **conceptualizations**, conceived in people's minds for a given purpose, rather than as observable objective realities. As a consequence, the primary focus of interpretive approaches is on the perceptions, values, beliefs and vested interest of the people involved in a **problem situation**. They recognize that in **human activity systems** different **stakeholders** have personal **world views** through which they perceive and interpret problem situations and issues **subjectively**, and that what they see differs, which may become sources of conflict. Furthermore, values and views are not free of the context of a problem, or even of how it is framed. Hence, there is a need to understand subjectively the points of view and positions taken by various stakeholders, or in other words, there is a need to externalize (at least partially) their world views and bring about a better mutual appreciation. This is in contrast to the objective approach of 'hard' MS/OR and **systems analysis**, and the functionalist approaches to systems thinking, such as **system dynamics** and Beer's **viable systems models**, where reality is largely assumed to be independent of the observer.

There is no coherent interpretive systems theory. Different interpretive systems methodologies, methods and techniques find their theoretical foundations mainly in phenomenology (the study of human consciousness and self-awareness in abstraction of any claim of existence), sociology, and organization theory, although some are also steeped in systems theory. The best-known approaches are Warfield's interactive management, Churchman's social system design, Mason and Mitroff's **strategic assumption surfacing and testing** (SAST), Ackoff's social system sciences or **interactive planning**, Checkland's **soft systems methodology** (SSM), and other **soft operational research** techniques and methods. They all share one basic tenet, namely that effective management, and particularly substantive change in an organization, requires the active involvement of all vested parties. They differ in how they bring this about. *H.G. Daellenbach*

Reference
Jackson, M.C. (2000) *Systems Approaches to Management*, New York: Kluwer/Plenum, ch. 7.

Inventories: costs, management, and control

Inventories are stocks of goods held in expectation of some future demand, either from external customers or from users inside the organization itself. If there exists no future demand, the goods are valueless and should be scrapped. Inventories come in most diverse forms: goods for sale, raw materials, goods in process (partially completed goods), goods in transit (pipeline stocks), supplies, spare parts, livestock held for fattening, cash, funds in bank accounts, water in reservoirs, blood in blood banks, etc.

There are various **costs** associated with inventories. Inventories tie up funds. Hence they incur an inventory *holding or carrying cost* (**opportunity cost**). They may become obsolete or spoil, losing value. They incur *handling costs*. They must be insured against loss or fire. Most of these costs vary more or less proportionately with the size of the inventory. Running out of stock may cause *shortage costs* (opportunity cost), such as lost profit and loss of goodwill, or a cost of expediting a stock replenishment. To keep shortage costs low, the organization may keep excess stock, called *safety stocks*. Replenishing inventories results in production *setup or ordering costs*, such as the cost of preparing the replenishment documents, fixed shipping and receiving costs, and the cost of setting up equipment for a production run, including the initial learning cost, and the equipment change-over cost between similar products. These are usually fixed, regardless of the size of the replenishment or production run.

Suppliers may offer single or multiple *quantity discounts* if an order exceeds certain quantities, called *price breaks*, to encourage placement of larger orders. These may apply to all units or only units in excess of the price break. These savings must be balanced with the increased inventory holding costs resulting from larger orders.

The major decisions in inventory management are:

- The timing of a replenishment, which depends on the replenishment **lead time** (i.e. the elapsed time between initiating a replenishment and the goods becoming available), the predicted demand or usage during the lead time, and the relative size of the holding and shortage costs.
- The size of the replenishment, which depends mainly on the relative size of the holding costs and the fixed setup or ordering cost, quantity discounts, and the predicted demand or usage. Other factors considered are product shelf life and risk of obsolescence.

For control purposes, goods are often grouped (e.g. by an **ABC classification** based on a decreasing annual dollar-volume of usage), with tight control for A-products, and loose control for C-products. Other management concerns are the frequency of reviewing the inventory positions, product assortment choice, disposal of excess stock or stock becoming obsolete, and acquisition and disposal of speculation stocks.

H.G. Daellenbach

Reference

Silver, E.A., Pyke, D. and Peterson, R. (1998) *Inventory Management and Production Planning and Scheduling*, New York: Wiley & Sons.

Inventory control models

Mathematical models for inventory control are based on the following major control aspects: Q = replenishment size, s = reorder point (triggering replenishment), S = order-up-to level, t = time between successive replenishments, and R = interval between successive stock level reviews. Models can be classified into six basic types. The first is **deterministic**, while the remaining five assume that the demand is a **random variable**.

1. **Economic order quantity** (EOQ) models are the most basic deterministic models. They minimize the sum of annual inventory holding costs, annual setup or ordering costs, and annual product costs. If the product cost per unit is constant, regardless of the order quantity, the optimal order quantity is given by the well-known square root formula:

$$EOQ = \sqrt{\frac{2 \times \text{annual demand} \times \text{fixed setup cost}}{\text{unit holding cost}}}$$

2. **Newsboy problem:** Assumes that there is only one opportunity to acquire stock to meet a random demand. It finds the optimal order quantity S which minimizes the sum of expected overage cost (loss incurred on stock left over) and expected underage cost (shortage cost).
3. (s, Q) model, R fixed: A multi-period EOQ model with a safety stock to protect against shortages during the replenishment **lead time**. It is often approximated by a two-step procedure: find Q^* using the EOQ model and then find s^* using a newsboy problem with the overage cost equal to the unit holding cost per cycle $t = Q^*/D$.
4. (s, S) model, R fixed: Whenever I = (stock on hand) + (goods on order) $\leq s$, initiate a replenishment of $Q = S - I$. It can be shown that this rule is an optimal policy for a fixed setup or ordering cost, and linear holding and shortage costs. Q is different for each replenishment.
5. (S, t) model, $R = T$: The multi-period version of (2). A replenishment of $Q = S - I$ is initiated after every interval of length t.
6. (s, S, R) model: This is the most comprehensive model, finding the jointly optimal values of all three decision variables. No analytic solutions exist.

Computations for solving models (4) to (6) are onerous and may not be justified by the possibly small gain (often less than 1 per cent) in savings achieved over the (s, Q) model. A number of approximate solution procedures have been developed for finding near-optimal solutions for the (s, S) model.

Some models restrict Q to be a multiple of a basic lot size, q. Others derive approximation for the best joint replenishments of several products. The general form of such policies is to set the replenishment cycle equal to $t_i = Rn_i$ for each product i, where R is the shortest review period (often equal to the replenishment cycle of the most frequently replenished product) and n_i is integer. Each product is replenished at every n_ith review period, resulting in different joint replenishments.

For non-stationary random demands, optimal Q's can only be determined using stochastic **dynamic programming**. For deterministic demands, the optimal policy can be found using the **dynamic economic order quantity** model for which the **Silver–Meal heuristic** yields excellent approximations.   *H.G. Daellenbach*

Reference

Silver, E.A., Pyke, D. and Peterson, R. (1998) *Inventory Management and Production Planning and Scheduling*, New York: Wiley.

Inverse transformation, see **random numbers**

Ishikawa diagram, see **seven tools of quality management**

ISO 9000

The ISO 9000 family of standards were released in 1987 with major revisions in 1994 and 2000. Produced by the International Organization for Standardization (ISO), a worldwide federation of national standards bodies, they provide generic principles for how companies should establish management practices and procedures for control of quality, consistency and reliability in manufacturing processes and service provision. The standards, however, are not product or service standards and alone do not guarantee quality.

The standards-based approach to quality assurance through a documented quality system provides:

- a framework for planning, control, verification and improvement of critical business processes;
- a means of systematically identifying and communicating current best practice;
- a systematic approach for generating evidence of past and current performance; and
- a mechanism for establishing customer confidence in an organization's ability to meet contractual obligations.

The ISO 9001 standard also provides a template to assess an organization's quality management system for the purpose of independent certification. They provide guidance on the aspects of an organization's operations that need to be incorporated and effectively managed within a quality system if the organization is to do business well, and obtain and maintain certification if appropriate. They do not have product, industry or technology-specific requirements. The series is intended to be relevant to all types of organizations, whether manufacturing or service. They do not tell an organization how to run its business.

The standards also require that the system is effectively documented through:

- A quality policy which publicly states the organization's intentions with regard to quality.

- A quality manual which gives the framework of the system identifying responsibilities, organizational structure, and the areas of operations to be procedurally controlled. The manual is the road map through the system.
- Procedures which give the framework for individual processes and identify the links between processes. Procedures are detailed road maps of important processes.
- Records which detail the results of activities which are part of the quality system.

The specifics of system documentation and operations are the responsibility of the individual organization. The standards' broad specification must be interpreted and applied to each organization to best meet the needs of that organization and its customers.

Much implementation of ISO 9000-based systems has been characterized by two common misconceptions. First, equating the system with documentation. Documentation is only a representation of the system. Secondly, the system has been seen as only applying to production. The first element of the standards is management responsibility which requires a public commitment to quality from senior management and their active involvement in the system: management practice is the intended focus of the standards.

The year 2000 revision of the standards simplifies the structure of the standards, increases the explicit customer focus of the quality system and increases the emphasis on systematic quality improvement. *D.J. Houston*

References

Voehl, F., Jackson, P. and Ashton, D. (1994) *ISO 9000: an implementation guide for small to mid-sized businesses*, Delray Beach, Florida: St Lucie Press.

ISO 9000: 2000 Quality Management Systems – Fundamentals and vocabulary.

ISO 9001: 2000 Quality Management Systems – Requirements.

Web site

International Organization for Standardization: http://www.iso.ch

Iteration, see algorithm

Jackson networks

Many systems, including communication and manufacturing systems, can be modelled as networks of queues (see **queueing networks**), where the output of one queue becomes part of the input to some subsequent service facility. One of the aspects of such systems we usually want to know is the state distribution – the numbers of customers at the various queues at a particular time. Normally this is a very complex problem, but for one class of networks the answer is very simple. Suppose we have a network where all external streams of customers arriving from outside the network are independent *Poisson processes*, service times are negative exponentially distributed, and upon leaving a queue a customer joins another queue or leaves the network with a fixed probability which is independent of the **state of the system**. The service rate at a particular queue can depend on the number of customers there, so you can have multiple servers in parallel, for example, but you cannot have finite capacity.

If these conditions are met, J.R. Jackson (1963) proved that the joint state distribution at a fixed point of time in **steady-state** has a very simple form, often called a product form, which can be written down as the product of the state distributions that we would get if we considered each queue as a separate independent system, each with an input Poisson process whose rate is equal to the rate at which customers flow through that queue.

This is an amazing result. It implies that at a fixed point in time the numbers of customers at each of the queues are independent of each other. Of course the state of a particular queue will influence that of any queue 'downstream' from it, but it will do so at a future point in time. This argument also shows why the easy product form solution does not give us easy results for the sojourn-time distribution – the time that a customer spends in particular parts of the network, since the time spent at successive queues may be dependent.

In 1967 Gordon and Newell showed that this product form extends to the case of a closed network, containing a finite number of customers who never leave. The problem with large closed networks is calculating the normalization constant which is required to make the distribution add to one. Closed network models have been proposed for systems such as time-sharing computers, where a finite number of jobs share a single processor.

While the assumptions for Jackson networks are restrictive, their unique (in the class of queueing networks) easy solutions mean they have been used to model many systems, possibly inappropriately. *D.C. McNickle*

References

Gordon, W.J. and Newell, G.F. (1967) 'Closed queueing systems with exponential servers', *Operations Research*, 15:254–265.

Jackson, J.R. (1963) 'Jobshop-like queueing systems', *Management Science*, 10:131–142.

Kelly, F.P., Zachary, S. and Ziedins, I. (eds) (1996) *Stochastic Networks: Theory and Applications*, Oxford: Clarendon.

Johnson's algorithm, see sequencing

Judgemental bootstrapping

Judgemental bootstrapping – also called *policy capturing* – is a type of **expert system** that infers the expert's model by examining predictions made by that person (the 'expert' may be a group). Judgemental bootstrapping – though not the name – was originally conceived and tested in the early 1900s in a study on the quality of next summer's corn crop (Hughes, 1917). In the 1960s, researchers in a variety of fields studied judgemental bootstrapping, initially unaware of each other's work until Dawes reviewed the research and coined the term 'bootstrapping'. The term suggests that forecasters can lift themselves by their own bootstraps.

To develop a bootstrapping model, an expert is presented with quantitative and qualitative information abut real or simulated situations and asked to make predictions about certain aspects of the situation. **Regression analysis** is then used to infer the rules the expert *appeared to use* in making these forecasts.

Bootstrapping models resemble *econometric models*, except that the dependent variable represents the expert's forecasts, rather than actual outcomes. For example, one could provide a manager with data about a sample of 20 new product plans that differ on price, advertising, and product design. One would then regress the data on the explanatory variables against the managers' predictions.

Although a judgemental bootstrapping model is not as comprehensive or flexible as an expert, it applies the expert's rules consistently; this improves the reliability of judgement. Consistency is useful for comparing policy alternatives and for improving accuracy. Judgemental bootstrapping can also help to identify judgemental biases, and can make the predictions by the best experts available for use by others with less expertise. When repetitive forecasts are needed, bootstrapping can reduce the costs of forecasting and provide forecasts rapidly.

Bootstrapping is useful when historical data on the variable to be forecast are lacking or of poor quality; otherwise, econometric models should be used. Bootstrapping is most appropriate for complex situations, where experts' judgements are unreliable but have some validity. If an expert's predictions are based on invalid cues, the bootstrapping model can be harmful because it does the wrong thing more consistently.

Studies from psychology, education, personnel, marketing and finance have compared bootstrapping forecasts with those made by experts using unaided judgement. Bootstrapping was more accurate for eight comparisons, less accurate in one, and there were two ties (Armstrong, 2001). *J.S. Armstrong*

References

Armstrong, J.S. (2001) 'Judgmental bootstrapping: Inferring experts' rules for forecasting', in J.S. Armstrong (ed.) *Principles of Forecasting*, Norwell, MA: Kluwer Academic.

Dawes, R.M. (1971) 'A case study of graduate admissions: Application of three principles of human decision making', *American Psychologist*, 26:180–188.

Hughes, H.D. (1917) 'An interesting seed corn experiment', *The Iowa Agriculturist*, 17:424–5, 428.

Just-in-time or JIT, Kanban

Just-in-time is a production system developed between 1965 and 1978 at Toyota Motor Company by the industrial engineer Taiichi Ohno which since then has spread worldwide.

JIT is a production system in which inventory is low, distances between operations are short, and each operation does not produce until an order is received. JIT is most appropriate for repetitive manufacturing and assembly systems. Synonyms for JIT include zero inventory, lean production, stockless production, material as needed, continuous flow manufacturing.

The goals of JIT are to:

- reduce work-in-process inventory by eliminating overproduction and make only what is needed only when it is needed;
- eliminate production steps that do not add value, such as transportation, waiting and inspections;
- reduce machine setup times to a negligible length;
- find production rates that coordinate processing between work centres;
- eliminate defects at all stages.

JIT is a *pull system*, i.e. products are requested in amounts needed by the user, rather than a *push system*, where the producer pushes quantities convenient for them to the user. JIT is usually implemented with *kanban*, which means card in Japanese. Kanbans are used for controlling when production at each work centre may occur. Production is always in quantities that fill exactly one or more standard containers. Each container carries one card. Therefore, for a given part the number of cards issued controls its maximum in-process inventory. The system works as follows: Empty containers are returned from a work centre CU that uses part X to the work centre CP that produces part X. As soon as centre CP has finished producing other parts which have been requested earlier, it will start producing part X. If there is no card for X, then there is no production. When centre CU is close to running out of part X, it will fetch a full container from centre CP.

Since each work centre produces several items, several cards are needed for each different part to guarantee that the using work centre never runs out. The aim is to keep the number of cards as low as possible. The normal procedure is to start with a conservatively high number. As experience increases, cards are gradually withdrawn until a further reduction would disrupt a smooth operation. This may take days or weeks. Hence, output rates need to be relatively stable over time.

JIT requires better quality to work, but also allows better quality. Small lots result in a short time between when an error is made and when it is discovered, so fewer bad parts are produced. Finding the cause of an error is easier as well. By contrast, large lots result in long delays before errors are discovered. Furthermore, because inventory takes up space, operations are farther apart and communication among workers is harder. Successful implementation of JIT requires low setup times. One of the important tasks for production engineering is to come up with product designs, processes and special equipment which result in very short setup times. So both continuous improvement and effective quality control are important attributes of JIT. Whenever possible, external parts suppliers and subcontractors, located close by, are integrated into the system. Those that are distant make daily overnight deliveries.

J.F. Raffensperger

References

Hall, R.W. (1981) *Driving the Productivity Machine*, American Production and Inventory Control Society, Inc.

Hay, E. (1988) *The Just-in-Time Breakthrough*, New York: Wiley.

Ohno, T. (1988) *Toyota Production System: Beyond Large-Scale Production*, Cambridge MA: Productivity Press.
Schonberger, R.J. (1982) *Japanese Manufacturing Techniques*, New York: Free Press.

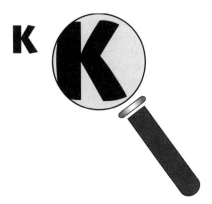

Kaizen

The development of quality management practices in Japan reflects the concept of Kaizen: continuous improvement in personal life, home life, social life and working life. When applied to the workplace, Kaizen means ongoing improvement involving everyone – top management, middle managers, and workers. Kaizen is a philosophy for organizations which sweeps in a wide range of processes and techniques. Masaaki Imai, who brought Kaizen to the attention of western managers in the 1980s, argues that the most important difference between Japanese and Western management concepts is Kaizen's process-oriented way of thinking versus the West's innovation and results-oriented thinking. While the two approaches are characterized as fundamentally different, they do have a common underlying focus – the reduction of quality-related costs.

Reliance on innovation or breakthrough has historically been more common in Western firms than attention to continuous improvement. An exclusive emphasis on results, and in particular short-term results, can lead to an attitude where 'the ends justify the means', and focus is on short-term profit over long-term survival. Kaizen's process orientation recognizes effort and contribution, not just achievement.

Critical elements of Kaizen include:

- Process orientation and results orientation: Both how the organization operates (process) and what it achieves (results) need to be kept in balanced perspective. Ongoing, systematic process improvements contribute to the organization's ability to improve its results. Results flow from a quality process.
- Defined roles of managers, supervisors, workers: Managers are constantly concerned with improvement of the whole organization. Supervisors are constantly concerned with maintaining and improving processes throughout their work group. Workers maintain systems and make immediate small improvements.
- A wide range of techniques: Anything that contributes to improvement is part of Kaizen. It incorporates: **total quality** control, **just-in-time**, *zero defects*, *robotics* and *automation*, **quality control circles** and small group activities, suggestion systems, total productive maintenance, cooperative labour–management relations, new product development.
- Improvement at all levels: Management-focused improvement, work-group-focused improvement, and individual-focused improvement.
- Gradual and constant change: Based on collectivism, group efforts and systematic, **systemic** approaches.

- Limited investment, but great effort to maintain it.
- Sharing, caring, commitment, communication, and leadership based on personal experience and conviction, not authority, age or task.

Many Western organizations in the 1980s and 1990s adopted one or more of the techniques included within Kaizen, but only a few can be found where all or most of the practices and principles are evident. The exceptions are often subsidiaries of Japanese companies. It is rare to find a Western organization for which the concept of Kaizen is an accepted, normal element of the organization's philosophy because the principles of Kaizen run counter to fundamental concepts of Western management.   *D.J. Houston*

Reference

Imai, M. (1986) *Kaizen The Key to Japan's Competitive Success*, New York: McGraw-Hill.
Imai, M. (1997) *Gemba Kaizen: A Commonsense, Low-Cost Approach to Management*, New York: McGraw-Hill.

Kalman filter

Forecasts are rarely perfect; instead they show what is likely to happen 'on average'. So it is a good practice to complement forecasts with measures of the forecast uncertainty. The most common measure of uncertainty is the variance. Such measures are particularly useful for decision making. For example, when determining the amount of stock to keep in a warehouse, it is necessary to consider being able to meet above normal levels of demand, not just the average demand. The amount of stock in the warehouse should be based on a measure of uncertainty such as the forecast variance.

The Kalman filter is an iterative computational algorithm designed to calculate forecasts and forecast variances for time series models. It can be applied to any time series model which can be written in *state space* form. Almost all of the standard time series models in common use can be written in this form. The Kalman filter is applied recursively through time to construct forecasts and forecast variances. Each step of the process allows the next observation to be forecast based on the previous observation and the forecast of the previous observation, i.e. each consecutive forecast is found by updating the previous forecast. The update rules for each forecast are weighted averages of the previous observation and the previous forecast error. These update rules resemble those of an allied approach to forecasting called **exponential smoothing**. The intriguing feature of the Kalman filter is that the weights in the update rules are chosen to ensure that the forecast variances are minimized. These weights, referred to collectively as the *Kalman gain*, play a similar role to the so-called *smoothing constants* in exponential smoothing.

The Kalman filter is important because it may be applied in real time. As each new value of the time series is observed, the forecast for the next observation can be computed. It has been widely used in engineering and the natural sciences, and to a lesser extent in economics and finance.   *R. Hyndman and R.D. Snyder*

Reference

Harvey, A.C. (1991) *Forecasting, Structural Time Series Models and the Kalman Filter*, Cambridge: Cambridge University Press.

Kanban, see just-in-time

Karesh–Kuhn–Tucker theory and conditions

Optimization is mostly concerned with finding the best solution to a decision maker's problem. A large class of optimization problems are constrained, so finding the **optimal solution** may not be as simple as finding maxima or minima of the objective function. A constraint may mean that the maximum of the objective function is actually outside the **feasible** region. The figure below shows contour plots for an objective function involving two variables, and a feasible region defined by the shaded quadrilateral ABCD.

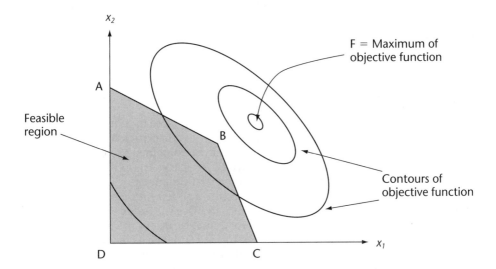

Figure: **A nonlinear optimization problem**

H.W. Kuhn and A.W. Tucker produced a set of conditions or equations that hold at the optimal solution to constrained problems. Originally named after Kuhn and Tucker, these conditions are now referred to as Karesh–Kuhn–Tucker (KKT) conditions, since Karesh independently arrived at the same set of conditions before Kuhn and Tucker.

The KKT conditions involve not only the decision variables of the original problem, but an additional variable associated with each constraint, known as a **Lagrange multiplier**. Each can be interpreted as a penalty for 'shifting' the solution from the unconstrained maximum of the objective function (point F) to a solution that satisfies the associated constraint.

Of course, there may be a number of solutions that will satisfy each constraint (any point along the sides of, or inside, ABCD), but we want the solution that will provide the best value of the objective function, i.e. the one that is 'closest' to the unconstrained maximum (point B in the figure). This is equivalent to minimizing the amount of 'shift' that the Lagrange multipliers produce. In this way, the KKT conditions represent a saddle point – they describe a solution (to a maximization) that minimizes the effect of the Lagrange multipliers that keep the solution feasible, but maximizes the profit from the original decision variables.

Specifically, the KKT conditions describe the following characteristics of the optimal solution to constrained problems:

- No improvement in the objective function can be obtained without violating at least one constraint, if that constraint is binding at the optimal solution.
- The Lagrange multipliers reflect the fact that constraints that are not binding at a particular solution are irrelevant. The Lagrange multipliers are equal to zero when the associated constraint is not binding (since the solution does not require 'shifting'), and positive when the constraint is binding.
- Feasibility, i.e. constraints are satisfied and variables are non-negative.

In the case of linear problems, the Lagrange multipliers are equal to the **shadow prices** or optimal values of the dual variables.

While the KKT conditions are necessary for a solution to be the optimal point to a constrained problem, they are not sufficient. In other words, a solution could satisfy the KKT conditions, yet possibly still not be the **global optimum** to the problem. However, they are both necessary and sufficient if the feasible region is **convex**.

S. Batstone

Reference

Winston, L.W. (1994) *Operations Research Applications and Algorithms*, Belmont, CA: Duxbury, pp. 691–701.

Knapsack problem

The knapsack problem gets its name from the situation faced by a person who has to decide which items from a long wish-list to pack into a backpack for a trip into the wilderness. The items packed should maximize their combined usefulness or value, while not exceeding the volumetric capacity of the pack. The problem appears in various disguises in many industrial and commercial situations.

Suppose an investor has the choice of investing a sum b into a subset of n different discrete investments. His or her objective is to place the funds so as to maximize some measure of performance, such as the total final wealth or the total **net present value**. Let c_j be the net present value of investment j and a_j be its required initial cash outflow. Any investment must be for the full initial value or none; fractional investments are not allowed. Let $x_j = 1$ if an investment j is chosen, and 0 otherwise. The problem can then be formulated as the following **integer program**:

$$\text{maximize } z = \sum_{j=1}^{n} c_j x_j$$

$$\text{subject to } \sum_{j=1}^{n} a_j x_j \le b$$

$$x_j = 0, \text{ or } 1, j = 1, \ldots, n.$$

The knapsack problem can be solved efficiently by **branch-and-bound methods**, taking advantage of the fact that all variables must be either 0 or 1. It can also be reformulated and solved by **dynamic programming**.

Knapsack problems are encountered as subproblems in more complex models of integer programming problems, where a whole series of similar, possibly small knapsack problems are solved repeatedly; hence, the need for efficient solution algorithms.

The knapsack problems without the integer restrictions are known as continuous knapsack problems. These are very easily solved by inspection. Observe that the ratio of c_j over a_j indicates the relative contribution of item j for one unit of the resource. A **greedy algorithm** can therefore be used, allocating items in order of decreasing values of the ratio until the last item picked exceeds the resource left unused. A fractional amount of that item is then added to meet the resource constraint as an equality. The addition of the integer restriction makes this process much harder.

R. Sridharan

References

Hillier, F. and Lieberman, G. (2001) *Introduction to Mathematical Programming*, 7th edn, New York: McGraw-Hill.

Winston, W.L. (1999) *Operations Research – Applications and Algorithms*, 3rd edn, Belmont, CA: Duxbury.

Knowledge-power

By conjoining the terms 'knowledge' and 'power' we suggest that 'knowing' in society is related to the exercise of power. People who insist that power affects the way in which knowledge is created, challenge the commonly held assumption that it is possible to develop a way of advancing knowledge of reality as it exists outside of ourselves. Let us take the example of the way that science is traditionally seen as advancing knowledge of reality. It can be argued (if one challenges this tradition) that the reference to scientifically generated 'results' enables scientists, and those using scientific results, to give authority to their claims against the claims of others. This way of giving authority to viewpoints can be regarded as an exercise of power, springing from the suggestion that the scientific community can use reason to develop knowledge of reality.

Likewise, when so-called debating mechanisms are used in society to arrive at conclusions about what is considered as realistic action, this too may become a form of power play (see also **dialogue**). People may try to get others to believe that they have been developing a kind of truth through the debating process. However, it should be remembered that debate may easily be distorted by, say, coercive forces. During debate the voices of some people may be given more credibility (status) than those of others. Coercion can operate in subtle ways. This can affect the way that issues become raised for discussion and treated during discussion. This means that any decisions reached about what is achievable 'in reality' again can be challenged if one is concerned with the manner in which power is being used to define the terms of debate (and to set agendas for discussion).

So those focusing on conjoining knowledge and power suggest that we must be on the lookout for ways in which power is used to define 'reasonable' knowing. Some of those who have expressed concern about the link between knowledge generation and power have suggested possible ways of addressing it. For instance, one way of addressing it is to try to develop more awareness in society about the fundamental fragility of anybody's claims to knowledge. By creating this awareness, one can open up more space to challenge those who try to give authority to their claims (at the expense of others). Another way of addressing knowledge–power relations is to set up arenas for people to develop their own self-reliance, so that they will be less dependent on the power of others to define 'the realities'. (See also **boundary critique**.) *N.R.A. Romm*

References

Fals-Borda, O. and Rahman, M.A. (eds) (1991) *Action and Knowledge: Breaking the Monopoly With Participatory Action Research*, New York: Intermediate Technology Pubs/Apex Press.

Flood, R.L. and Romm, N.R.A. (1996) *Diversity Management: Triple Loop Learning*, Chichester: Wiley.

Gordon, C. (ed) (1980) *Power/Knowledge: Selected Interviews and Other Writings 1972–1977/Michel Foucault*, New York: Pantheon Books.

Habermas, J. (1993) *Justification and Application*, Cambridge: Polity Press.

McKay, V.I. and Romm, N.R.A. (1992) *People's Education in Theoretical Perspective*, Cape Town: Longman.

McLagan, P. (1998) *Management and Morality*, London: Sage.

L

Lagrange multipliers, see **shadow prices, Karesh–Kuhn–Tucker theory**

Lead times

In general, lead time is the elapsed time before an action becomes effective. It is the time lag between initiating an activity and its effects being felt or the activity being completed. Lead times are commonly encountered in strategic planning, project management, **production planning and control**, and **inventory control**. It basically means that any planning has to take lead times into account. Actions have to be initiated well ahead of the time their effects are to be realized.

Lead times are also common in **system dynamics** modelling in the form of *transport lags* in **feedback loops**. An example of a transport lag is the time it takes for the water temperature to change after you adjust the valves in the shower.

In project management (see **critical path methods** or **PERT**) the times of various tasks have to include lead times, either as part of the task time itself or as a separate activity that is a precedent to the actual task. The latter alternative is preferred, since it clearly shows the composition of the total time and allows action to be focused on either if need be.

Production scheduling depends on accurate specification of lead times for various operations. Traditionally, production lead times cover not only the actual time of production or of an operation, including the time needed to set up the equipment or to change over from one product to another, but also the average waiting time prior to the start of the activity. Some modern methods of production planning and scheduling, such as **JIT** and **synchronous manufacturing**, have as their goal to reduce both these times through tight coordination of activities, e.g. by eliminating most or all of the waiting time between operations and scheduling consecutive operations to overlap, whenever possible, as well as scheduling small lots, and reduction of setup times.

In inventory control, accurate estimates of the procurement lead times are needed in order to provide for appropriate protection, so-called *safety stocks*, against possible shortages. The lead time should include not only the time between initiating a replenishment and having the goods available for use, but also the time between successive reviews of the stock levels, since during that time no action is taken to initiate a replenishment. For example, if stocks are reviewed only once each week, then one week needs to be added to the procurement lead time to account for this period of lack of remedial action.

Leadership

The notion of leadership has greatly fascinated scholars and practitioners in management. There are also management writers who find the notion of leadership almost repulsive. The well-known management guru Peter Drucker claims that leadership is all hype. On the other side, there are equally well-known management thinkers like Mary Parker Follet, who observed: 'The person who influences me most is not he who does great deeds but he who makes me feel I can do great deeds.'

It is important to distinguish between management and leadership. It is possible to be a good manager without being a leader, but the reverse is not true. If management is doing things right then leadership is doing the right thing. According to Stephen Covey, management is climbing the ladder of success. Leadership determines whether the ladder is leaning against the right wall. Second, it is important to distinguish between *transactional leadership* and *transformational leadership*. Transactional leadership is more similar to management. It is the leadership exercised in day-to-day situations. This is the type of leadership that can be taught as a technique or skill. Transformational leadership, as the name implies, is concerned with bringing about change. This is not a skill or technique that can easily be taught. While individuals can learn how to improve their transformational leadership abilities, they cannot be trained to become transformational leaders.

The most important lesson from leadership research in recent years is that transformational leadership is rooted, not in charisma, but in character. Integrity is probably the most important attribute of such a leader. Integrity means consistency in thought, word and deed. While qualities of character and integrity cannot be taught, they can be acquired by discipline and self-development. In his book *Principle-Centered Leadership*, Stephen Covey identifies eight positive qualities of principle-centred leaders. They:

- are continually learning
- are service-oriented
- radiate positive energy
- believe in other people
- lead balanced lives
- see life as an adventure
- are synergistic
- exercise physically, mentally, emotionally and spiritually for self-renewal.

According to Covey, there are three barriers that inhibit us from developing these positive qualities. These are 'appetites and passions', 'pride and pretension' and 'aspiration and ambition'. These forces lower our credibility as principle-centred leaders. When we lose credibility, we are unable to create or promote trust. Therefore, these restraining forces are like diseases that have to be overcome through appropriate antidotes. The antidote for appetites and passions is self-discipline and self-denial. The cure for pride and pretension is character and competence. The medication for aspiration and ambition is working for noble purposes by providing service to others.   *V. Nilakant*

References
Covey, S.R. (1990) *The Seven Habits of Highly Effective People*, New York: Simon & Schuster.
Covey, S.R. (1991) *Principle-Centered Leadership*, New York: Simon & Schuster.
Follet, M.P. (1918) *The New State*, New York: Longmans, Green & Co.

Line balancing, assembly line balancing

An assembly line is a manufacturing mode where a product is produced or assembled on a fixed sequence of consecutive workstations, as shown in the figure below. The typical example is the assembly of cars. A workstation is a place along the assembly line where a worker or an unattended machine, e.g. a robot, performs a particular operation consisting of one or more tasks. A task is a unit of work that for technological, safety or managerial reasons cannot be subdivided into smaller parts. There are strict precedence relationships between most tasks, although some tasks may be slotted between alternative pairs of several consecutive tasks. The most common assembly line is a conveyor that moves past a series of workstations at uniform rate, forcing each workstation to complete its task in a maximum amount of time, called the *cycle time*. The aim is to assign one or several consecutive tasks to each workstation such that all workstations perform the same amount of work. Hence no workstation remains periodically idle, waiting for the preceding station to finish its operation. This will maximize the output of the line for a given speed. Such a perfectly balanced line is rarely achieved. The slowest workstation will determine the output rate for the line.

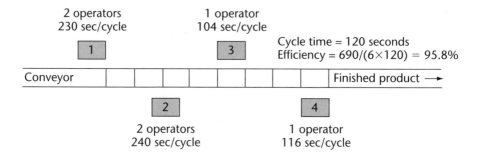

Figure: **Assembly line**

The following algorithm generates good if not optimal workstation assignments:

Step 1: Order the tasks needed to complete one unit of the product.

Step 2: Draw a precedence diagram – a flow chart that shows which task must follow which other task or tasks. This will show which combinations of task assignments are feasible.

Step 3: Determine the desired cycle time, using the following formula:

$$\text{Cycle time} = \frac{\text{Production time per day}}{\text{Output target per day in units}}$$

(Note, the actually achieved cycle time may be higher.)

Step 4: Assign tasks to workstations so that the precedence relationship among the tasks is respected and the total operating time at each workstation is close to, but does not exceed, the cycle time. A number of heuristics can be used to assign tasks to workstations, such as selecting the longest imminent task first (to fit the tough task first), or the task that has the most successors on the precedence diagram (to prevent problems of assigning those later).

Once the first trial line has been done, the efficiency of a line balance is calculated as

$$\text{Efficiency} = \frac{\text{Sum of task times}}{(\text{Number of workers}) \times (\text{Cycle time})}$$

Line balance efficiency is used to determine the sensitivity of the line to changes in the production rate and workstation assignments. If the efficiency is considered too low, say below 90 per cent, improvements may be sought by interchanging some tasks between consecutive stations.

Large-scale line-balancing problems are often solved by computers. COMSOAL (Computer Method for Sequencing Operations for Assembly Lines) is a software package that is widely used for large line-balancing problems. *P. Venkateswarlu*

Reference
Stevenson, W.J. (1996) *Production/Operations Management*, 5th edn, Homewood, IL: Irwin, pp. 272–279.

Linear programming

Linear programming (LP), as the name implies, deals with constrained optimization problems where all relationships are linear. The constraints may represent limits on the available resources, output requirements, quality standards, relationships between the decision variables, and so on.

Consider the following simplified example. A firm produces two models of bookshelves, consisting of a tubular frame and painted shelves, involving the following operations: shelf making, painting and assembly. The basic model requires 36 minutes of operator time to make shelves and 22.5 minutes for assembly. Its material cost is $43, and it sells for $143. The standard model requires 50 minutes for making shelves, 25 minutes for assembly, costs $60 in materials and sells for $180. The daily capacity is 23 hours for making shelves, 15 hours for assembly and at most 32 units can be spray-painted. Marketing considerations dictate that the number of basic units produced should be no more than three times the number of standard units. The firm would like to know the product mix that maximizes the difference between total revenue and total material costs. (Labour costs can be ignored, since they are fixed.)

Formulated as an LP, this problem results in the following expressions:

Objective function:	maximize 100 BASIC +	120 STANDARD
	subject to	
Shelving constraint:	36 BASIC +	50 STANDARD ≤ 1380 (minutes)
Assembly constraint:	22.5 BASIC +	25 STANDARD ≤ 900 (minutes)
Painting constraint:	BASIC +	STANDARD ≤ 32
Basic/Standard relationship:	BASIC ≤	3 STANDARD
Nonnegativity conditions:	BASIC ≥ 0,	STANDARD ≥ 0

Note that all relationships are linear. The figure below shows these relationships graphically. The objective function and each constraint (as an equality) can be graphed as a straight line. The shaded area is called the **feasible** region. It represents all combinations of the two decision variables that satisfy all constraints. It is a polygon. The two broken lines correspond to those combinations of BASIC and STANDARD that generate a value for the objective function of $2400 and $3536.71, respectively. As the two variables assume combinations that increase the value of the objective function, the broken line moves parallel up and to the right. To maximize it, we want to push it as far up as possible, while still having at least one point in common with the feasible region. That occurs at point C for the value of $3536.71, which corresponds to an output of $15\frac{5}{7}$ units of basic and $16\frac{2}{7}$ units of standard per day. These have to be interpreted as averages, with partially finished units carried

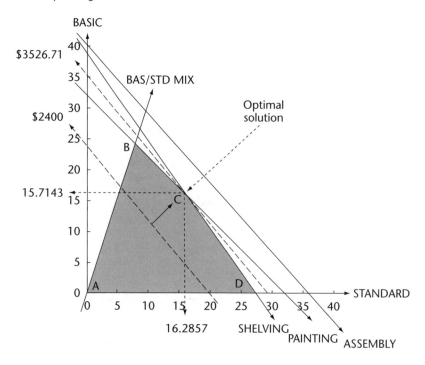

Figure: **Graphical representation of LP**

forward from day to day. Interestingly, the assembly constraint has no point in common with the feasible region. It is so-called *redundant*.

Real-life problems have thousands of variables and constraints. The optimal solution is found by a computationally efficient **algorithm**, such as the **simplex method**, developed in 1947 by the US mathematician, G. Dantzig.

In a mid-1980s a survey of Fortune 500 firms reported that 85 per cent had used LP.
A.D. Tsoularis

Reference
Any text book on MS/OR techniques has one or more chapters on LP and the simplex method.

Little's queueing formulas

Named after J.D.C. Little, a past president of the Operations Research Society of America (now INFORMS), these **queueing** formulas relate

- the mean number of customers in a queue L_q, to the mean time that a customer spends waiting in the queue before service, W_q, by

$$L_q = \lambda W_q$$

and

- the mean number of customers in the queueing system (this includes any in service) L, to the mean time that a customer spends in the system, W, by

$$L = \lambda W$$

where λ is the arrival rate of customers.

Take the second formula, for example. Consider the queue of ships waiting at, or being serviced, at a port. We want to work out the total cost of ship time per day. Suppose that it costs $8000 per day to operate a ship while it is in port (either at, or waiting for a berth). Thus the average total cost of ship time per day is given by ($8000) times the mean number of ships in port, L. On the other hand you could argue that each ship spends an average of W days in port. On any given day, we expect on average λ ships to arrive, each of which is going to spend W days in port. Hence the additional cost, considering only that day's average number of arrivals, is ($8000)$\lambda W$. But this must also be the average daily cost of all ships in port. Both arguments are right, and equating the answers and cancelling out the $8000 gives $L = \lambda W$. A proof of $L_q = \lambda W_q$ can be done similarly.

Notice that the proof of these formulas does not make any assumptions about the probability distribution of arrivals or service times, or about the number of servers and (under some restrictions) the service discipline.

The value of the formulas is that often one side of them is easier to calculate or estimate than the other. For theoretical models it is often easier to find a formula for L or L_q, and then use Little's formulas to calculate W or W_q, whereas it is often easier to estimate W or W_q if you are directly observing the queue. For estimates over a finite time period the formulas are only approximately true, although if the observation period starts and ends with an empty system they do hold exactly. Since $W = W_q$ + [mean service time], knowing any one of L, L_q, W or W_q means that you know the other three.

D.C. McNickle

References

Gross, D. and Harris, C. (1998) *Fundamentals of Queueing Theory*, 3rd edn, New York: Wiley.
Little, J.D.C. (1961) 'A proof of the queueing formula $L = \lambda W$', *Operations Research*, 9:383–387.

Living systems theory

This theory suggests that all living systems share a number of similar characteristic processes and subsystems. The theory conceptualizes seven different levels of living systems, namely cell, organ, organism, group, organization, society, and supranational system. The theory postulates that any living system must have access to the following 20 subsystems in order to carry out the characteristic processes of living:

1. Reproducer: Replicates the system and contributes to the system's evolution.
2. Boundary: Defines the spatial limits of the system and protects it from the environment.
3. Ingestor: Brings in matter and energy from outside.
4. Distributor: Distributes the inputs and the outputs of the various subsystems.
5. Converter: Converts the form of inputs and outputs, e.g. matter to energy.
6. Producer: Synthesizes materials and provides energy.
7. Matter–energy storage: Stores matter and energy and retrieves when required.
8. Extruder: Transmits matter and energy out of the system.
9. Motor: Moves various parts within the system.
10. Supporter: Maintains a set of spatial relationships.
11. Input transducer: Brings in information markers from the environment and transforms these to markers that can be interpreted within the system.
12. Internal transducer: Receives internal information markers from one subsystem and transforms these to a form that can be interpreted by another subsystem.
13. Channel and net: Provides a route for the various information markers.
14. Timer: Transmits time-related states to the decider subsystem.
15. Decoder: Transforms received (public) information to a (private) code understood within the system.
16. Associator: Helps the system learn by relating and associating certain streams of information.
17. Memory: Stores and retrieves information.

18. Decider: Receives information inputs and generates outputs that guide, control, and coordinate the system's functioning.
19. Encoder: Transforms private information to a public information code.
20. Output transducer: Transforms internal information markers to an external form.

The theory has been applied successfully to diagnose various malfunctions in living systems, including businesses and other organizations, with the aim of improving or designing, and redesigning systems that are effective in achieving their goals and aims.   *D.P. Dash*

References

Dean, E.B. (1997) *Living Systems Theory from the Perspective of Competitive Advantage*, available online [(]http://akao.larc.nasa.gov/dfc/lst.html[)].
de Geus, A. (1997) *The Living Company: Growth, Learning, and Longevity in Business*, London: Nicholas Brealey Publishing.
Miller, J.G. (1978) *Living Systems*, New York: McGraw-Hill.

Local optimum, see optimization

Logistics, supply chain management

According to the Supply Chain Council web site, 'the supply chain encompasses every effort involved in producing and delivering a final product or service, from the supplier's suppliers to the customer's customers. Supply chain management includes managing supply and demand, sourcing raw materials and parts, manufacture and assembly, warehousing and inventory control and tracking, order entry and order management, distribution across all channels, and delivery to the customer.'

The US Council for Logistics Management (CLM) defines logistics as 'that part of the supply chain process that plans, implements, and controls the efficient, cost-effective flow and storage of goods, services, and related information from the point of origin to the point of consumption in order to meet customers' requirements'. Thus, in the CLM's perspective, logistics is subsumed within supply chain management. Two aspects of the definition are especially noteworthy: the management of

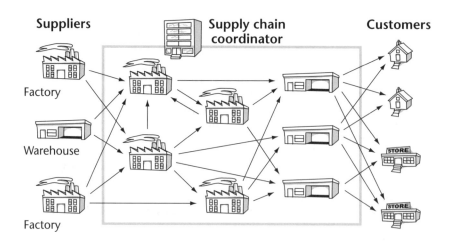

Figure: **Supply chain**

not just material flows, but information flows as well, and the emphasis on both efficiency (with regard to cost) and effectiveness (with regard to meeting adequate levels of customer service).

The figure below depicts the supply chain graphically.

Coyle, Bardi and Langley (1996) identify three phases in the historical evolution of logistics/supply chain management: physical distribution management, integrated logistics management, and supply chain management. The focus in physical distribution management was to systematically manage a set of interrelated activities pertaining to the outbound distribution of finished goods from manufacturing plants with reference to a systems viewpoint. Closely related to the systems viewpoint is the 'total cost concept', which means that rather than seeking to minimize the costs of individual activities (transportation, warehousing, etc.), one should seek to minimize the total cost, given the interactions between the various activities.

Integrated logistics management arose from the recognition of additional opportunities for cost savings, such as backloading to reduce dead mileage of empty vehicles, through the integration of the inbound procurement of raw materials to manufacturing plants, with the outbound distribution of finished goods. In this respect, supply chain management can be thought of as the integrative management of the end-to-end supply chain that encompasses all the firms that are involved in ensuring that the final customer receives the right product at the right time, place and cost and in the right condition.

Ellram and Cooper (1993) lucidly clarify how supply chain management departs from traditional systems with regard to pipeline (goods-in-transit) coordination of the management of inventories; the seamless nature of inventory flows; the focus on landed cost to the end-consumer; the sharing of information and risk, based on trust, between trading partners; and a supply chain team approach towards planning.

Johnson and Pyke (2000) present 12 content areas in supply chain management, several of which could quite easily be subsumed under logistics management. They include location, **transportation**, warehousing, material handling, **inventory management** and **forecasting**, sourcing and supplier management, service and after-sales support, reverse logistics and green issues, and global issues (e.g. cross-border sourcing and distribution). The possible exceptions are outsourcing and strategic alliances, metrics and incentives, marketing and channel restructuring, and product design and new product introduction. The first three speak broadly to partnerships, which are critical to supply chain management, and to the coordination and sharing of information between supply chain partners. The last topic reflects the heightened importance of supply chain management in the face of ever-shortening product life cycles. *J. Sankaran*

References

Coyle, J.J., Bardi, E.J., and Langley Jr., C.J. (1996) *The Management of Business Logistics*, 6th edn, Minneapolis/St. Paul: West Publishing Company, pp. 2–23.

Ellram, L.M. and Cooper, M.C. (1993) 'Characteristics of supply chain management and the implications for purchasing and logistics strategy', *International Journal of Logistics Management*, 4(2):1–10.

Johnson, M.E., and Pyke, D.F. (2000) 'A framework for teaching supply chain management', *Production and Operations Management*, 9(1)(Spring):2–18.

Manpower planning, see **applications of MS/OR to crew scheduling**

Marginal and incremental analysis

Marginal and incremental analysis are basic tools of economics and their principles are used in a number of MS/OR techniques. They are used for determining the optimal activity level of an operation. The underlying idea is that a change from a given position is desirable if a small (or marginal) change in the activity level produces an improvement in the performance measure. The basic principle of marginal analysis states that (unless prevented by the constraint) such small changes should continue until no further improvements in the performance measure occur. Consider the following simplified example for finding the best volume of irrigation water used by a farmer: Line 2 lists the total water charges by the irrigation board, and line 3 gives the total net value (expected revenue minus all other costs) of the crop produced.

	0	1	2	3	4	5	6
1. Water units used (units of 10 000 m³)	0	1	2	3	4	5	6
2. Total water charges (£)	0	1200	1400	1600	1800	2100	2500
3. Total net value of crop (£)	4000	6000	6800	7200	7500	7700	7800
4. Net profit (£)	4000	4800	5400	5600	5700	5600	5300
5. Marginal net profit (£)		800	600	200	100	−100	−300
6. Marginal water charge (£) = $MC(Q)$		1200	200	200	200	300	400
7. Marginal net crop value (£) = $MR(Q)$		2000	800	400	300	200	100

The net profit (line 4) reaches its maximum for a water usage of four units. Marginal analysis reaches the same conclusion by looking at the marginal net profit (line 5: difference between successive values in line 4). It is positive for increased water usage of up to a level of 4 units. After that, further increases have a negative result.

A variant of marginal analysis compares the marginal increase in costs, $MC(Q^*)$ (line 6), with the marginal revenue, $MR(Q)$ (line 7). As long as marginal costs are less than marginal revenue, it pays to increase the activity level – water usage in our case. This property is expressed as follows: The optimal level of activity Q^* occurs where

$$MR(Q^*) \geq MC(Q^*) \text{ and } MR(Q^*+1) < MC(Q^*+1)$$

Marginal analysis assumes that changes can be infinitesimally small. Hence, marginal costs and marginal revenues may be expressed as the first derivatives of the corresponding total cost and total revenue functions. The optimal activity level occurs where the derivatives are equal.

Incremental analysis assumes that changes can only be made in discrete amounts, which may not be equal. The principle of marginal analysis is then applied to these consecutive discrete amounts (as in the above example), i.e. as long as the incremental cost is less than the incremental revenue, the increase in activity level is advantageous.   *H.G. Daellenbach*

Reference
Daellenbach, H.G. and McNickle, D.C. (2001) *Systems Thinking and OR/MS Methods,* Christchurch: REA, ch. 2.

Marginal costs, see costs, type of

Marketing, see applications of MS/OR in marketing

Markov chains, Markov processes, Markovian decision processes

Markov chains model the probabilistic behaviour of **discrete event systems**, i.e. systems where there are n possible states, such as n possible stock levels in an inventory system, or up to n machines requiring service from an operator in a machine shop. Consider the following simple example. The market for a product is shared by three brands. Consumers have mixed loyalties and tend to switch brands. The table below shows the probabilities of staying with the same brand or switching to one of the other brands at consecutive purchases:

Percent switching		to brand		
		A	B	C
	A	0.6	0.1	0.3
from brand	B	0.1	0.4	0.5
	C	0.2	0.2	0.6

This is a *Markov chain* with three states that describe the purchasing behaviour of the average consumer. The numbers in the table represent the transition probabilities p_{ij} of going from state i to state j in one transition. In the example, each purchase corresponds to a transition. Note that these transition probabilities depend only on the present state the system is in and not on how the system got into that state. This is a fundamental property of Markov chains.

In a Markov chain, the state of the system is observed only at the time a transition occurs, such as, for instance, at the end of each day in an inventory system or at each new purchase in this brand-switching example.

It is interesting to determine the long-run behaviour or the **steady-state** of such systems. The steady state is the fraction of time the system is in each state in the long run, or the probabilities of finding the system in each state at a random point after a large number of transitions. For *regular chains*, i.e. chains where it is possible to get directly or indirectly from all states to all states without cycling, the steady-state probabilities are independent of the starting state of the process. In the brand-switching example, we should expect to find the system on average 29.8% in state A, 21.3% in state B and 48.9% in state C. These numbers can also be interpreted as the shares of the market for each brand.

In a *Markov process*, the state of the system is a function of time, where time is a continuous variable and the state transitions can occur at any point in time. The behaviour in a waiting line or **queueing system** is of that sort. The number of customers in the system is the state variable. The state changes when a customer arrives or a service is completed. These events can occur at any point in time.

A.D. Tsoularis

References

Hillier, F.S. and Lieberman, G.J. (2001) *Introduction to Operations Research*, 7th edn, New York: McGraw-Hill, ch. 16.

Ross, S.M. (2000) *Introduction to Probability Models*, San Diego, CA: Academic Press, ch. 4.

Markov decision processes

A **Markov chain** where in each **state** there is a choice of possible actions with different transition probabilities and different costs or benefits is called a Markov decision process. The problem is to choose a policy, i.e. an action for each state, so as to maximize either the long-run **discounted** benefits or the average benefit per transition.

Consider the following example: A taxi operator works the San Francisco–Oakland area. Past experience indicates the following pattern of calls depending on whether she cruises in each city or always returns to the main bus terminal:

Location	Action	Percentage of calls going to		Expected
		San Francisco	Oakland	net profit
San Francisco	Cruise	60	40	$14.80
	Bus terminal	50	50	$15.20
Oakland	Cruise	25	75	$7.50
	Bus terminal	40	60	$12.00

What is her best policy?

In 1960 R.A. Howard at MIT showed how this problem can be solved via a special **algorithm** based on **dynamic programming** – named *approximation in policy space* – for an objective of either maximizing the expected discounted net profit over an infinite **planning horizon**, or maximizing the average net profit per trip. The algorithm consists of four steps:

1. In each state, select an initial policy (e.g. in each state select the action that has the highest immediate benefit).
2. Evaluation routine: Assume that this policy is followed forever and compute the expected return (discounted or per trip) for each state, by solving a set of linear equations.

3. Improvement routine: The decision maker now decides to postpone using that policy until after the next trip and instead in each state selects for the next trip the action that maximizes the sum of the immediate return plus the expected return over the states that result from that action.
4. Stopping rule: If the evaluation routine returns the same policy, this is the optimal policy, otherwise return to Step 2.

Applied to our example, starting with an initial policy of always going to the bus terminal in each city, the algorithm converges in two iterations to the optimal policy of {cruising in San Francisco; bus terminal in Oakland}, giving an average return per trip of $13.60, an improvement of $0.36 over the initial policy.

When maximizing the discounted expected returns over an infinite planning horizon, the optimal policy may change as the discount rate changes. Howard also showed that the problem can be reformulated and solved as a **linear program**.

H.G. Daellenbach

References
Hillier, F.S. and Lieberman, G.J. (2001) *Introduction to Operations Research*, 7th edn, New York: McGraw-Hill, ch. 21.
Puterman, M.L. (1994) *Markov Decision Processes: Discrete Dynamic Programming*, New York: Wiley-Interscience.

Matching

As the name suggests, matching involves the process of pairing objects or people together to achieve some desired goal. Matching problems arise in many applications, such as personnel assignment, machine scheduling and inventory planning. One of the most common situations is the **assignment problem**, i.e. assigning people to jobs in order to maximize a measure of total suitability, under the restriction that each person is assigned to only one job, and every job is assigned to only one person.

The most convenient way of representing a matching problem is by a **network**, in which the nodes represent the objects to be matched, and the arcs represent the possible matchings allowed. In this way, finding a matching is equivalent to finding a set of arcs, no two of which meet at the same node.

The personnel assignment problem is an example of *weighted bipartite matching*, i.e. in which the objects being matched come from two groups, and the value of the matching is based upon the weights. It can be solved very efficiently by special versions of the network **simplex method** or by the **Hungarian method**.

The assignment problem seeks an optimal weighted matching; there are two other forms of objective: *maximum cardinality* and *bottleneck matching*. An example of the first is where a hostel manager wishes to assign pairs of roommates to rooms on the basis of (known) 'compatibility'. Finding the maximum number of compatible pairs will minimize the number of rooms required. It turns out that bipartite maximum cardinality matching problems are equivalent to **maximum flow problems**. Bottleneck (or maximin) bipartite matching problems seek a maximum cardinality matching for which the minimum arc-weight in the matching is maximal. For instance, suppose that n workers are to be assigned to n workstations on an assembly line, and that the potential work rate of each worker at each station is known. Since the rate at which the line can operate is determined by the rate of the slowest worker, the assignment of workers to workstations that maximizes the production rate is of the required bottleneck form. Algorithms for bottleneck matching problems are also computationally efficient.

In more general situations, however, there may be more than two aspects to be matched. For instance, the assignment problem matching may involve constructing teams of workers, with team leaders assigned to each team; 'compatibility' in the hostel problem may be defined separately in terms of religion, culture and social interests. These *non-bipartite matching problems* are significantly harder to solve than their bipartite counterparts, but are still efficiently solvable in polynomial time.

J.W. Giffin

Reference

Evans J.R. and Minieka, E. (1992), *Optimization Algorithms for Networks and Graphs*, 2nd edn, New York: Marcel Dekker.

Mathematical modelling, mathematical models

The OR/MS methodology consists of three major phases: *problem formulation* (where the problem is identified and formulated within its wider context), *mathematical modelling* (where the problem is analysed in mathematical terms), and *implementation* (where the recommendations derived are implemented).

Mathematical modelling itself consists of four major steps:

1. Translating the (explicitly defined or implicitly conceived) **system** corresponding to the problem formulated into a quantitative model: The relationships between inputs, components, and outputs are expressed in the form of functions, equations, inequalities, and logical relations. Some models may contain tens of thousands of expressions. For decision problems, a mathematical model will contain at least one performance measure or objective function. Some expressions may be constraints that restrict the values of decision variables.
2. Exploring the **feasible solution space**: The model is manipulated to observe its behaviour as a function of the decision variables, often in view of finding the **optimal solution**, i.e. the set of values for the decision variables that optimizes the performance measure.
3. Verify and validate the model: **Verification** checks whether the logic of the model and all its inputs are correct. **Validation** establishes whether the model is a sufficiently close approximation to reality at an appropriate level of resolution and provides answers relevant for decision making. While verification is a matter of attention to detail, validation involves subjective judgements on the part of the analyst and the decision maker and careful review of the **boundary judgements** implied by the system underlying the model.
4. Analyse the sensitivity of the solution: This is a systematic exploration of how the answers from the model, such as the optimal solution, are affected by changes in critical inputs, e.g. resource availabilities, cost and revenue factors. Thorough **sensitivity analysis** often provides the most valuable insights for informed decision making and risk assessment.

MS/OR uses many so-called *general-purpose models*. Each type has a well-defined form in terms of its mathematical properties. For example, if all relationships are linear and all variables nonnegative, the model is a **linear program**. Powerful computational solution methods or **algorithms**, usually incorporated into computer software, facilitate and speed up the exploration of the solution space and **sensitivity analysis**.

J.D.C. Little (1970) lists desirable properties for mathematical models:

- Simplicity: Simple models are more easily understood.
- Completeness and parsimony: The model should include all significant aspects, and only those.
- Ease of manipulation: Solutions should be obtained with a reasonable amount of computations.

- Adaptiveness: It should be easy to update the model to reflect non-structural changes in relationships and changes in inputs.
- Ease of interface: The model should be user-friendly in terms of input requirements and output interpretation.

A further important property is:

- Appropriateness and relevance: The model should give answers and insights that are relevant and appropriate for decision making within the time frame required.

H.G. Daellenbach

References

Daellenbach, H.G. (2001) *Systems Thinking and Decision Making*, Christchurch: REA, chs. 5–8.
Little J.D.C. (1970) 'Models and managers: concept of decision calculus', *Management Science*, 16(8):B469–471.
William, H.P. (1999) *Model Building in Mathematical Programming*, New York: Wiley.

Mathematical programming, nonlinear programming

Mathematical programming is a catch-all term for **optimization** problems with the following characteristics:

- There are n decision variables, usually restricted to assume nonnegative values.
- The objective is to maximize or minimize a function of these decision variables.
- The combinations of values for the decision variables that are **feasible** are restricted by m constraints.

If the objective function and the constraints are all linear expressions of the decision variables, then the problem is a **linear program** for which highly efficient solution techniques exist (the **simplex method, interior-point methods**). **Integer programming** problems are similar to linear programs, except that some or all of the decision variables are restricted to assume integer values only, e.g. only 0 or 1, or 0, 1, 2, 3, . . . etc. These problems are solved by so-called **branch-and-bound methods**, which are computationally more demanding, since they usually involve solving a large sequence of linear programming problems. Another subclass, called **quadratic programming** problems, allows the objective function to include quadratic terms of the decision variables (i.e. squares of decision variables or the product of two decision variables), with all constraints restricted to be linear. Again highly efficient solution methods exist to solve them.

Another subclass of problems can be broken into stages and solved by **dynamic programming**, provided the objective function lends itself to being expressed as a recursive relationship between consecutive stages as a function of a **state variable**. The latter could, for instance, correspond to the amount of a resource left to be allocated at that stage. The overall solution is found by systematically, stage by stage, solving many simple optimization problems, one for each possible value of the state variable.

Those problems that make no assumptions as to the form of the objective function and the constraints are simply referred to as *nonlinear programming problems* or NLPs. Unfortunately, no single optimization procedure can solve all NLPs efficiently. The famous **Karesh–Kuhn–Tucker conditions** state what conditions a solution has to satisfy to be an **optimum**, but it may not be a **global optimum**. To be a global optimum, the objective function and the **feasible** region both have to be well behaved (i.e. the feasible region has to be **convex**, and the objective function **concave** for a maximization problem). A large number of solution methods have been proposed over the years, the most famous ones being **gradient methods** and **barrier and penalty methods**.

Since the development of **algebraic modelling languages**, such as GAMS, AMPL or AIMMS, which greatly facilitate both the formulation and solution of NLPs, much interest in the more esoteric NLP solution methods has waned. The processing power of today's computers overcomes most of the computational difficulties that were faced in the 1950s, 60s and 70s. *H.G. Daellenbach*

References

Hillier, F.S. and Lieberman, G.L. (2001) *Introduction to Operations Research*, 7th edn, New York: McGraw-Hill, ch. 13.

Nemhauser, G.L., Rinnooy Kan, A.H.G. and Todd, M.J. (eds) (1989) *Handbooks in Operations Research and Management Science, 1: Optimization*, Amsterdam: Elsevier. Authorative overview, advanced treatment.

Maximum flow problem, max-flow/min-cut theorem, Ford–Fulkerson algorithm

The maximum flow problem (MFP) seeks a **feasible solution** that sends a maximum amount of flow from a single source node to a single sink node in a **network** of arcs with limited capacities along each arc. Applications include sending fluid (water or oil) in a pipeline network, messages in a telecommunications network, electricity in an electrical network or traffic on a road network. The MFP is often found as a subproblem in much larger problems.

The MFP can be formulated as a **linear program** and solved by the **simplex method**. There are, however, algorithms specifically designed to take advantage of the network structure, such as the *augmenting path, preflow-push* or *capacity scaling* algorithms. The following example describes the simplest of these methods, the augmented path algorithm by Ford–Fulkerson.

Node 1 is the source and node 6 is the sink. (Multiple sources or multiple sinks are linked to a supersource or supersink, respectively, with infinite capacity.) The number shown beside each arc is its maximum flow capacity (ignore the numbers in brackets). The idea behind the augmenting path approach is to iteratively find a connected sequence of arcs (a path, best described by its node sequence) from the source to the sink which allows the total flow in the network to be increased, i.e. augmented. We start with a flow of zero. At the first iteration, we arbitrarily pick path 1–2–4–6, which allows a maximum flow of 2 – the lowest capacity on any of its arcs. This leaves residual capacities of

$$r_{12} = 0, r_{13} = 9, r_{24} = 2, r_{25} = 5, r_{35} = 3, r_{46} = 0, \text{ and } r_{56} = 6.$$

At the second iteration, we pick 1–3–5–6, which allows a flow of 3, augmenting the total flow to 5. The residual capacities are updated to

$$r_{12} = 0, r_{13} = 6, r_{24} = 2, r_{25} = 5, r_{35} = 0, r_{46} = 0, \text{ and } r_{56} = 3.$$

The algorithm also allows flow to be, so to speak, 'pushed back' on any arc. So at iteration 3, we send 2 units via path 1–3–4–2–5–6. This implies that 2 units are pushed

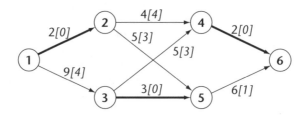

Figure: **The maximum flow problem**

back from node 4 to node 2 (which previously went from 2 to 4), increasing the total flow to 7. As shown in the brackets in the figure, the updated residual capacities are now

$$r_{12} = 0, r_{13} = 4, r_{24} = 4, r_{25} = 3, r_{35} = 0, r_{46} = 0, \text{ and } r_{56} = 1.$$

Extra flow can now be pushed from the source via node 3 only to node 4, i.e. not through to the sink. Have we found the optimal solution?

The so-called 'max-flow/min-cut theorem' states that the value of the maximum flow is equal to the capacity of the minimum 'cut', which is the set of arcs whose deletion from the network would cut the sink off from the source. For our example, the cut turns out to be (1,2), (3,5) and (4,6) – the heavy lines, for which $u_{12} + u_{35} + u_{46} = 2 + 3 + 2 = 7$; so optimality is confirmed. The final solution is: 2 via 1–2–5–6, 3 via 1–3–5–6, and 2 via 1–3–4–6. *J.W. Giffin*

Reference

Winston, W.L. (1994) *Operations Research: Applications and Algorithms*, 3rd edn, Belmont, CA: Duxbury, ch. 8.

Means–end chain, see cause-and-effect thinking

Medicine, see applications of MS/OR in medicine

Metagame analysis

A strictly rational approach to decision making can lead to 'dilemmas' casting doubt on the very idea of rationality. The best-known comes from **game theory**, the *prisoners' dilemma* – a situation in which two prisoners are held as suspects of a crime and are unable to communicate with each other. Each is offered leniency if he confesses and turns evidence against his partner, but will get a maximum sentence if the partner confesses while he does not. Although both realize that if neither confesses they can only be charged with a lesser offence, the only rational choice for either is to confess, since this gives a better outcome whatever the other does. However, if they both do so, they get heavier sentences than if they had remained silent. Hence both are worse off if each makes a rational choice.

Game-theoretic dilemmas are typical of situations in which parties are interdependent, yet have differing objectives. Metagame theory 'solves' the prisoner's dilemma by asking how rational players who are able to completely predict each other's choices will choose. Von Neumann and Morgenstern tackled this question in game theory, concluding that a rational solution corresponds to an equilibrium in which each player optimizes against the other's (predicted) choice. Metagame analysis goes further by modelling the possibility that players may not only know each other's choices, but also their reactions to such knowledge, and so on, ad infinitum. A result is that dilemmas of cooperation are 'solved'. Players who know not only what each other will do, but how each other will react to such knowledge, are strictly rational to cooperate and trust each other. 'General' metagames extend the approach to consider partial or incomplete knowledge.

Metagame analysis was developed in the 1960s under a contract with the US Arms Control and Disarmament Agency for use on arms control problems, including the negotiations which led to the first Strategic Arms Limitation Talks (SALT) agreement. It has been used for other international conflict situations and has been applied to resolve conflicts on policy and management issues in business organizations.

Experience with applications of metagame analysis led in the early 1990s to the ideas of **drama theory** and the associated technique of *confrontation analysis*. *P. Bennett*

References

Howard, N. (1990) 'The manager as politician and general: the metagame approach to analysing cooperation and conflict', and 'The CONAN play: a case study illustrating the process of metagame analysis', in J. Rosenhead (ed.) *Rational Analysis for a Problematic World*, Chichester: Wiley, pp. 239–282. A good starting point.

Howard, N. (1971) *Paradoxes of rationality: theory of metagames and political behavior*, Cambridge, MA: MIT Press.

Von Neumann, J. and Morgenstern, O. (1959) *Theory of Games and Economic Behavior*, Princeton, NJ: University of Princeton Press.

Military OR, see applications of MS/OR in military services

Mind maps

When you think about something, a phenomenon, an issue or a **problem situation**, a host of thoughts are evoked in your mind: things, aspects and concepts, including fears and aims, data and facts, possible actions and reactions by yourself or other people or entities involved and their consequences, both planned and unplanned, desirable and undesirable that result from such actions, and the wider context or environment of it all. A mind map is all this or a judiciously chosen subset put down on paper in words, slogans or sentences. The things are arranged in a meaningful way by showing them closely related in clusters, by links that connect things that are related, and by arrows that indicate causal relationships between items shown. No other conventions are used. You may introduce your own, such as solid lines for strong relationships, broken lines for weak ones, or enclosing items that form a cluster inside a circle or 'cloud'. The diagram below is a mind map of what gets evoked in my mind when I think about whether or not I should go to work by bicycle today.

Mind maps can easily be used to capture and consolidate the thoughts and ideas of several people, borrowing rules of *brainstorming*. An effective way to go about it is to write down each item on a 'post-it' as it is evoked. These are initially stuck on a

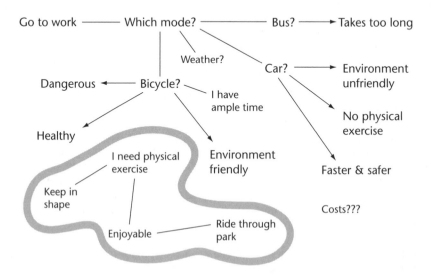

Figure: **Mind map**
(*Source*: H.G. Daellenbach (2001) *Systems Thinking and Decision Making*, Christchurch: REA, p. 62)

whiteboard or big flip chart. As more and more concepts and relationships are elicited from the participants, the 'post-its' are rearranged and appropriately connected by lines or arrows. This is an effective way to help bring about a better appreciation of the problem situation. Differences in viewpoints or disagreements between **stakeholders** can also easily be depicted by showing alternative configurations side by side.

In contrast to a formal written description of a situation, which by necessity has to be sequential, a mind map, similarly to a **rich picture diagram**, shows the situation as a whole in much of its complexity at a glance, so to speak. It can be 'read' in any direction with all aspects remaining 'present' for instant reference. The information contained in it can be processed in parallel, while a verbal description can only be processed serially. It is thus a much more effective and potent vehicle to present a problem situation of complex issues and to use as a basis for communication than a formal write-up.   *H.G. Daellenbach*

Minimax criterion, maximin criterion

A housing complex, which consists of 46 townhouses and two 10–storey apartment blocks of 30 units each, has to allocate its 120 above-ground parking spaces one space to each apartment, leaving 14 unassigned for visitors. The parking spaces are scattered around the various buildings in small groups. The aim is to allocate the spaces to apartments in a fair way, such that no apartment occupant will have to walk excessively far. What **criterion** would be suitable to get a fair allocation? Obviously, one could try to minimize the sum of the distances between apartments and allocated space. But that could easily lead to some occupants having to walk very far. How about trying to make the maximum distance any occupant has to walk as small as possible? This is what the *minimax criterion* does.

If d_{ij} is the distance from apartment i to space j, and $x_{ij} = 1$ if apartment i is allocated to space j and 0 otherwise, the minimax criterion will find an allocation such that

$$\text{minimum [maximum } d_{ij}\, x_{ij}]$$
$$\text{all } i, j$$

and no apartment is allocated to more than one space
and no space is allocated to more than one apartment.

Minimax problems occur in many contexts. When selecting a site for a fire station, one might wish to minimize the maximum distance to any part of the city. When scheduling items of work through parallel machines, the objective might be to minimize the time until all the work is completed, which will be the maximum time of completion for any item. Faced with alternative choices of action that have different costs or losses, depending on which of several events will eventuate, the decision maker may wish to protect her- or himself against the worst possible outcome, i.e. select the alternative that minimizes the maximum cost or loss, or the one that has the smallest largest regret if the wrong choice has been made (the so-called *minimax regret criterion*). In some forms of fitting a model to a set of data, the best model may be the one that minimizes the largest difference between theory and observation. In **game theory** a rational player will usually select the minimax action or strategy.

Related to the minimax criterion is the maximin criterion, where, as the name implies, one wishes to maximize the minimum value of some benefit. This might be useful, for example, in locating a building which is considered unwelcome (such as a waste disposal site) and making it as far away as possible from towns, or in selecting the alternative that maximizes the lowest return, regardless of which event will occur.

One of the difficulties with minimax (and maximin) problems is that the function which is largest for one set of decision variables need not be the largest for all sets. This means that the objective function is not smooth or well behaved (in a mathematical sense) as is usual in **optimization** problems.   *D.K. Smith*

References

Stahl, S. (1997) *Gentle Introduction to Game Theory*, American Mathematical Society.

William, H.P. (1999) *Model Building in Mathematical Programming*, New York: Wiley.

Minimum spanning trees

Many managerial problems in areas such as design of transportation systems, information systems and project scheduling have been successfully solved with the aid of network models and their analysis. A **network** is a type of **graph**, consisting of a series of nodes linked or connected by arcs. If the links have a length (such as the distance between the connected nodes), a cost or a (transport or flow) capacity associated with them, then the graph is a network. For example, a **transportation problem** can be represented as a network, where the nodes represent the sources and destinations of goods and the arcs represent possible shipping routes, each with its unit transport cost.

The minimum spanning tree (MST) problem is a network that involves using the arcs to reach all nodes in such a way that the total length of all the arcs used is minimized, where the length of an arc can represent distance, time, cost, and so on.

A spanning tree for a network of n nodes is a set of $n-1$ arcs that connect (directly or indirectly) every node to every other node. A minimum spanning tree is a spanning tree with minimal total length of the connecting arcs. The following simple **greedy algorithm** identifies a minimum spanning tree:

Step 1: Select any node and connect it to its closest neighbour. Break ties arbitrarily.
Step 2: Identify the unconnected node that is closest to one of the connected nodes. Break ties arbitrarily.
Step 3: Add this new node to the set of connected nodes. If all nodes have been connected, the minimum spanning tree has been found. Otherwise, return to Step 2.

The MST problem is one of a few MS/OR problems in which one will not be misled in one's pursuit for an optimal solution by being greedy at each intermediate step of its solution procedure.   *B. Chen*

References

Cook, W.J. Cunningham, W.H., Pulleyblank, W.R. and Schrijver, A. (1998) *Combinatorial Optimization*, New York: Wiley, ch. 2.

Papadimitriou, C.H. and Steiglitz, K. (1982) *Combinatorial Optimization: Algorithms and Complexity*, Englewood Cliffs, NJ: Prentice-Hall, ch. 12.

Monte-Carlo simulation

Monte-Carlo **simulation** is a technique to generate frequency distributions for the outcome of a complex process that is affected by various random phenomena. Consider the following situation: An insurance company has to provide cover for any late completion penalties that could be incurred on a large construction project. The project consists of many separate components and tasks that have to be undertaken in a fixed order, some concurrently, some in sequence. For example, the ground has to be excavated before the concrete for the basement can be poured. This done, the casings for the basement walls can be built, and so on. For each week that the completion of the project is late, the construction company will be fined $10,000. In order to fix the premium, the insurance company must assess the risk it runs. A frequency distribution of the size of late penalties would provide the needed information. Assume that it is possible to specify for each component and each task a probability distribution of how long it will take. Using **random numbers**, we can then simulate an execution time for each component and each task. This in turn will allow us to determine an overall completion time for the project. Repeating such simulations 1 000 times will produce a frequency distribution of completion times. From this a frequency distribution of late penalties can easily be derived. This is an example of Monte-Carlo simulations.

The American statistician E.L. de Forest may have been the first to use this technique during the Manhattan Project in World War II, and it was named 'Monte-Carlo' by two other mathematicians, Ulam and von Neumann, who were also associated with that project. However, the idea can be traced back to the French naturalist Georges Louis Leclerc, Comte de Buffon (1707–1788). He attempted to calculate the number π by throwing pins randomly onto a tiled floor, and then counting the proportion of pins that fell inside a circle of all those that landed in the smallest square circumscribing the circle. The idea remained forgotten until the invention of electronic computers, which made it possible to execute the large number of trials that the method requires in order to get sufficiently reliable results.

Monte-Carlo simulation has found numerous uses in such diverse fields as chemistry, physics, economics, and business and industry. Obviously, generating frequency distributions for complex industrial, financial or marketing processes still remains its major application. **Risk analysis** is an extension of the technique, and is used extensively to assess the risk profile of investment projects that can be affected by a variety of uncertainties over the lifetime of the project. More recent applications are in econometrics for finding the distribution of error terms in complex econometric models. In finance, banking and foreign exchange, Monte-Carlo simulation is being used to assess risks as well as set prices for stock and currency instruments, such as options and derivatives.   *N.C. Georgantzas*

References
Fishman, G. (1996) *Monte Carlo: Concepts, algorithms and applications*, New York: Springer Verlag.
Gentle, J.E. (1985) 'Monte Carlo methods', in S. Kotz, N.L. Johnson and C.B. Read CB, (eds) *Encyclopedia of Statistical Sciences*, 5:280–287, New York: Wiley.

Multiattribute utility analysis

Multiattribute utility analysis or MAUA, a tool pioneered by Keeney and Raiffa, is an extension of single-attribute **utility** analysis (as used in statistical **decision analysis**) to several objectives. Prominent applications are blood bank management (objectives: minimize blood shortages and blood outdating), the siting of airports (objectives: minimize travel time, flight path lengths, noise, disruption to residents, operating costs and construction costs), the siting of nuclear power stations and nuclear waste storage facilities, land use for joint recreation and production purposes, etc.

Consider the blood bank example. The outcome of any policy is measured by two **attributes**, i.e. the amount of blood shortages and the amount of blood outdated. These outcomes are, however, random. Let $x_j^{(n)}$ and $y_j^{(n)}$ denote the value of the two attributes for outcome j if policy n is followed. That outcome occurs with probability $p_j^{(n)}(x_j^{(n)}, y_j^{(n)})$. With each pair $(x_j^{(n)}, y_j^{(n)})$, the decision maker associates a utility $U_j^{(n)}(x_j^{(n)}, y_j^{(n)})$ that measures the outcome's personal (subjective) worth. In parallel with single-attribute decision theory, the best policy is the one that maximizes the expected utility, i.e.

$$\text{maximize } \sum_j^J p_j^{(n)}(x_j^{(n)}, y_j^{(n)}) U_j^{(n)}(x_j^{(n)}, y_j^{(n)}).$$
$$\text{all policies } n$$

Without strong simplifying assumptions, the assessment of U is extremely difficult and beyond the ability of most decision makers. The strongest simplification is to assume that the utility of each attribute is independent of what value the other attribute assumes. As a consequence, the multiattribute utility function U is equal to the weighted sum of two single attribute utility functions, i.e.

$$U_j^{(n)}(x_j^{(n)}, y_j^{(n)}) = w_1 u_1^{(n)}(x_j^{(n)}) + w_2 u_2^{(n)}(y_j^{(n)})$$

This is the approach used by the **multiattribute value function** method. A less restrictive assumption is mutual utility independence. Attribute 1 is utility independent of attribute 2 if the preference ranking of the possible outcomes for attribute 1 remains unchanged, in relative terms, for all possible values of attribute 2. (Note the proviso 'in relative terms', in contrast to the assumption of absolute independence earlier.) If this holds for both attributes, the utility function U for a small number of objectives (2 to 4) can be assessed by requiring the decision maker to answer a few indifference statements, in the form of 50–50 **reference lotteries**.

The major criticisms of MAUA are the difficulties in assessing multiattribute utility functions that are a reliable measure of the decision maker's preferences, and the highly subjective nature of the results produced. Furthermore, for the type of problem where such a sophisticated method seems appropriate and has been applied, e.g. public works projects, the decision-making body consists of the 'public', often represented by several elected or appointed public officials. This raises the question of how 'public' preferences can be captured. Is it valid to aggregate the preferences of the public officials, and how is such an aggregation to be made?   *H.G. Daellenbach*

References

Goodwin, P. and Wright, G. (1999) *Decision Analysis for Management Judgement*, Chichester: Wiley.

Keeney, R.L. and Raiffa, H. (1976) *Decisions with Multiple Objectives*, New York: Wiley.

Multiattribute value functions and V.I.S.A.

Several **multicriteria decision making** methods or aids are founded on the use of an *aggregative multiattribute value function* (MAVF). The most commonly used form is a simple weighted sum:

$$V_i = \sum_{\text{all } j} w_j v_{ij}$$

where V_i is the overall value derived for alternative i from the aggregation of partial preference functions v_{ij}. The latter measure the performance of alternative i on objective j. w_j is a weight or scaling factor. It defines acceptable *trade-offs* between **objectives**, essentially specifying how much a decision maker would be willing to give up on one objective in order to achieve an improvement on another. Rigorous procedures, involving the decision makers, must be followed to define the scales for the partial preference functions and to elicit acceptable trade-offs which relate to these scales. Unfortunately, some textbooks propose simplistic procedures which do not adopt the necessary rigour and can easily be shown to be flawed.

The use of such a value function calls first for the determination of the alternatives to be evaluated and the objectives, or criteria, against which they are to be evaluated. This initial phase of problem structuring is very important – an art informed both by the theoretical assumptions on which the use of a MAVF lies and guidelines derived

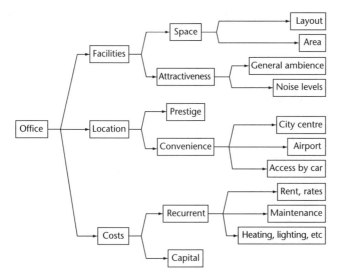

Figure: **Value tree**

from the extensive experience of practitioners. The objectives are usually structured as a **hierarchical** (value) **tree**, such as depicted in the figure above.

The set of objectives must be

- mutually preference independent, e.g. in terms of the figure shown, preference statements about one objective (say layout) can be made without concern for the level of any other objective (say prestige);
- complete, capturing all important factors; and
- usable, i.e. meaningful and measurable, requiring a manageable number of judgements or comparisons.

The alternatives must be well defined. They may be 'given', as when evaluating bids to provide for a service, build a facility or develop a system, or they may emerge during the problem structuring process, such as a strategy development session.

The whole process, including problem structuring leading to the definition of alternatives and a value tree, elicitation of preferences, and the interactive construction and use of a model, may take place in one or more **decision conferencing** workshops, led by a facilitator skilled in both the use of the analytic technique and in managing group processes.

In this task, the facilitator may make use of interactive graphics software, such as V.I.S.A. (for Visual Interactive Sensitivity Analysis). V.I.S.A. combines simple visual displays, facilitating exploration of the implications of **uncertainties** about stated preferences, trade-offs, and different priorities.

Although MAVF provides an overall score for each alternative, this should not be taken as 'the answer'. The real power of MAVF is in the provision of a structure for thinking through an issue in all its detail, as a vehicle for testing intuition and exploring different perspectives, and in the learning derived from this process. Its simplicity and ease of use greatly facilitate this.   *V. Belton*

References
Belton, V. (1985) 'The use of a simple multiple criteria model to assist selection from a short-list', *JORS*, 36:265–274.

Goodwin, P. and Wright, G. (1999) *Decision Analysis for Management Judgement*, Chichester: Wiley. Easy treatment with cases.

Keeney, R.L. and Raiffa, H. (1976) *Decisions with Multiple Objectives*, New York: Wiley. Advanced treatment of multiattribute utility theory.

Web site

www.decision-conferencing.com/index.html

V.I.S.A. web-page: www.multi-criteria.com/visa

Multicriteria decision making

For many, if not most, decision problems the decision maker may wish to pursue several, possibly conflicting or incommensurable **objectives**. Rarely will the same decision choice simultaneously 'optimize' the achievement level for all of them. Therefore, the best or most preferred decision must be a compromise, balancing the achievement levels over the various objectives. Multicriteria decision making or MCDM is a collection of techniques and methods to deal formally with multiple objectives.

The traditional OR/MS approach to multiple objectives is to identify the most important objective, which is used for a single objective optimization, and substitute minimum performance targets or constraints for other objectives. The irony of this approach is that the most important objective becomes subordinated to the constraints safeguarding the secondary objectives. Furthermore, the target levels for the secondary objectives, often chosen arbitrarily, may in fact lead to inferior or **dominated solutions**, i.e. there may exist solutions with better performance on some objectives without lowering the performance on the other objectives. MCDM methods are attempts to overcome the shortcomings of this *ad hoc* approach. They also recognize that one of the most difficult aspects of MCDM is the externalization of the decision maker's preference structure. Often, that preference structure is determined interactively as an integral part of the method (see **interactive multicriteria decision making**).

Goal programming, a version of **linear programming**, developed in the late 1950s, was one of the first methods attempting to deal with multiple objectives. Since the 1970s, a large number of MCDM methods and algorithms have been proposed. Aggregate value function methods assume that achievement levels between objectives can be traded off and that the overall score of a decision choice is a weighted function of the outcome scores of the individual objectives. Some find the most preferred solution based on importance weights provided by the decision maker. Others determine the weights by presenting the decision maker with a small number of trade-off questions calling for answers in numerical form or of the yes–no–inconclusive type. In either case, the search is restricted to **efficient solutions**, i.e. solutions that are not dominated. *Outranking methods* assume that the decision maker is unwilling or unable to define trade-offs between objectives. Any ranking of alternative decision choices has to be done on pairwise comparisons, either on a holistic basis or based on the two sets of outcome scores over all objectives. The comparison will indicate that one alternative is preferred to the other, that the decision maker is indifferent between the two, or that the comparison is inconclusive. The end result of the process is a complete or partial ranking of the alternatives.

H.G. Daellenbach

References

Belton, V. (1990) 'Multiple criteria decision analysis' in L.C. Hendry and R.W. Egles (eds) *Operational Research Tutorial Papers 1990*, Birmingham: OR Soc., pp. 53–102.

Gal, T., Stewart, T.J. and Hanne, T. (eds) (1999) *Multicriteria Decision Making: Advances in MDM Models, Algorithms, Theory and Applications*, Dordrecht: Kluwer.

Multimethodology

In the past 20 years many new 'soft' management science methods have been developed, such as the **soft systems methodology** (SSM), **strategic option development and analysis** (SODA), **strategic choice approach** (SCA), etc. These are also called **problem structuring methods** or PSMs (Rosenhead and Mingers, 2001). This led to a new problem – deciding which method(s) to use in a particular situation. One approach is to select a particular method – be it hard or soft – on the assumption that a problem can be matched to that particular method. However, a better approach is multimethodology (Mingers and Brocklesby, 1997; Mingers and Gill, 1997), i.e. deliberately seek to combine together a range of methods, perhaps both soft and hard, in order to match the richness of the problem situation and to deal effectively with the different stages or aspects of a project.

Any real-world **problem situation** will be a complex mix of the material, the social and the personal. Material or physical characteristics can be modelled using traditional MS/OR techniques, but social conventions, politics and power, and personal beliefs and values need quite different, qualitative approaches. Equally, a real project goes through several stages – understanding and appreciating the situation, analysing information, assessing different options, and acting to bring about change. The various methodologies can be more or less useful at these different phases. These are two strong arguments for combining together different PSMs.

In use, multimethodology is a creative process of design, based on competence in a range of methods. Each project or intervention is seen as a unique situation (although of course having features in common with others) for which a particular combination of methods, or parts of methods, needs to be constructed. This is an ongoing process throughout the project, as events occur and the situation evolves. The general process can be described as follows:

Reflection:

- Review the current situation.
- Determine which areas of the problem situation currently need addressing.

Design:

- Understand what methods or techniques could possibly be useful.
- Choose the most appropriate to use in relation to the project context as a whole.

More guidance about multimethodology design, both for practical interventions and for research, can be found in Mingers (2000), while Ormerod (1995, 1998) has documented a number of sophisticated multimethodology projects.   *J. Mingers*

References

Mingers, J. (2000) 'Variety is the spice of life: combining soft and hard OR/MS methods', *International Transactions in Operational Research*, 7:673–691.

Mingers, J. and Brocklesby, J. (1997) 'Multimethodology: towards a framework for mixing methodologies', *Omega*, 25(5):489–509.

Mingers, J. and Gill, A. (eds) (1997) *Multimethodology: Theory and Practice of Combining Management Science Methodologies*, Chichester: Wiley.

Ormerod, R. (1995) 'Putting soft OR methods to work: information systems strategy development at Sainsbury's', *Journal of the Operational Research Society*, 46(3):277–293.

Ormerod, R. (1998) 'Putting soft OR methods to work: the case of the business improvement project at Powergen', *European Journal of Operational Research*, 118:1–29.

Rosenhead, J. and Mingers, L. (eds) (2001) *Rational Analysis for a Problematic World Revisited*, Chichester: Wiley.

Multiple objective mathematical programming (MOMP)

Traditional **mathematical programming** has a single objective function and results in one or several **optimal solutions** all having the same objective function value. When there are multiple **objectives**, it is only by coincidence that the same solution optimizes each and every objective. The best solution is a compromise which depends on the relative importance of the various objectives. For simplicity, assume that there are two objectives and three decision variables $\{x_1, x_2, x_3\}$. Then the corresponding MOMP problem is

$$\max_{x_1, x_2, x_3} f_1(x_1, x_2, x_3)$$

$$\max_{x_1, x_2, x_3} f_2(x_1, x_2, x_3)$$

$$\text{subject to } \{x_1, x_2, x_3\} \text{ in } X$$

where X is the set of **feasible solutions** that satisfy all constraints on the decision variables.

MOMP solution methods restrict their search to the set of so-called *Pareto optimal* or **efficient solutions**. A solution is efficient if there exists no other solution which achieves objective values that are no worse on all objectives and better for at least one.

Two approaches for finding the best compromise solution are used, both based on assigning to each objective an importance weight w_k. The first maximizes the sum of the weighted objectives. The other, often called a *Tchebycheff formulation*, seeks to minimize the maximum weighted deviation from the 'ideal solution'. In terms of our two-objective example above, the ideal solution is defined as $\{U_1, U_2\}$, where $U_1 = \max f_1(x_1, x_2, x_3)$, and $U_2 = \max f_2(x_1, x_2, x_3)$. Then the weighted-sum formulation is

$$\max w_1 f_1(x_1, x_2, x_3) + w_2 f_2(x_1, x_2, x_3), \text{ subject to } \{x_1, x_2, x_3\} \text{ in } X$$

and the Tchebycheff formulation is

$$\min y, \text{ subject to } y \geq w_1 (U_1 - f_1(x_1, x_2, x_3))$$

$$\text{and } y \geq w_2 (U_2 - f_2(x_1, x_2, x_3))$$

$$\text{and } \{x_1, x_2, x_3\} \text{ in } X$$

Many different solution methods exist for MOMP problems. One way to classify them is according to when the preferences of the decision maker are elicited. In the first category, *a priori articulation* of preferences, all preference information is obtained from the decision maker up front before the problem is solved. The problem here is that decision makers may not always have a good understanding of their preferences at the start or they adjust their preferences as they learn more about the **feasible solution space**. The second category, which comprises most solution methods, is *progressive articulation* of preferences. These solution methods interact with the decision maker, soliciting some preference information at each iteration with the goal that the sequence of solutions will converge to a final, preferred solution (see **interactive multicriteria decision methods**). The third category is *a posteriori articulation* of preferences. In these solution methods, all efficient solutions are generated and presented to the decision maker to choose the best one. A problem here is that there may be thousands of efficient solutions, most of which will be of little or no interest to the decision maker.   *J.T. Buchanan*

References

Miettinen, K. (1999) *Nonlinear Multiobjective Optimization*, Boston: Kluwer.

Steuer, R.E. (1986) *Multiple Criteria Optimization: Theory, Computation and Application*, New York: Wiley.

White, D.J. (1990) 'A bibliography on the applications of mathematical programming multiple-objective methods', *JORS*, 41(8):669–691.

N

Narrow system of interest, see hierarchy of systems, systems concepts

Nash equilibrium

John Nash, in a series of seminal papers on game theory in the early 1950s, extended the notion of the **minimax solution** of **zero-sum games** and the dominant strategy equilibrium solution to games like *prisoners' dilemma* by introducing a broader concept of equilibrium, subsequently named after him as the *Nash equilibrium*.

Nash's concept of equilibrium can be best illustrated through an example. Consider two companies, A and B, each having available three possible marketing policies for their products. The market shares are shown below for each combination of strategies:

Market share percentage [company A; company B]		Strategies for company B		
		B1	B2	B3
Strategies for company A	A1	25; 25	50; 30	50; 20
	A2	30; 50	15; 15	30; 20
	A3	20; 50	20; 30	10; 10

This is not a zero-sum game, so the minimax solution does not apply. There is also no single dominant strategy equilibrium. This is the situation Nash considers. If there is a set of strategies with the property that a unilateral departure from it would make the player worse off, this set of strategies and the corresponding *payoffs* constitute the Nash equilibrium.

The Nash equilibrium is a fairly simple idea: Given a strategy chosen by a participant, the other participant chooses his/her best strategy. This is exactly what happens in this game. If company A chooses marketing policy A1, company B should respond with policy B2, as this gives it a 30 per cent share, leaving A 50 per cent of the market. If company A chooses A2, company B should select B1 as this gives it a 50 per cent share, leaving A 30 per cent of the market. In both cases the two companies together capture 80 per cent of the market. The worst both companies can do is to adopt simultaneously policy 2, which allocates 15 per cent of the market to each. Obviously this is something either wants to avoid. Also, neither company will consider its **dominated** policy 3, as this only offers a 20 per cent share at best. Our example turns out to have two Nash equilibria: (A1=50%, B2=30%) and (A2=30%, B1=50%). If either one eventuates, both companies would be worse off by changing policy.

Naturally, if this game is played many times, A and B may wish to cooperate and equally share the market, by synchronized switching of policies. Also, the financially stronger company can force the weaker one to go to its preferred policy, by punishing the weaker until it switches.　　　　　　　　　　　　　　*A.D. Tsoularis*

References
Colman, A.M. (1995) *Game Theory and its Applications in the Social and Biological Sciences*, Oxford: Butterworth-Heinemann. Excellent introduction for the novice with real-life examples.
Nash, J.F. (1996) *Essays on Game Theory*, Cheltenham: Edward Elgar Publishing Company. A collection of Nash's papers.

Natural resources, see applications of MS/OR to environmental issues

Negotiation theory

Negotiation is a process whereby parties with conflicting aims establish the terms on which they will cooperate. For example, a vendor may wish to sell a car for $5000, and a purchaser to buy it for $4500. Negotiation is the process of how they influence each other's behaviour to arrive at an agreed purchase price.

The negotiation process consists of a number of stages. The first stage is preparation, which is usually regarded as being crucial to achieving a successful outcome. In this stage each party determines their own needs, assesses the other party's needs, identifies items which are potentially tradable, and plans their concessions. They should also consider process issues such as their negotiating style, timing, location and agenda. Once negotiations begin, debate between the parties occurs. In this stage the parties outline their positions and engage in behaviours as diverse as argument and joint problem solving. At some point, if agreement is to be reached, the parties need to start making proposals to make concessions on tradable items, usually in return for concessions by the other party. The extent of each party's concessions is dependent on their skill as a negotiator and the power they bring to the negotiating table. The final stage in the process is the post-negotiation phase where the parties record their agreement and, if necessary, restore the relationship between them.

Negotiation has become a discipline in its own right. A number of centres (e.g. at Harvard University) dedicated to research and education on negotiations have been set up, and specialist journals such as the *Negotiations Journal* and the *Journal of Conflict Resolution* started. Research into negotiations is being conducted in a number of related disciplines such as:

- social psychology, where the focus has been on how negotiators frame their objectives and expected outcomes (Rubin and Brown, 1975)
- economics, where researchers have been applying the tools of negotiation analysis to cooperative coalitions and the *prisoners' dilemma* to understand pricing and market behaviour (Porter, 1985)
- **game theory**, which suggests that negotiations can be distinguished by the payoff structure, number of parties, size of parties, and concession rules (Raiffa, 1982)
- management, where the importance of negotiating is being increasingly recognized as part of the complex interactions managers are involved in (Lax and Sebenius, 1986)

- **soft operational research methods**, such as **strategic options development and analysis**, and the game theory offshoots **metagames, hypergames,** and **drama theory**.

I. Brooks

References
Fisher, R. and Ury, W. (1981) *Getting to Yes*, Boston MA: Houghton-Mifflin.
Lax, D. and Sebenius, J. (1986) *The Manager as Negotiator*, New York: Academic Press.
Porter, M. (1985) *Competitive Advantage*, New York: Free Press.
Raiffa, H. (1982) *The Art and Science of Negotiation*, Cambridge MA: Belknap Press.
Rubin, J. and Brown, B. (1975) *The Social Psychology of Bargaining and Negotiation*, New York: Academic Press.

Net present values, see discounted cash flows

Networks

The term network is generally used to denote a set of nodes linked together by arcs. Nodes and arcs can assume a wide variety of meanings. Nodes could be concepts of a theory or operations of a job and arcs logical or precedence relations between them. Nodes could be geographical locations and arcs road or air links between them, e.g. for scheduling pick-ups and deliveries of goods or finding an itinerary to visit all locations, the so-called **travelling salesperson problem**. Arcs may be links that allow material flow, such as through pipelines, cables, over rail tracks, or even roads, forming *network flow problems* (see **maximum flow problem**). Nodes may represent college courses and arcs forbidden pairings of courses (because they share some students), the so-called **timetabling** or *graph colouring* problem.

The network diagram below consisting of four nodes and five arcs. In this example, the links are directed, indicating that travel or material flow can go only in the direction of the arrows. The number attached to each arc could represent a penalty for traversing that arc, or the maximum carrying capacity of that arc. Such diagrams are often very useful for capturing and displaying the structure of a system or a model.

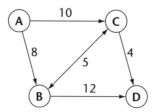

Figure: **The graph of a network**

Many important network problems, such as the travelling salesperson or the graph colouring problem, are extremely difficult to solve if the number of nodes and arcs is large. Rather than try to find optimal solutions, **heuristic methods** are used to find good solutions.

Network flows have directed arcs between nodes, indicating the permissible direction of flow. There are some nodes which have all links directed away from them, called *source nodes*, while other nodes have all links directed towards them, called *sink nodes*. Source nodes may have limited supply capacities and sink nodes minimum requirements for materials. The figure shown can be viewed as a network flow, where node A is the source node and node D the sink node, with maximum carrying capacity along all arcs.

Network flow problems can be formulated as **linear programs**. A variable is associated with each arc, measuring the flow on the arc. Each node gives rise to a conservation of flow constraint. The total flow into a node must equal the total flow out. Sources have supply constraints, sinks requirement constraints. Each unit of flow over an arc incurs a cost. The objective is to find a flow that minimizes total costs.

Network flow problems have the special property that if all source supply, sink requirement and arc capacities are integer, the optimal solution will also be integer. Efficient **algorithms** solve network flow problems. They are used to model many MS/OR applications, such as in communication and traffic networks, for power generation and transmission, product distribution, etc. *S. Dye*

Reference

Winston, W.L. (1994) *Operations Research Applications and Algorithms*, Belmont, CA: Duxbury, ch. 8.

Neural networks

A neural network (NN) is a network of many interconnected simple processing units, called neurons, which usually carry numeric data, as opposed to symbolic data, encoded by any of various means. The units operate only on the inputs they receive via the connections. Each connection is associated with a weighting factor which the network adjusts to adapt to changes in the environment.

Some NNs are models of biological neural networks and some are not, but historically, much of the inspiration for the field of NNs came from the desire to produce artificial systems capable of sophisticated computations and human-type inferences similar to those that the human brain routinely performs. It is hoped that such an endeavour will eventually shed some light on how the human brain actually works.

Most NNs have some sort of training rule or learning algorithm whereby the weights of connections are adjusted on the basis of new available data. In other words, NNs learn from examples and exhibit some capability for generalization beyond the training data.

The original ideas of NNs began in the 1940s, but kindled interest in the MS/OR community only in the 1980s. However, the number of useful applications have been few. Examples are time series recognition and prediction and pattern and speech recognition. *A.D. Tsoularis*

References

Aleksander, I. and Morton, H. (1990) *An Introduction to Neural Computing*, London: Chapman & Hall. Aimed at beginners; first author is a well-known British researcher in the field.

Dodd, N. (1992) 'Introduction to neural networks', in M.E. Mortimer, (ed.) *Operational Research Tutorial Papers 1992*, Birmingham: Operational Research Society. More advanced.

Newsboy problem

The newsboy (or Christmas tree) problem determines the ordering or purchasing quantity for a product when there is only a single occasion to do so, and the demand for the product is a **random variable** with a known probability distribution. It gets its name from the daily problem faced by a newspaper vendor of how many copies to get to meet each day's demand. If the vendor acquires too many, a large number may remain unsold which have to be discarded. If he or she acquires too few, potential sales are lost. Analogous situations are faced for many products, such as magazines, fast food items with a one-day shelf life, fashion goods where the manufacturers only make one run, fad products, Christmas trees, flowers, spare parts ordered at the same time as specialized equipment, etc. Stock left over, called overages, at the end of the period may have little or no value, with the initial purchase cost not recoverable, and may even incur a disposal cost. These costs are called *overage costs*. Shortages of goods result in lost profits – an **opportunity cost**. Both are assumed to be proportional to the overage or shortage.

Marginal analysis shows that the optimal order quantity S^* occurs where the marginal expected overage cost for the S^*th unit is just equal to or larger than the marginal expected underage cost. This yields the following expressions: Find S^* so that

$$P(x<S^*) < [c_u/(c_u+c_o)] \le P(x \le S^*) \text{ for a discrete random variable,}$$

and

$$F(S^*) = [c_u/(c_u+c_o)] \text{ for a continuous random variable,}$$

where $P(x \le S)$ and $F(S)$ are the probability that the demand x is less than or equal to S.

Below is a simple example for a discrete distribution. The overage cost is $c_o = \$1$, the underage cost $c_u = \$9$. Hence the ratio $c_u/(c_u+c_o) = 9/(9+1) = 0.9$. The probabilities for the possible levels of demand are shown in the row $p(x)$.

S	0	1	2	3	4
$p(x)$	0.4	0.3	0.15	0.1	0.05
$P(x \le S)$	0.4	0.7	0.85	0.95	1

The value of 0.9 falls between an S of 2 and 3. Hence the optimal order quantity is $S^* = 3$.

Adaptations of this model are used to determine *safety stock* protection for multi-period **inventory models**.

H.G. Daellenbach

Reference

Silver, E.A., Pyke, D. and Peterson, R. (1998) *Inventory Management and Production Planning and Scheduling*, New York: Wiley, ch. 10.

Nonbasic variables, see basic feasible solutions

Non-convex, non-concave, see convex

Nonlinear programming, see mathematical programming

NP-hard, computational complexity

In the world of **optimization**, some problems are intrinsically 'easy' to solve, while others are intrinsically 'hard', where easy and hard refer to the computational requirements to find the **optimal solution**. Before attempting a solution, this would be useful to know. We may go for a good rather than the optimal solution for a hard problem.

The field of computational complexity endeavours to formally categorize the computational requirements of both **algorithms** (i.e. solution procedures) and problems. This categorization, however, uses the so-called *recognition* version of a problem, for which only a 'yes' or 'no' answer is possible. For example, for the **travelling salesperson problem** (TSP) the recognition version is: 'Is there a tour visiting all n cities whose length is at most K?'

'Easy' problems are those for which algorithms have been devised to solve every instance 'efficiently', i.e. the number of steps the algorithm takes can be no more than a *polynomial* function of the size of a problem instance. For the TSP, size is the number of cities, for a **linear program** (LP) it is the number of constraints. Such problems are said to be solvable in polynomial time, and belong to *class P*. They include LP, but not TSP.

'Hard' problems belong to *class NP* (for 'non-deterministic polynomial'). Class NP consists of all recognition problems for which it is possible to verify, in polynomial time, that there is a 'yes' answer to the recognition problem. A solution method for finding such an answer is not required; only the efficient verification of the answer's correctness, such as the solution to a TSP instance forming a tour through the *n* cities. Therefore, class NP contains problems (such as the TSP) where it is easy to check a solution that may have been very hard to find, whereas class P contains problems whose solutions are easy to find.

A recognition problem, Q, is said to be *NP-complete* if it belongs to class NP, and every problem in class NP may be efficiently transformed to it, i.e. any instance of any other problem in class NP can be converted, in polynomial time, to an instance of Q such that both instances have the same yes-or-no answer. NP-complete problems all seem to be intrinsically very difficult, and no efficient algorithms have been found for any of them. But neither has it been proved that efficient algorithms for these problems do not exist. Furthermore, if any efficient algorithm could be found that solved one NP-complete problem, then there would be efficient algorithms for all of them.

The term *NP-hard* refers to any problem that is at least as hard as any problem in class NP. Any optimization problem whose recognition version is NP-complete will be NP-hard, since solving the optimization version requires solving a sequence of recognition versions. Note that NP-complete problems are both NP-hard and in class NP.   *J.W. Giffin*

Reference
Wilf, H.S. (1986) *Algorithms and Complexity*, Englewood Cliffs, NJ: Prentice-Hall.

Objectives, objective function, see criteria

Objectivity, subjectivity

The dictionary definition of objectivity is 'the expression or interpretation of facts or conditions as perceived without distortion by personal feelings or prejudices' – in other words 'independent of the observer', whereas subjectivity is 'a personal view of something as perceived by a person and not independent of the mind'. Subjectivity allows that different people view or interpret the same thing in different ways. Subjectivity is affected by the person's state of mind, his or her values, background, education and experience (see **world view**).

Objectivity is an unreachable ideal. We can never step outside our own mind. What we see, how we interpret things only reflects our perceptions. Knowledge of 'objective reality' always escapes us. Naturally, this brings up an age-old debate of whether 'reality' exists. It is largely of theoretical interest, since we can never know reality. We can only work from our perceptions.

That does not preclude that on many things, most people agree. We may all see the same thing for relatively simple situations, e.g. the colour red at the traffic light, except if you are severely colour blind. But even the meaning of a £20 note is subjective – a small fortune for a beggar, not worth the effort for a billionaire. Whose perception is objective?

The only operational meaning for objectivity is what R.L. Ackoff, a systems thinker, calls 'the social product of an open interaction of a wide variety of individual subjectivities' – a sort of *consensual subjectivity*. So wide consensus of interpretations on many things is not excluded. Modern scientific knowledge is based on that. But, even that may be only a temporary consensus. The most famous example comes from physics. Newton's laws of dynamics were accepted as the 'truth' for over three centuries and still are the most widely used principles in mechanics. Yet Einstein showed that, when considering motions with a velocity comparable to the speed of light, Newton's laws break down. Maybe Einstein should have the last word on this controversy: 'The only justification for our concepts is that they serve to represent the complex of our experiences; beyond this, they have no legitimacy.'

The issue of subjectivity is important to **systems thinking**. The interpretation of what is a problem or even its context, i.e. the **problem situation**, may be different for the various **stakeholders**, and the **system** envisaged to analyse it is a **conceptualization** by the analyst; it is personal and hence subjective, nor is there any claim that this system exists or should exist in reality. A system description is simply a convenient way to organize our thoughts.

H.G. Daellenbach

References
Ackoff, R.L. (1977) 'Optimization + objectivity = opt out', *EJOR*, 1:1–7.
Ackoff, R.L. (1979) 'The future of operational research is past', *JORS*, 30(2):93–104, but especially pp. 102–3.

Object-oriented simulation

An object-oriented simulation (OOS) is a computer **simulation** which uses *software objects* for modelling the system or operation of interest. Software objects are *data structures* which encapsulate a state (the value of its data or attributes) and a behaviour (its operations or methods). Using software objects to model the **system** of interest is intuitively appealing as the world can be regarded as composed of interacting physical objects. For example, a real-world vehicle might be modelled by a software vehicle object with attributes, including weight, position and speed, and actions, including 'start moving', 'stop moving' and 'change direction'. The start moving action would be executed by a message sent by another object in the simulation (e.g. a traffic light object) or alternatively by a human user who controls the simulation via a graphical user interface. The position attribute of the vehicle would then be incremented as simulated time advances. Many software objects representing real-world objects (termed *domain objects*) may interact with each other in various ways as simulated time advances.

A distinctive feature of an OOS is that software objects may be used not only to model the world, but to perform all the other actions which occur in a simulation run. For example, all simulations use simulated time, and in an OOS this could be an attribute of a special object termed a 'simulation controller' and advanced by this object, either in fixed or variable increments. At each time increment messages would be sent by the simulation controller object to the domain objects updating their attributes. Other special objects (termed *application objects*) may be used in an OOS to generate **random numbers**, collect statistics on the results of the simulation, and produce graphical output to a screen.

There are many advantages of object-oriented software engineering (of which OOS is a subset). General object-oriented analysis and design methodologies, such as the *unified modelling language* (UML), are applicable to OOS, and there are also simulation-specific analysis and design methodologies such as *discrete event system specification* (DEVS). Implementation of an OOS can be via general-purpose object-oriented programming languages such as C++, or special OOS languages such as MODSIM III. Although OOS can be used for almost every type of simulation application, it is particularly well suited for larger, more complex and **advanced distributed simulations**, including Web-based simulations running over the Internet.

R. Watson

References
Hill, D.R.C. (1996) *Object-Oriented Analysis and Simulation*, Harlow, UK: Addison-Wesley.
Levasseur, J. (ed) (1998) Special Issue of *Simulation*, 70:6, on 'Object-Oriented Simulation', San Diego: Society for Computer Simulation International.
Zobrist, G.W. and Leonard, J.V. (1998) *Object-Oriented Simulation: Reusability, Adaptability, Maintainability*, New York: IEEE Press.

Oil and petroleum industries, see applications of MS/OR to oil industry

Open systems, see **closed systems, general systems theory**

Operating characteristics curve, see **acceptance sampling**

Operational gaming, see **gaming**

Operational research/management science, 'hard' MS/OR

Operational research, operations research (US terminology), or OR has its origins in interdisciplinary operational planning teams of scientists, attached to WWII allied military commands. As OR was adopted by governmental agencies, business and industry in the 1950s and the first university courses were offered in the USA and the UK, OR became a discipline in its own right, losing its interdisciplinary flavour. It also became increasingly mathematical, spawning numerous powerful mathematical tools, techniques and solution algorithms. A 1960s definition would say: 'OR is the application of **scientific methods** to build mathematical models and derive optimal solutions for decision problems in government, business, industry, and agriculture.' The emphasis was firmly on **mathematical modelling** and **optimization**.

This 'hard' view of MS/OR presupposes that the problem is largely devoid of human aspects, but instead is clearly defined, with strong systems boundaries, known objectives, known decision choices, known constraints, and well-structured systems relationships that can be translated into tractable mathematical models. All data are readily available, and decision makers can enforce adoption of solutions. The management of natural resources in forestry, oil, gas, and electric power industries, various scheduling problems in the airline, railroad, ocean shipping, road transport, construction, and manufacturing and service industries, and the military have seen numerous successful applications of hard MS/OR.

In practice, many important strategic problems violate some or all of these conditions. Attempts in the 1950s to broaden OR led to the concept of 'Management Sciences' (MS). However, by the mid-1960s, OR and MS were again seen as one discipline. This led to the search for new approaches to decision making, better able to cope with the practical difficulties of real-life, messy problem situations involving human aspects. They became known as **soft OR** methodologies or **problem structuring methods**, and now all go together with OR under the umbrella of the Management Sciences.

A year 2000, more realistic view of MS/OR recognizes that the methodology is not a version of the scientific method of the natural sciences. MS/OR projects are largely unique and reproducibility – one of the cornerstones of science – is absent. MS/OR projects are undertaken to improve decision making, rather than further scientific knowledge. But most important, MS/OR involves considerable **subjectivity** on the part of the analyst in terms of **boundary judgements** about system. Hence, any solution is optimal only with respect to that choice. In fact, most operations researchers would agree that the aim of MS/OR is not to produce optimal solutions, but to provide valuable insights that lead to more informed and therefore better decision making. *H.G. Daellenbach*

Reference

Daellenbach, H.G. (2001) *Systems Thinking and Decision Making,* Christchurch: REA, ch. 12.

Operations management

An operations manager has the job of running the part of a company (or department) where goods and/or services are produced. Operations managers use many resources. As an example, the cook in your family is practising operations management by using food, energy, recipes, time, equipment (stoves, refrigerators, pots and pans) and people to produce the meals that you eat. Commercial examples of operations management include making cars in a plant, performing surgery in a hospital, providing loans and monthly statements at a bank, making and serving meals at a restaurant, growing food on a farm, and providing air travel arrangements.

Generally, the largest number of jobs and highest resource use are found in the operations area of the company. Hence, operations managers pay attention to how **efficiently** they produce the goods and services needed. They are also responsible for the quality, timing, and (sometimes) the place where goods and services are needed. The job is so important in many firms that they have a chief operating officer (COO) at the most senior levels in the company.

A key operations objective is to match supply with demand. For example, skis and air planes to ski resorts are needed in winter, and bicycles with bike trails in summer. The skis and bicycles can be stored until the appropriate time, while the air travel and trail use cannot be. This distinction between goods and services affects the resources available for managing the operations. In both cases, however, the operations manager is responsible for providing either inventory or capacity to meet the forecast demand.

To carry out their responsibilities, operations managers are concerned with a vast range of decision making, from short-term tactical decisions to long-term strategic planning. Short-term tactical activities cover **supply chain management**, i.e. managing the flow of materials from the supplier through the plant(s) to the customers. At the strategic level, they are involved in long-term demand **forecasting**, capacity planning (buildings, equipment, people), and **facility location**, **facility layout**, process and technology selection for the next several years. They also help determine the qualitative focus of operations needed to achieve the company's objectives. For example, should the firm produce high quality product, respond quickly to changes, or produce at low cost? Very general tools like spreadsheet models are used for this activity.

With these decisions made, operations managers plan the output and inventory levels (or capacity and service levels) for the next several months. To carry out these plans, they make specific schedules for people, equipment and so on. Both planning and scheduling are often done with the aid of **mathematical models**, such as **linear programming**. Finally, operations managers are concerned with making sure the plans are met in the face of changes in demand, quality problems, absent people, and so forth. For this they use overtime, reschedule customers, provide alternative means of production and use other general problem-solving skills.   *D.C. Whybark*

Reference

Chase, R.A., Aquilano, N. and Jacobs, R.J. (2000) *Production and Operations Management*, 8th edn, New York: McGraw-Hill/Irwin.

Opportunity cost, see costs, type of

OPT, see Drum–Buffer–Rope, OPT, synchronous manufacuring

Optimal solution, see optimization

Optimization, suboptimization, optimal solution, global optimum, local optima

Consider the following type of problem: The relationships underlying some phenomenon, or the operation of a **system**, or a purely abstract mathematical

conceptualization are formulated as a **mathematical model**, consisting of an objective function, expressed in terms of decision variables, where the latter may be subject to constraints. The aim is to find values for the decision variables that optimize the value of the objective function, i.e. either maximize or minimize it, whichever is relevant, while at the same time satisfying all constraints on the decision variables. The set of values of the decision variables for which the objective function assumes its optimum is called the optimal solution.

If the mathematical structure of the model is well behaved (e.g. **convex feasible region**, an objective function that is concave for maximization, and convex for minimization), the optimum is unique. For example, for a two-decision-variable problem, if the objective function has the shape of an upside-down bowl, then its maximum value is unique. However, if its shape is such that it has several peaks of different heights (such as a sandcastle), then there are several optima, each corresponding to a different set of decision variable values. Each such peak is referred to as a *local optimum*. The peak or local optimum which has the highest value of the objective function of all local optima is called the *global optimum*. These concepts readily generalize to situations with more than two decision variables, as well as minimization problems.

When related to systems, optimization means finding the best or the globally optimal mode of operation of the system as a whole. *Suboptimization*, on the other hand, only refers to the best operation of part of a system, i.e. a **subsystem**. Its optimum is only best for the subsystem, but is not necessarily part of the global optimum of the system as a whole. It may not even be a local optimum of the system.

In practice, most systems' optimizations are only suboptimizations, since the choice of what is part of the **narrow system of interest** and what is part of the **wider system of interest** or in the environment involves (possibly arbitrary) **boundary judgements**.

Furthermore, if there are **multiple objectives** (which may be conflicting and incommensurable), each objective is likely to assume its optimal value for a different set of values of decision variables. Hence, there is no global optimum. What will be judged as the 'most preferred' solution can only be a compromise which balances (or trades off) the achievement level between different objectives. There is little meaning in talking about optimization of such problems.

Outputs, see systems concepts

Outranking methods, see ELECTRE, multicriteria decision making

P

Parametric programming, post-optimal analysis

An optimal solution should be analysed for **sensitivity** or **robustness** to changes in parameters, such as objective function coefficients or the right-hand side (RHS) of constraints. In principle, such so-called *post-optimality analysis* can be performed on any form of **optimization** problem, but here the discussion is restricted to **linear programs** (LPs).

The most common form of post-optimality analysis is sensitivity analysis, which finds the range over which individual parameters can be varied such that the same variables keep positive (though changed) values. Parametric programming extends this idea by systematically covering the full range of values that a particular parameter can take, as it increases from zero, while the solution moves through a sequence of so-called **basic feasible solutions**. (A basic feasible solution corresponds to a corner point of the feasible region.) For changes in objective function coefficients, moves (called *pivots*) from one basic solution to the next are obtained by using the **simplex algorithm**; for changes in the RHS of constraints, the **dual simplex algorithm** is used. The sequence of values of the objective function forms a piecewise-linear function of the parameter being changed.

The process is illustrated below for simultaneous changes to the RHS. Suppose the RHS of constraint i is changed from b_i to $b_i + \alpha_i t$, where α_i is the chosen rate of change of t in constraint i, and t is gradually increased from zero. The procedure is as follows:

1. Solve the LP with $t = 0$ and identify the optimal basic solution, B. Set $t > 0$ and compute the values of the non-zero variables as functions of t.
2. Let t increase from zero and find the maximum of t for which the current basic feasible solution is replaced by another basic feasible solution (i.e. the optimal solution has moved to an adjacent corner point of the feasible region). Perform one iteration of the dual simplex method to generate the new B.
3. Repeat Step 2 until there are no further basic solution changes (or the problem becomes infeasible).

Example: maximize $z = x_1 + x_2$

subject to $x_1 + x_2 + x_3 = 6 - t$ $(\alpha_1 = -1)$

$-x_1 + 2x_2 + x_4 = 6 + t$ $(\alpha_2 = 1)$

all variables ≥ 0

When $t = 0$, the optimal basic variable values are $x_1 = 2$, $x_2 = 4$ with $z = 14$. Introducing $t > 0$, yields $x_1 = 2-t$ and $x_2 = 4$, with $z = 14-t$. When t is increased to 2, x_1 becomes 0 and is replaced by x_4. In terms of t, the new values are $x_2 = 6-t$ and $x_4 = -6+3t$, with $z = 18-3t = 18-3(2) = 12$. When $t = 6$, x_2 goes to 0 and $z = 0$. Increasing t beyond 6 allows no more feasible solutions. The values of z can now be graphed as t ranges from 0 to 6, to visually track the effect on the objective function of the changing constraint values, as shown in the figure.   *J.W. Giffin*

Reference

Hillier, F.S. and Lieberman, G.J. (2001) *Introduction to Operations Research*, 7th edn, New York: McGraw-Hill.

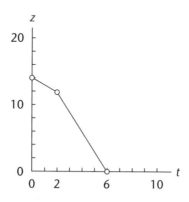

Figure: **Parametric programming: change in objective function**

Pareto analysis, see **ABC classification**

Pareto charts, see **seven tools of quality management**

Pareto optimality, see **dominance**

Penalty methods, see **barrier methods**

PERT: Project evaluation and review technique

PERT is a technique for monitoring and controlling complex projects that consist of a large number of individual tasks that have to be executed in a certain order, similar to the **critical path method** (CPM). Developed by the US Navy, in collaboration with Booz, Allen & Hamilton, it extends the ideas of CPM to include probabilistic task times and allows time–cost trade-offs.

Like CPM, it determines those tasks which, if delayed, will delay completion of the entire project – the so-called *critical tasks* that form the *critical path*. Unlike CPM, task times are not specified as a fixed length, but by three numbers: an optimistic time, t_{opt}, the most likely time or its mode, t_{mod}, and a pessimistic time, t_{pes}, as shown in the table below for a simplified house construction project, where task A is laying the foundations and erecting the exterior walls, task B refers to having the windows and doors made, task C putting on the roof and finishing the exterior, task D building the interior, and task E installing windows and doors and finishing the interior.

Table: **Time and cost data for building a house**

Task code	Optimistic time	Mode	Pessimistic time	Crash time	Regular cost*	Crash cost*
A	1 week	2 weeks	3 weeks	1 week	2000	3000
B	5	7	9	5	7000	9000
C	3	4	5	3	4000	5000
D	2	3	4	2	3000	4000
E	1	2	3	1	2000	3000

* Excluding the cost of material.

The critical path is determined by the same method as for CPM, except that the average time t for each task and its standard deviation s are approximated by a particular form of the *Beta probability distribution*:

$$t = \frac{t_{opt} + 4t_{mod} + t_{pes}}{6}, \qquad \sigma = \frac{t_{pes} + t_{opt}}{6}$$

The expected length of the critical path is given by the sum of the average task times on the critical path and the standard deviation as the square root of the sum of the squared task standard deviations. PERT uses this to make statements about the approximate probability that the project can be completed in less than a given time T, based on a normal approximation. For example, for the house construction project, the critical path consists of tasks B and E, with an expected path length of 9 weeks and a standard deviation of about 0.75 weeks. The probability that the house is finished within 10 weeks is about 0.9.

This is only an approximation, since it ignores the possibility that one or more non-critical tasks can become critical if their times are somewhat larger than their averages. **Monte-Carlo simulation** can be used to find more accurate completion time distributions.

The PERT method also considers *time–cost trade-offs*. By using additional labour or more costly equipment, the average task times may be reduced down to a minimum level, called the *crash time*. As the task times of critical tasks are reduced, previously non-critical tasks may become critical. Hence, time reductions may become progressively more expensive. The figure below shows the time–cost trade-off for our example. If reduction in total completion time also saves fixed daily project costs, then there is an optimal schedule of crashing task times that minimizes total project costs.

N.C. Georgantzas

References

Brassard, M. (1989) *The Memory Jogger Plus+*, Methuen, MA: GOAL/QPC.
Turner, J.R. (1993) *The Handbook of Project-based Management*, New York, NY: McGraw-Hill.

Web sites

Mind Tools: http://www.mindtools.com/critpath.html
Naval Postgraduate School: http://vislab-www.nps.navy.mil/me/calvano/asnesem/tsld104.htm

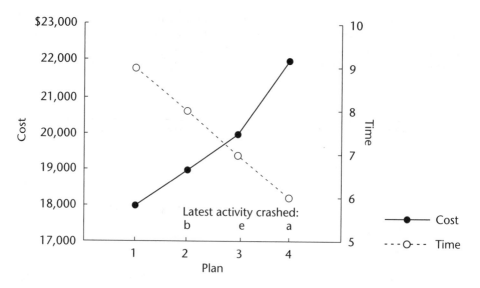

Figure: **PERT's time–cost trade-off**

Planning horizon, rolling planning horizon

Multi-period decision problems involve making a set of decisions in each period, where decisions in later periods may be conditional upon decisions in earlier periods. For some problems, the length of the planning horizon is predefined. For instance, a multistage investment project has a known estimated lifetime, which will serve as the planning horizon. In other cases, the problem is part of an ongoing operation or activity with no predicted known endpoint. This is the case for the planning of daily, weekly or monthly operations of a factory or service, such as the generation of electricity or the production schedule of a plant. No natural choice of planning horizon imposes itself.

Furthermore, the decision maker is in most instances primarily interested in the best set of decisions to make in the first period of the planning horizon. Implementation of decisions for later periods is postponed until the appropriate time arrives. This is particularly the case in the presence of **uncertainty** about what will happen in the future. The main reason for extending the planning horizon beyond the first period is that the best set of decisions in the first period is not independent of the exogenous future events and the decision options available in subsequent periods. In most instances, this dependence becomes weaker the farther away the future is. The length of planning horizon chosen should be such that what happens beyond the planning horizon has no or only a negligible effect on what is the best set of decisions for the first period.

Once events in the first period of the planning horizon have been observed, the problem is resolved, using as input the latest information about the **state of the system** and the uncertain future events. A common approach is to shift the planning period forward by one period, i.e. the previously first period (now past) is dropped, and a new period is added at the end, keeping the planning horizon the same length, as shown in the figure below. This is known as a *rolling planning horizon*.

This approach recognizes that information about the future becomes progressively less reliable the farther away it is. It also says that decisions should be based on the known current **state of the system**, rather than some projected state, and on the

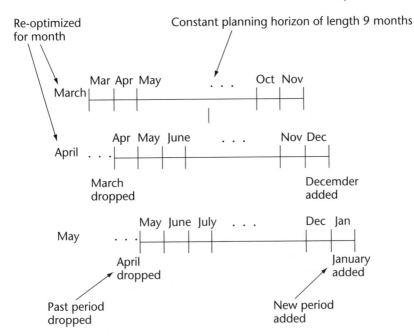

Figure: **Rolling planning horizon**
(*Source:* H.G. Daellenbach (2001) *Systems Thinking and Decision Making*, Christchurch: REA, p. 296)

most recent information about what the future might bring, rather than be based on possible outdated information.   *H.G. Daellenbach*

Portfolio selection, portfolio theory

Pension funds, trust funds, insurance companies, banks and private individuals invest funds in portfolios of tradable securities, such as company shares, bonds, treasury bills, foreign currencies, share options, even bank accounts. They are usually concerned with two aspects: the return on their investment and the risk involved. Return over a given period is measured by the ratio of [price change of the security + dividends or interest] divided by [initial price]. Risk refers to the level of uncertainty or variation in this ratio. Return and risk tend to vary in the same direction. Treasury bills are seen as riskless and offer low return, while speculative shares are seen as high potential earners, but also of high risk. Most investors are looking for a balance between return and risk. It can be shown that skilful *diversification* (i.e. mixing securities whose returns tend to move in opposite directions) increases the return while holding risk constant, or decreases risk while holding the return constant. Hence it is possible to diversify away the unsystematic risk associated with individual securities, leaving only the systematic market risk. Mathematical portfolio selection models exploit this property.

The first portfolio selection model was formulated in 1952 by Markowitz. It starts from the premise that the investor is willing to trade off returns against risk (e.g. accept reduced return for reduced risk), and wishes to maximize the weighted difference between the two. If λ is the weight given to returns and 1 the weight given to risk, measured by the total variance of the portfolio, we get the following model:

$$\text{Maximize } \left[\lambda \sum_{\text{all } i} r_i x_i - \sum_{\text{all } i} \sum_{\text{all } j} x_i x_j c_{ij} \right]$$

$$\sum_{\text{all } i} x_i = 1$$

$$x_i \geq 0, \quad \text{all } i$$

where the x_i is the fraction of funds invested in security i and c_{if} is the covariance between security i and j (if $i=j$ it is the variance). The constraint says that all funds have to be invested. This is a **quadratic programming problem** since it involves products of the variables. The figure below depicts the problem graphically. The enclosed area contains all possible portfolios. The solid line is the set of all so-called *efficient* portfolios, i.e. portfolios for which it is impossible to find another portfolio with a higher return without an increase in risk, or with a lower risk without a decrease in return. They form the **efficient frontier**. The straight lines represent various constant values of the objective function. The value of λ determines the slope and hence where the optimal solution occurs. λ also corresponds to the trade-off the investor is willing to make between return and risk. As λ increases, the optimal solution shifts down and to the left along the efficient frontier, reducing return and risk.

In practice, adaptations of the Markowitz model are usually implemented, the most famous ones by W. Sharpe in 1963. The returns of all securities are related only via an index, such as a stock exchange index. Risk is now measured only in terms of the variances of securities and the index. This reduces the amount of data needed dramatically. Sharpe later reformulated this problem as a **linear program**.

H.G. Daellenbach

References

Markowitz, H. (1959) *Portfolio Selection*, New York: Wiley.
Markowitz, H. (2000) *Mean-Variance Analysis in Portfolio Choice and Capital Markets*, New York, McGraw-Hill.
Sharpe, W.F. (1970) *Portfolio Selection and Capital Markets*, New York: McGraw-Hill.

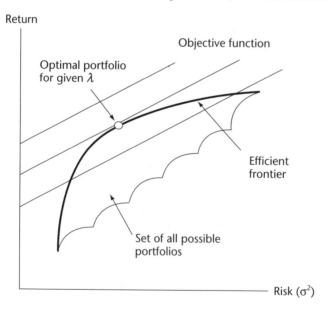

Figure: **Return/risk outcomes of all feasible portfolios**

Posterior probability, see **Bayesian decision analysis**

Post-optimal analysis, see **parametric programming**

Preemptive goal programming, see **goal programming**

Preventive maintenance

Preventive maintenance (PM) is a maintenance philosophy aimed at reducing the likelihood of unexpected plant failures occurring. PM originated as a practice mainly in mechanical plant that is subject to an ageing, damage or wear-out mechanism that increases the probability of failure over time and with use.

The traditional approach of preventive maintenance is to schedule specified maintenance operations that will replace, refurbish or overhaul certain parts of equipment at fixed intervals. The intervals are determined in a conservative manner, usually based on either design estimates or on operating experience. The objective of PM is to pre-empt failure. The use of this philosophy will result in the replacement of many serviceable components with new components, just because of the fixed-time nature of the process. The very act of disturbing working equipment, showing no signs of distress, also often results in maintenance-induced failures. The economic efficacy of this approach, used as a general policy, is therefore questionable. The exceptions are certain industries with continuous operation, process-based and often seasonal in nature, where preventive maintenance may be the only option, because plant can only be accessed during narrow time windows. In such cases extensive project management effort using **critical path scheduling** is required to perform the set workload within the available time.

Reliability prediction, as a basis for determining preventive maintenance intervals, has been proven by Patrick O'Connor of British Aerospace to be overly theoretical and largely devoid of reality to act as a basis for determining PM policies. Only in limited cases, such as in the case of well-defined load and fatigue cycles, or where degradation rates are predictable and constant, will it be of any use, and even then PM may still be less economical than other, more suitable, approaches.

PM nowadays must be viewed as an archaic approach for determining maintenance actions and task schedule intervals. It is ineffective and not applicable to modern multi-celled equipment consisting of elements that show no changes in their inherent failure probability with use or over time. A more modern approach, called **reliability-centred maintenance**, has largely taken its place.   *P. Beukman*

References

Levitt, J. (1997) *The Handbook of Maintenance Management*, New York: Industrial Press Inc.

Kelly, A. (1997) *Maintenance Strategy – Business-centred Maintenance*, Oxford: Butterworth Heinemann.

Moubray, J. (1997) *RCM II, Reliability-centred Maintenance*, 2nd edn, Oxford: Butterworth Heinemann.

Price decomposition, see **decomposition**

Prior probability, see **Bayesian decision analysis**

Prisoners' dilemma, see **cooperative games**

Problem situation

The problem situation is the context and situational framework within which a given issue or problem, that may have triggered a project, is embedded. The diagram below depicts the various factors that are relevant for describing the problem situation. It includes all aspects that may help in understanding the **systemic** relationships that affect the problem. This should cover not simply the 'hard' facts, such as the physical structure, resources, activities, data on which most **stakeholders** of the problem agree, but also 'soft' facts, such as **world views**, conflicts and dilemmas, **uncertainties**, and interpersonal relationships, all of which may influence what is culturally feasible within the organization.

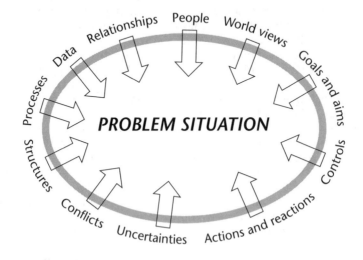

Figure: **Aspects defining the problem situation**
(*Source:* H.G. Daellenbach 2001, p. 59)

Furthermore, the description should not only restrict itself to those aspects that are seen as directly affecting the problem, but also aspects that may have indirect effects, via other aspects, and that includes other issues that may be related to it. R.L. Ackoff, one of the early systems thinkers, says that problems do not come alone but in 'messes' of interconnected issues. Interfering with one may affect others and vice versa.

   Rich pictures or **mind maps** are ideal tools to represent the problem situation, while *brainstorming* and **cognitive maps** are effective means to capture it.

   The problem situation should not be depicted in a **system** format. This is unhelpful at this initial stage of a project. It may impose a structure onto the problem that is not appropriate and may send the project down the wrong alley. At this stage, it is important to keep a completely open mind and not prejudge the situation. It is also clear that different stakeholders will see the problem situation and hence the problem differently, which gives rise to conflicts that may have to be resolved before being able to analyse the problem itself. **Problem structuring methods** may be able to bring about sufficient consensus to let the analysis go ahead.   *H.G. Daellenbach*

Reference
Daellenbach, H.G. (2001) *Systems Thinking and Decision Making*, Christchurch: REA, pp. 58–68.

Problem structuring methods

Problem structuring methods (PSMs) are **soft operational research** (soft OR) approaches to dealing with messy, often ill-structured, complex issues, where quantitative approaches (or traditional 'hard OR') fails to fit the situation. Typical problem situations involve multiple **stakeholders** with possibly different values and different perceptions of the issue. PMSs often call for open discussion, debate and **negotiation**. They are generally based on either subjectivist or social constructionist paradigms: those that say either human **subjectivity** or the normative construction of human affairs should be the priority for exploration. All such paradigms accept the validity of different **world views**, pay attention to how perceptions change when stakeholders gain a greater appreciation of other people's views, and aim to bring about some commitment to action.

The name PSMs was coined by J. Rosenhead in 1989 to break down the 'hard OR/soft OR' dichotomy and to make clear that the two modes of OR have different purposes and are therefore equally useful. Furthermore, the name reflects the fact that messy complex situations first need to be structured before any attempt can be made to look at action plans. In fact, for several PSMs, the structuring aspect dominates, with action plans 'emerging' as part of the convergence of views and positions.

The majority of PSMs were developed through **action research** and/or consulting practice. Some, such as Checkland's **soft systems methodology**, are firmly based on **systems thinking** and **systems concepts**. Several methods, such as **metagame analysis, hypergame analysis, gaming**, and the further development into **drama theory**, originate from the concepts of **game theory**. However, these abandon game theory's rigorous axiomatic foundation and retain only its outer form, i.e. two or more *players* with opposing views, but seeking means to bring about cooperation. Some approaches, such as **cognitive mapping**, have radically subjectivist psychological roots. The majority use *ad hoc* processes that have evolved through consulting practice and have proven to be successful for certain situations.

All PMSs have one thing in common. They start out seeking to attain a reasonably comprehensive view of the issue(s) within the context, although most recognize that true comprehensiveness is impossible. This initial analysis is then structured in various ways, such as bringing **uncertainties** about values, choices and the environment to the fore, and/or identifying clusters of highly connected aspects. The main aim at this stage is to gain a shared understanding of the issues, which may or may not bring about a convergence of views. The aim is not necessarily consensus, but a commitment to action, which is taken in the light of greater understanding of the views of others than might have been available without the use of a PSM.

A significant minority of PSMs use specialized software to aid in the structuring process and/or the exploration of choices. Most require a facilitator, trained in the method, with good interpersonal and negotiation skills. Only a few are designed for implementation without the presence of a facilitator.

G. Midgley and H.G. Daellenbach

References

Rosenhead, J. (ed.) (1989) *Rational Analysis for a Problematic World*, Chichester: Wiley.

Rosenhead, J. and Mingers, J. (eds) (2000) *Rational Analysis for a Problematic World Revisited*, Chichester: Wiley.

Process of MS/OR

MS/OR is primarily a practical discipline and profession carried out in organizational settings. The process of MS/OR refers to the way that MS/OR techniques are brought to bear upon managerial and business situations by MS/OR practitioners. A defining characteristic of this process is the construction and use of a model, or representation, of the situation. Examining this process reveals that such work is much more than the straightforward application of simply mathematical and computer methods. The work is creative and varied and requires practitioners to display a variety of social and interpersonal skills as well as having highly developed technical abilities.

The benefits of understanding the process of MS/OR are several: it enables practitioners to reflect upon the nature of their work and improve their practice; students of the discipline can be made aware of the complex nature of the activity prior to entering the profession; and it offers a framework within which general issues of concern can be addressed.

There are two common approaches to understanding and exploring the MS/OR process. One approach seeks to formulate a general statement, or methodology, of how the models constructed using MS/OR techniques are built, validated and used in the situations they aim to represent. These approaches emphasize the technical aspects of MS/OR work and highlight the difficulties faced in constructing and validating such models in dynamic organizational settings. Problems of data collection, model evaluation and interpretation of results are practical issues that have to be addressed by a practitioner in any piece of analysis. Methodologies have the advantage that they are easily understood by students of the discipline and can provide a useful entry into the realities of MS/OR work. A main disadvantage is that they simplify the work, often to an unrealistic degree, and do not formally acknowledge its social aspects.

A second approach examines in detail particular pieces of MS/OR work and seeks to develop an understanding of how experienced practitioners carry out their work. Such examples of good practice can be translated into a rich description of practical MS/OR and insights into its nature generated. This type of approach emphasizes the social and creative character of MS/OR and shows how practitioners work with managers and other organizational actors to achieve a desirable result. The role adopted by analysts, the way they gain support from and involve relevant individuals, the management of projects, and the embedding of technical work within this set of social processes are all critical to the success of an MS/OR project. This approach is valuable for broadening the understanding of MS/OR work and provides a balance to the technical emphasis that the methodological descriptions of MS/OR present.

P. Keys

References

Eden, C. and Radford, J. (1990) *Tackling Strategic Problems*, London: Sage.
Fortuin, L., van Beek, P and van Wassenhove, L. (1996) *OR at wORk*, London: Taylor and Francis.
Keys, P. (1995) *Understanding the Process of Operational Research: Collected Readings*, Chichester: Wiley.
Keys, P. (1997) 'Approaches to understanding the process of OR', *Omega*, 25:1–13.

Producer's risk, see acceptance sampling

Production planning and control

The production planning and control function determines and manages the resources (material, people, machines, etc.) needed to produce a company's goods and services. As an example, consider baking a birthday cake for a party next Wednesday at 2:00 pm. You know you will need to mix the ingredients before you bake the cake in the morning. You need to make frosting, and the cake will need to cool before you can finish it. On Tuesday, you will check on the ingredients in case you need more of something. You know you cannot use the oven for anything else Wednesday morning, nor can you do much else.

These are all elements of the production planning process. You use the recipe to determine the ingredients, check your inventory for availability and purchase ahead of time anything you need. You schedule the stove and yourself for Wednesday morning. You just made a production plan. On Wednesday morning, when you start to mix the cake, you find that the eggs you were going to use are hard-boiled. You borrow from the neighbour, race to the store or call a friend for help. Making these adjustments is the control part of production planning and control.

Imagine, now, doing this in a bakery where each day there are cakes, cookies and many other products to be mixed, baked, decorated and packaged. All the ingredients must be purchased and the bakers, mixers and ovens all must be scheduled to produce the number and variety of goods required. For your birthday cake you could do this in your head, almost intuitively. For more complicated situations, companies use computer programs that do *material requirements planning* (MRP). These programs determine when, how much and where (at mixing, the oven or packaging) material is required and what capacity (bakers, ovens, and packaging people) is needed to complete everything on time.

All this is derived from a master schedule – a statement of the desired future output of products and services. It is based on known customer orders, forecasts of additional demand and any reserve inventory or capacity that is desired. An experienced manager is usually responsible for developing the *master production schedule* and keeping it synchronized with the overall production plans of the company.

Companies use several techniques for meeting the master schedule. They prioritize activities at each work area (e.g. always mix the cakes before the cookies). They check progress and expedite products that are behind schedule, authorize overtime, and so forth. When things go really badly, changes in the master production schedule are made and customers are informed of possible late deliveries. This way, the production planning and control activities match the output with future demands.

D.C. Whybark

Reference

Vollmann, T.E., Berry, W.L. and Whybark, D.C. (1997) *Manufacturing Planning and Control Systems*, 4th edn, Homewood, IL: McGraw-Hill/Irwin.

Public policy analysis

A policy, according to the *Oxford English Dictionary*, is 'a course or principle of action adopted or proposed by a government, party, business, or individual'. Like strategies, programmes, plans and similar concepts, policies provide general directives rather than detailed instructions for action. The specific function of policies is to provide normative orientation – guiding values and ends – for the elaboration of strategies, programmes and plans, which are more concerned with the selection of appropriate means for achieving those ends.

Policy analysis is a field of professional practice that is concerned with the scientific analysis of the contents and consequences of policies, particularly in public sector management and planning. The field emerged after World War II from the confluence of four parent disciplines: (1) management science and systems engineering; (2) welfare economics; (3) the political and administrative sciences; and (4) the empirical social sciences. Lerner and Lasswell (1951) are generally considered the fathers of the field, although its major concepts and tools were developed in the subsequent decades.

The object-domain of public policy analysis comprehends all stages of public policy-making, from policy formulation to policy implementation and to policy review, as shown in the table below.

It is obvious that this broad scope requires an equally broad theoretical and methodological basis. Policy analysis clearly needs to be an interdisciplinary field. Drawing on its four mentioned parent disciplines, some of its major tools are (1) **decision analysis, sensitivity analysis** and **systems analysis**; (2) **cost–benefit analysis**; (3) implementation research, financial auditing and reporting tools; and (4) evaluation research, environmental and social impact assessment. Additional

Table: **The object-domain of public policy analysis**

I – POLICY FORMULATION
1 – Policy problems
2 – Policy objectives
3 – Policy contents (action plans and resources)
II – POLICY IMPLEMENTATION
4 – Policy decisions and legislation
5 – Policy implementation
6 – Policy outcomes and impacts
III – POLICY REVIEW
7 – Policy monitoring
8 – Policy evaluation
9 – Policy reporting

tools include project management; social indicators; management information and reporting systems; **total quality management**; technology assessment; **forecasting**; **scenario analysis**; creativity techniques; idealized design; and many others.

A major methodological development in the 1960s was the *planning–programming–budgeting system* (PPBS). PPBS aimed to integrate the formulation of policy objectives with the processes of budgeting, in an effort to ensure a systematic allocation of financial resources to political priorities. The approach was hailed as a breakthrough in rational policy-making; policy analysis for a while became almost synonymous with PPBS. In the USA, PPBS was declared mandatory for all federal agencies under the Johnson Administration in 1965, but it never fulfilled its promises. Its conceptual framework relied all too one-sidedly on a utilitarian economic concept of rationality, and consequently could not deal satisfactorily with the core problem of public policy-making, of how to integrate differing views and conflicting interests into a meaningful concept of 'rational' action.

Subsequent major developments were, in the 1970s, the rise of *policy and program evaluation* to a major field of policy analysis, and more recently, in the 1990s, the development of the so-called *new public management*, which once again tries to achieve an integrated approach to policy planning, programming and budgeting.

The current state of public policy analysis is characterized by an increasing variety of tools. These range from financial controlling approaches such as *benchmarking* (performance comparison with best practices) and output-oriented performance indicator systems for public services (e.g. for hospitals) to new forms of participatory inquiry and design, such as citizen surveys, citizen reports, search conferences and others.

Despite its impressive development, the field's methodological foundations still appear deficient regarding the role of conflicting views and value judgements, power, civil society and democracy in 'rational' policy-making. More than other applied disciplines such as **action research, community OR** and management consulting, policy analysis has remained tied to its origin in 'hard' conceptions of systems analysis. It has hardly begun to incorporate the principles of soft and **critical systems thinking**, yet their concern for interpretive, critical and emancipatory issues, for **dialogue**, discursive rationality and **boundary critique**, appears relevant indeed. It would seem that public policy analysis has much to gain from a systematic revision of its conceptual foundations along these lines (see also **critical systems heuristics**).

W. Ulrich

References

Bardach, E. (2000) *A Practical Guide for Policy Analysis,* Chatham, NJ: Chatham House.

Dror, Y. (1968) *Policy Making Re-examined,* San Francisco: Chandler.

Dror, Y. (1971 *Design for Policy Sciences,* New York: American Elsevier.

Friedman, J. (1987) *Planning in the Public Domain: From Knowledge to Action,* Princeton: Princeton University Press. Critical historical overview.

Lasswell, H.D. (1971) *A Preview of Policy Sciences,* New York: American Elsevier.

Lerner, D. and Lasswell, H.D. (eds) (1951) *The Policy Sciences: Recent Developments in Scope and Method,* Stanford: Stanford University Press.

Quade, E.S. and Boucher, W.I. (eds) (1968) *Systems Analysis and Policy Planning: Applications in Defense,* New York: American Elsevier.

Quadratic programming

Quadratic programming problems (QPs) are a special case of **nonlinear programming problems**, where the objective function is a quadratic function and all constraints are linear. They can be solved by much simpler techniques than those needed for general nonlinear programs.

A function is said to be quadratic if it is the sum of terms that involve only squared variables or products of two variables, or variables themselves. The following expressions are an example of a QP:

$$\text{Maximize } 8x + 11y + 3z - 0.01\,x^2 - 0.01\,y^2 - 0.01xy$$

$$\text{subject to } x + y + z \leq 100$$

$$x, y, z \geq 0$$

One method for solving general nonlinear programs is to find a solution to the **Karesh–Kuhn–Tucker (KKT) conditions**. These conditions are a system of equations characterizing the **optimal solution**. If the KKT conditions are all linear, they are very easy to solve using a variant of the **simplex method** for **linear programming**. In 1956, P. Wolfe took advantage of this fact in developing what is now known as Wolfe's method for solving QPs.

Deriving the KKT conditions for a QP results in a set of mostly linear expressions, except for the **complementary slackness** (CS) conditions. The CS conditions state that either the slack variable in a constraint must be zero (i.e. the original inequality is binding), or the **dual variable** associated with the constraint is zero, but not both may be positive at the same time. Wolfe's method deals with this by removing the CS conditions from the system of KKT equations, and introducing a rule into the simplex method which achieves the same thing: at no iteration will a variable be introduced into the solution if its complementary variable is already in the solution. If the objective function is well behaved (i.e. **convex**), Wolfe's method is guaranteed to find the optimal solution to the QP. In 1959 E.M.L. Beale developed a more general method for solving QPs.

One of the first and most famous examples of a QP is the **portfolio selection** problem, introduced by Harry Markowitz in 1952. The investor wants to invest a given amount of funds into risky securities, such as company shares. The objective is to balance return against risk, expressed mathematically as the weighted difference between expected returns and risk. If risk is measured by the variance of total return, the objective function becomes quadratic since the variance of a sum of random variables involves products of two variables and squared variables. The example shown above is a portfolio selection problem for two risky investments, X and Y, and a riskless investment, Z. *S. Batstone*

Reference
Winston, W.L. (1994) *Operations Research Applications and Algorithms*, 3rd edn, Belmont CA: Duxbury, ch. 12.

Quality control circles

Quality control circles (QCCs) were developed in Japan in the 1960s as a group learning mechanism for foremen and workers. QCCs were one part of a widespread quality effort in Japanese industry developed cooperatively by major companies, the Japanese Union of Scientists and Engineers (JUSE) and the Japanese Standards Association. Kaoru Ishikawa, a leading figure in the Japanese QCC movement, identified ten key ideas of the approach:

- voluntarism by participants in the organization's QCC program, not coercion from above to participate
- self-development as the focus for QCC activities rather than problem solving
- group activity
- mutual development and benefit for members
- for any individual QCC, participation by all employees in the particular work group
- utilization of QC techniques and the basic **seven tools of total quality control**
- focus on workplace-related activities
- vitality and continuity based in collective understanding of opportunities to improve the workplace
- originality and creativity
- awareness of quality, of problems and of improvement.

Japanese QCCs apply a rigorous problem-solving approach to workplace problems in the context of providing development and learning opportunities for workgroups, with benefits accruing to team members individually and collectively and to the organization. Activities of QCCs are structured around Shewhart's 'Do, Study (or Check), Act cycle', statistical thinking and analysis of data and the use of *story boards* to present the work of the QCC to management.

During the late 1970s and 1980s QCCs spread to many countries. Western companies, impressed by the success of QCC activities in Japan, adopted quality circles (note that the term control tended to disappear from the title) as the latest solution to their quality problems. Some organizations achieved great success; others invested large amounts of money and time in forming and training quality circles for limited return.

Reviews of the experience with QCCs in the West have concluded that the mixed results reflected the approach taken by management to their use. The lack of success has been attributed to their overuse and misuse, with inadequate conceptualization and implementation. QCCs often were isolated from management, and implemented mechanically with separate infrastructure established to support QCC-type activities. The circles were seen as one-off problem-solving groups rather than part of an organizational fabric for continuous improvement. In the less successful cases, organizations adopted QCCs, the visible symbol of an organization-wide approach, as a quick fix for quality problems without changing the broader context of managerial attitudes, behaviours and priorities. In successful initiatives quality circles were seen not as the sole solution to all of the quality problems, but as one technique contributing to an organization-wide quality strategy.

In the 1990s the term was often replaced by titles such as quality improvement teams for groups which apply the QCC tools and frameworks to solving quality problems.   *D.J. Houston*

References
Ishikawa, K. (1986) *What is Total Quality Control? The Japanese Way,* Englewood Cliffs, NJ: Prentice-Hall

Lawler III, E.E., Mohrman, S. and Ledford, G.E. Jr (1992) *Employee Involvement and Total Quality Management,* San Francisco: Jossey Bass.

QUALY, see cost-effectiveness analysis

Queueing and waiting lines

Queueing or waiting in lines is an everyday experience. We wait for elevators, we wait at checkout counters or the library until it is our turn to be served. But queueing is also commonplace in industrial situations. Parts wait at machines for processing, aircraft wait for clearance to use the runway, ships wait for unloading or loading. Waiting has economic consequences. Each hour a ship waits may cost several thousand dollars. Service could be speeded up by having more container cranes, but that is costly too, and some may remain idle much of the time. There are thus costs to be balanced against each other. The first mathematical queueing models, dealing with old-fashioned telephone exchanges, were developed in 1917 by the Danish engineer Erlang – more than three decades before the name *operational research* was coined.

The basic structure of a queueing process is as follows: Customers arrive randomly at a service facility, requesting service. In the *multi-channel* case, one or several servers work in parallel, such as the checkout counters at a supermarket. In the *multi-phase* case, several servers work in series, each performing different tasks, such as in a surgery where patients first see a nurse, then a doctor, and may return again to the nurse. If no server is free, arriving customers join a queue and wait for their turn to be served. There may be a single queue feeding all servers, i.e. the servers are pooled, as is the case in most banks, or there is a separate queue for each server, as for checkout counters in supermarkets. The usual queue discipline for serving customers is on a first-come/first-served basis, but the queue discipline may be different, e.g. priority service based on urgency. Customer who have received service leave the system. The customer population may be modelled as infinitely large, such as for a supermarket, or finite, such as a limited number of machines that may require the service of an operator. The figure below depicts a typical waiting line structure and its inputs.

The A/B/m notation is used to denote various types of single-phase mathematical queueing models. 'A' denotes the arrival process, e.g. *D* for **deterministic** with fixed interarrival times, or *M* (short for 'Markovian') for negative exponentially distributed interarrival times. 'B' denotes the service time distribution, e.g. *G* for any general distribution (not exponential), such as Normal, and 'm' is the number of servers or parallel channels. Unless specified otherwise, the queue discipline is assumed to be first-come/first-served, and the number of spaces in the queue is assumed unlimited.

We are usually interested in the long-run behaviour of the system, i.e. when it has reached an equilibrium or **steady-state**. Let μ be the average rate of customer arrivals, μ the average rate at which customers are served and σ the standard deviation of the service time. Unless $\lambda < \mu$, a queueing system will never approach steady-state conditions. The queue will grow indefinitely. For an $M/G/1$ model, the important service performance measures are:

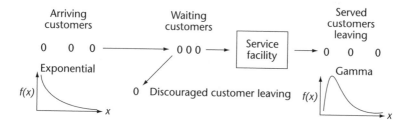

Figure 1: Queueing process

- The average number of customers in the queue, L_q:

$$L_q = \frac{\lambda^2\sigma^2 + (\frac{\lambda}{\mu})^2}{2(1 - \frac{\lambda}{\mu})} \text{ provided } \frac{\lambda}{\mu} < 1$$

For an $M/M/1$ queue this simplifies to $(\lambda/\mu)[\lambda/(\mu - \lambda)]$.

- The average waiting time per customer, W_q:

By the well-known Little's law

$$W_q = L_q/\mu$$

- The fraction of time the server is busy or the traffic intensity or the probability of finding the server busy at a random point in time:

$$P(\text{server busy}) = \lambda/\mu$$

For an $M/M/1$ queue it is also possible to state the probability of finding n customers waiting at any random point in time. Multi-channel queueing models show that pooling of customers into a single queue improves systems performance. Unfortunately, for most other queueing models the expressions lack exact analytic solutions and performance statistics can only be obtained by simulation.

N.C. Georgantzas

References

Gross, D. and Harris, C.M. (1998) *Fundamentals of Queueing Theory*, 3rd edn, New York: Wiley.
Hillier, F. and Lieberman, G. (2001) *Introduction to Operations Research*, 7th edn, New York: McGraw-Hill, chs. 16 & 17.
Wolff, R.W. (1989) *Stochastic Modeling and the Theory of Queues*, Englewood Cliffs, NJ: Prentice Hall.

Queueing networks

Many systems, such as computer, communication and manufacturing systems, can be modelled as networks of queues, where the output of one queue becomes part of the input to some subsequent service facility. We would like the *state distribution* – the probability distribution of the numbers of customers at the various queues at a particular time, and the *sojourn-time distribution* – the probability distribution of the time taken to traverse from one queue to some other. The difficulty is that usually queue outputs are very complex. For example, the output of an M/GI/1 queue is at best a **Markov renewal process** with an infinite number of possible states. When this becomes part of the input to a subsequent server we have a yet more complex model for which there are very few exact practical results available, and so on.

With very few exceptions, state distributions can only be determined where these have a product form, like those for **Jackson networks**. Product form is closely connected with the queues in the network exhibiting various forms of time-reversibility, where future departures are independent of the times of previous arrivals. Even if a state distribution can be found, this does not give the sojourn-time distribution, as the sojourn times of a single customer at a sequence of queues are usually dependent, unless other conditions, such as customers cannot overtake each other, are also imposed.

There are some extensions to the Jackson network model which still give product form solutions. For example, customers can be created and destroyed at service facilities. However, recent research on queueing networks other than Jackson networks has had to focus on less explicit areas. These include:

- Conditions for stability of more general models: it is often possible to check that the system does have a stable limit, even if we cannot find it, and possibly to give some limiting results.
- Approximations: especially when traffic is heavy the state of many systems can be modelled by assuming the state space is continuous rather than discrete. The state can then often be described in terms of reflected Brownian motions.

D.C. McNickle

References
Much of the research on these systems appears in the journals *Queueing Systems* and the *Journal of Applied Probability*.

Queueing Systems (1998) vol. 30(1,2), Special issue on 'Diffusion approximations of queueing networks'.

Queueing Systems (1999) vol. 32(1–3), Special issue on 'Approximations and stability in stochastic networks'.

Random numbers, random variates, random number generators

In computer **simulation** we often require long sequences of numbers that mimic the properties of certain independent **random variables**. We refer to them as *random variates*. For example, we may need a sequence of random variates that appear to be independent samples from a particular normal or a negative exponential distribution to model the variation of a particular random event. This is done by a two-stage process. The first stage generates a random variate for a uniform distribution between zero and one. These are what is commonly called *random numbers*. The second stage is to convert these random numbers into random variates from the desired distribution. There are formulas or algorithms to do this for most common distributions. For example, the formula $-\alpha \ln(1 - u)$ will convert a uniform random number u into one which comes from a negative exponential distribution with mean α. This is an example of the most common conversion method – the *inverse transform method*. For discrete distributions, the inverse transform method uses look-up tables for the cumulative distribution.

The random numbers themselves are produced by a numerical procedure, called a *random number generator*. Most generators in common use are of the *linear congruential* or *multiplicative congruential* class. A linear congruential generator produces the next random number, Z_i from the previous number by $Z_i = (aZ_{i-1} + c)(\mathrm{mod}\ m)$ where a, c and m are integer constants. So Z_i is the remainder left over after dividing $(aZ_{i-1} + c)$ by m. The speed of the generator relies essentially on the fact that a digital computer can only store a certain number of binary digits (16 or 32) to represent a number. Choosing m appropriately (e.g. $m = 2^{32}$) allows the division operation to be avoided, because Z_i is given by the digits which will overflow the storage. When converted into fractions between zero and one these numbers can often pass statistical tests for independence of each other, lack of short-run serial correlation and coming from a uniform distribution between zero and one.

It is important to recognize that these numbers are generated by a formula; hence once the initial number, called a *seed*, is given, the future sequence of numbers is exactly determined. For this reason the sequences produced are sometimes referred to as *pseudo-random* numbers. In fact, a linear congruential generator like the one above will begin to repeat the same sequence after generating $2^{31} - 2$ random numbers. While this is quite long enough for most practical simulations, it could well be too short for applications that seek to determine theoretical properties. However, this property is highly useful. It allows testing of different operating rules in a simulation using the same sequence of random events. This eliminates one source of variability and therefore produces a more reliable comparison. Another useful property of many random number generators is the ability to jump out to a certain number in the sequence (e.g. to the 100 000th number), called an *offset*. Offset and seed must not be confused.

D.C. McNickle

References

Law, A.M. and Kelton,W.D. (2000) *Simulation Modelling and Analysis*, 3rd edn, New York: McGraw-Hill. Extensive list of formulas to convert random numbers into random variates.

L'Ecuyer, P. (1998) 'Random number generation', in J. Banks (ed.) *Handbook on Simulation*, New York: Wiley, pp. 93–137.

Random variables and their probability distributions

A variable that represents the yet unknown numeric value of a random event is called a random variable. Although its value is not known yet, it will be one of the values in the range of possible values. For a continuous random variable, that range could be any real number from plus infinity to minus infinity, as for a normal random variable, or any nonnegative value, as for a negative exponential random variable, while for a discrete random variable it could be a set of nonnegative integer values, 0, 1, 2, ..., N, as for a binomial random variable. Once the result of the event is known, it is not a random variable any longer, but a constant, i.e. it is one of its possible numeric outcomes.

If its probability distribution is known, then prior to knowing the result of the event, we can assign a probability that a given discrete random variable will assume a given value or that a continuous random variable will fall into a specified interval. We can also determine the probability $P(X \leq x)$ that the random variable X will assume a value of less than or equal to x or $1 - P(X \leq x)$ that it will assume a value larger than x.

The expected value or mean of a random variable represents its centre of gravity or mass and is given by

$$\mu = \int_a^b x f(x) dx \quad \text{for } x \text{ continuous, } a \leq x \leq b,$$

$$\mu = \sum_{i=a}^{i=b} x_i p(x_i) \quad \text{for } x_i \text{ discrete, } x_i = a, a+1, a+2, ..., b$$

where $f(x)$ is the probability density or mass function and $p(x_i)$ the probability that $X = x_i$.

The normal distribution is a suitable probability law for situations where the random variable is the sum of a large number of small independent events, such as the daily demand for a product being the sum of individual purchases by many customers. It is completely defined by its mean and its standard deviation. The Poisson distribution is also a sum of independent rare events with the proviso that each event assumes either the value 0 or 1. It is completely defined by its mean, the standard deviation being simply the square root of the mean.

H.G. Daellenbach

Reference

Daellenbach, H.G. and McNickle, D.C. (2001) *Systems Thinking and OR/MS Methods*, Christchurch: REA, pp. 127–135. Elementary introduction.

Recourse programming

Recourse programming or stochastic *linear programming problems with recourse* deal with decisions where some outcomes are random variables with known probability distributions. They assume that some firm decisions have to be made before the outcomes of uncertain events are known, and provide for recourse actions to be taken for each possible event. The concept of a stage is used to cover all decisions made with the same level of information.

As an example, consider a hydro-power generating company which must choose some or all of several contracts for the firm supply of power at various prices in several future periods. The inflow of water into its reservoir is a random variable, hence the amount of water available for power generation is also random. If the amount of water available during any period is insufficient to meet the firm supply contracts, the company's possible recourse action consists of making up the shortfall in power by buying it from other generators at the predicted spot prices, which increase as the shortfall rises, or possibly pay a heavy shortage penalty. The selection of the firm contracts is the set of unconditional first-stage decisions. The recourse actions are the decisions to be taken at subsequent stages.

When the probability distributions are discrete, the problem structure can be illustrated using an *event tree*, as shown in the figure below. The nodes of the tree correspond to decision points and the arcs correspond to possible outcomes of the random variables. All nodes at the same level in the tree are in the same stage, with the root node being Stage 1. The diagram succinctly shows the information that may be used at any stage. A path from the root node to a leaf node is called a *scenario* and represents one possible sequence of random events.

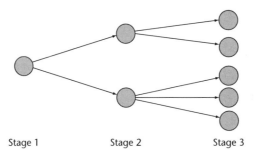

Stage 1   Stage 2   Stage 3

Figure: **Event tree**

For discrete probability distributions or a discretized approximation of a continuous distribution, it is possible to translate the problem into a deterministic equivalent **linear program**, albeit a possibly very large one. Many algorithms for solving recourse problems capitalize on the structure of the problem and decompose it into several smaller interconnected subproblems. One common difficulty is keeping the approximate event tree relatively small. Keeping track of the structure of an event tree and the corresponding data can be very difficult, especially considering that each node of the event tree may itself be a large linear program. *S. Dye*

References

Kall, P. and Wallace, S.W. (1994) *Stochastic Programming*, Chichester: Wiley.

Ruszczyński, A. (1997) 'Decomposition methods in stochastic programming', *Mathematical Programming*, 79:333–353.

Wets, R. and Ziemba, W. (1999) *Stochastic Programming. State of the art 1998. Annals of Operations Research*, vol. 85, Bussum, Netherlands: JC Baltzer Science. Extensive bibliography on stochastic programming in the preface.

Reduced costs, see simplex method

Reductionist thinking, see cause-and-effect thinking

Reference lottery, see utility

Regression analysis

MS/OR analysts regularly must determine if a given variable is affected by other variables which would allow its value to be predicted from the values of these other variables. Regression is a statistical technique for analysing the numeric relationships between a given variable, Y, called the *dependent variable*, and one or more other variables, $X_1, X_2, ..., X_n$, called the *independent or explanatory variables*. The inputs used are the corresponding sets of observations on both the dependent and the independent variables. The output is a *regression equation* that expresses the predicted value $\hat{y}$ of Y as a linear function of given x_i values of X's of the form

$$\hat{y} = a + b_1 x_1 + b_2 x_2 + ... + b_n x_n$$

where a and the b_i's are called the regression coefficients. They are estimated from the observations. The figure below demonstrates this dependence for the case of two variables.

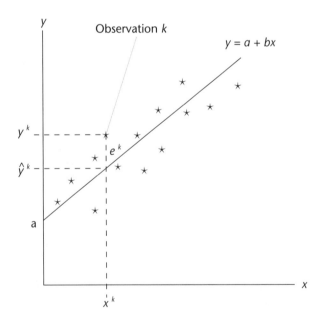

Figure: **Regression equation**

The figure shows a set of observations on X and Y. Higher values of X tend to be associated with higher values of Y. The straight line, $y = a + bx$, through the data captures this positive relationship. The line has intercept a and slope b. Note that no observations may, in fact, fall on the line. So the line is not a perfect predictor of Y. For example, the kth observation for X, marked in the figure, results in a prediction of $\hat{y}^k$. The difference between the actual and the predicted Y, $e^k = y^k - \hat{y}^k$, is the amount by which the prediction is 'in error' and is called the *error term*. However, no 'error' is involved. It simply means that there are factors, other than X, that affect Y which are unaccounted for by the regression equation.

The regression equation is the straight line that minimizes the sum of the squared error terms, hence the term *ordinary least squares regression*. By squaring the error terms large errors are penalized proportionately more than small errors.

Although a least squares regression line can be calculated for any set of data, the statistical regression model makes the important assumption that the error terms are independently distributed normal random variables. Only if this assumption holds is it possible to make statistical inferences upon the coefficients derived, such as whether they deviate significantly from zero, or make interval estimates on them and on the prediction. If the relationships are not linear, regression may be applied on suitable transformations of either the dependent variable or of one or more of the independent variables. Finding the 'best' regression equation is a matter of relying upon underlying economic, social or scientific theory, as well as that of exploring various combinations of explanatory variables and transformations. There are powerful statistical software packages, such as SPSS, SAS or Minitab, that simplify and speed up such analysis.   *S. Stray*

References

Dielman, T.E. (2001) *Applied Regression Analysis for Business and Economics*, 3rd edn, Pacific Grove, CA: Duxbury.

Draper, N. and Smith, H. (1998) *Applied Regression Analysis*, 3rd edn, New York: Wiley.

Smith, G. (1998) *Introduction to Statistical Reasoning*, New York: McGraw-Hill.

Reliability-centred maintenance

Reliability-centred maintenance (RCM) is a maintenance philosophy aimed at using the severity of a failure consequence as the basis for making decisions on what approach to take for maintaining plant and equipment. The term RCM and the approach, as well as the treatment of hidden failures, were defined by Nowlan and Heap (1978), based on earlier work done by the Maintenance Steering Groups in 1968 within the airline industry in response to the diverse and complex demands placed on maintenance by the advent of wide-body transport aircraft and on the maintenance and safety of nuclear power plants. With the availability of affordable computer technology, RCM has now become cost-effective for most industries. Moubray (1997) has refined the philosophy, now called RCM II, to be more generally applicable.

RCM has replaced traditional **preventive maintenance** (PM), which schedules maintenance or equipment and replacement of parts at fixed intervals. PM is ineffective, uneconomical, and inappropriate for modern multi-celled equipment consisting of elements that show no change of inherent failure probability with use or over time.

RCM starts out by analysing the consequences of failures. Failure consequences are defined as *safety-critical, operational- or economical-critical*, or as *hidden*. The next step is to assign applicable and effective maintenance tasks that are based on the particular failure consequences. For safety-critical failure consequences, redesign of the item is mandatory if no applicable or effective maintenance task can be found. For operational- or economic-critical consequences, task selection is based on cost-effectiveness. By matching maintenance task types to failure consequence, no single approach to task type is prescribed. Only where hard-time limits are dictated by equipment that shows degradation with use is PM-type fixed scheduled overhaul or replacement utilized.

The real strength of RCM lies in not requiring any maintenance work on items where the condition of that item does not indicate a potential failure mode or imminent failure. In cases where the occurrence of a failure has no safety, operational or economic impact, or where standby provides continuation of functionality, that item will be allowed to fail before any action is taken. The RCM philosophy results in maximizing the useful life of many items by generally avoiding premature replacement of serviceable components.

RCM therefore offers a balanced portfolio of maintenance tasks that are based upon failure consequence, rather than elapsed time. This has resulted in overall improvements in plant reliability as well as a reduction in maintenance costs. RCM is eminently suited as a maintenance philosophy for structural elements, electrical and mechanical systems, and electronic components. *P. Beukman*

References

Kelly, A. (1997) *Maintenance Strategy – Business-centred Maintenance*, Oxford: Butterworth Heinemann.

Levitt, J. (1997) *The Handbook of Maintenance Management*, New York: Industrial Press Inc.

Moubray, J. (1997) *RCM II, Reliability-centred Maintenance*, 2nd edn, Oxford: Butterworth Heinemann.

Nowlan, F.S. and Heap, H.F. (1978) *Reliability Centred Maintenance*, United Airlines.

Reliability theory

Most equipment or production components used in industrial processes and subjected to intermittent or continued use may fail suddenly. Their manner of failing varies depending on the component. A fuse fails suddenly – one moment it works perfectly and the next moment is not functioning at all, whereas a steel beam under heavy load gradually weakens over time and then fails or becomes unsafe at some point in time. The consequences of failure may be economic (e.g. lost production, damage) or endanger safety with possible injury or loss of life. It is therefore important to be able to predict when a failure might occur in the expectation of intervening prior to the component failing.

Unfortunately, identical components may fail at different points in time. Their failure can only be predicted in probabilistic terms, e.g. we may be able to determine the probability that a component fails or loses its ability to perform within a given length of time t, defined as $P(T \le t)$, for all values of t.

$R(t) = P(T > t) = 1 - P(T \le t)$ is called a component's *reliability function*. Associated with a component's time to failure is its *hazard function* or instantaneous failure rate. This rate may remain constant, regardless of how long the component has been used; it may increase with use – the more common case, or it may decrease with use (at least initially, as was the case with aircraft piston engines (the reason why these engines were run-in prior to mounting on the aircraft). For constant or decreasing failure rates, it never pays to replace an item prior to failure.

The two prominent questions asked when dealing with reliability problems are:

1. What type of failure probability distribution or law best fits a component's failure pattern, as observed in past, or as assumed to hold for future failures? The normal, exponential and Weibull probability distributions, depicted in Figure 1 below, are commonly used approximations. For example, the failure law for a component whose failure rate is constant is given by the negative exponential distribution.

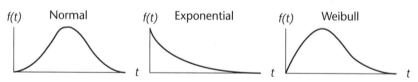

Figure 1: **Common failure laws**

2. When two or more components with known failure laws are combined as a system in series or in parallel, as shown in Figure 2, what is the failure law of their system?

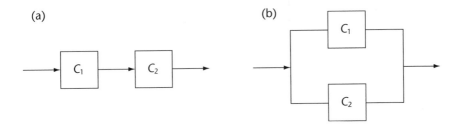

Figure 2: **Independent components/items connected (a) in series and (b) in parallel**

The second question is difficult to answer for large systems. If two components, arranged in series, function independently of each other, as depicted in system (a) of Figure 2, the reliability function of the system as a whole is equal to the product of the two individual reliability functions, i.e. $R_1(t) R_2(t)$. The reliability of the system as a whole is less than the reliability of any of its parts. Conversely, a system of two independently functioning components, operating in parallel, as shown in (b), is more reliable than either of its components, i.e. the system fails only if both components fail. Its reliability function is equal to $R_1(t) + R_2(t) - R_1(t)R_2(t)$.

N.C. Georgantzas

References
Blanchard, B. (1981) *Logistics Engineering and Management*, Englewood Cliffs, NJ: Prentice-Hall.
Lloyd, D. and Lipow, M. (1982) *Reliability: Management, methods and mathematics*, Englewood Cliffs, NJ: Prentice-Hall.
O'Connor, P. (1981) *Practical Reliability Engineering*, New York, NY: Wiley.

Repairperson problem

Consider a production system that consists of M identical machines, M_1, M_2, ..., M_M. Each machine may stop at a random point in time and require the attention of an operator or repairperson. The reasons for the stoppage may be machine breakdown or failure, the machine may run out of raw material, or there occurs some fault in the production process, such as the rupture of a thread on a loom. There is a pool of S_N operators for servicing the machines. If a machine stops and an operator is free, the machine is serviced immediately. If all operators are busy, the machine waits its turn for the first operator to become free (a so-called first-come/first-served service rule). This process is depicted in the figure below. The arrows show the **activity cycle** of the servers, being called from the pool by a machine stoppage and returning to the pool after completing the service. Under normal conditions, some of the operators may be idle from time to time.

There are intangible costs associated with stoppages, such as lost production time or loss of safety. Each operator in the pool has a cost, such as wages. The more operators there are available, the higher their total cost, but the less likely a machine will have to wait for service, the shorter the total waiting time over all machines during a given time period, and consequently the lower the cost of stoppages, and vice versa. There is thus the possibility for a trade-off between these two costs. For a given number of machines and failure and service times it is possible to determine the optimal number of operators so as to minimize the overall total cost.

For simple failure and service time distributions, the repairperson problem can be formulated as a **queueing model** with a finite population of customers and N servers.

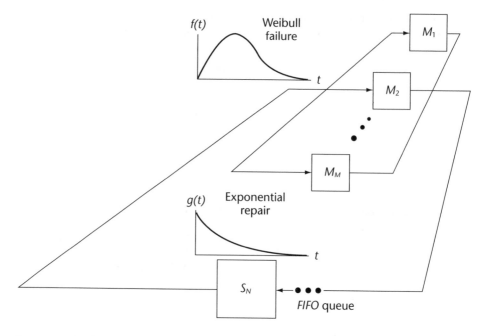

Figure: **An M-machine × N-server maintenance float system**

Situations with more complex and more realistic failure and service time distributions can be analysed by **simulation** models.

When it is crucial to keep critical facilities, such as air defence or civil defence systems at a high level of availability, then additional servers and/or additional machines may have to be used to improve service reliability. In other situations, major improvements in performance or reduction in total costs may only be achieved by reducing or eliminating the causes of failure. *N.C. Georgantzas*

References

Barlow, R.E. and Proschan, F. (1981) *Statistical Theory of Reliability and Life Testing*, New York: Holt, Rinehart and Winston.

Madu, C.N. and Georgantzas, N.C. (1988) 'Waiting line effects in analytical maintenance float policy', *Decision Sciences*, 19(3): 521–534.

Web sites

Engineering statistics handbook: http://www.itl.nist.gov/div898/handbook/index.htm
INFORMS Online: http://www.informs.org/Conf/WA96/TALKS/MD29.1.html

Replacement problem

The performance of many types of equipment (machines, vehicles, installation) tends to deteriorate with age. Ageing equipment becomes increasingly prone to breakdowns, maintenance, overhaul and repair costs increase, and the quality and volume of output tend to decrease. As a result, total operating costs increase, while the potential for profit contribution decreases. Furthermore, the resale or salvage value of the equipment also drops as the equipment gets older, requiring a larger net outlay when it is replaced. At some point in time it becomes more cost-effective or more profitable to replace the old equipment with new similar equipment or more modern technology. The replacement problem finds the optimal time to replace ageing equipment so as to maximize the long-run present value of net profits generated by the equipment, given by revenues minus costs, all expressed in terms of a given base year (thereby excluding the effect of inflation).

The procedure consists of two separate phases. The first determines the optimal replacement age of the proposed new equipment (usually with another unit having the same operating, cost and profit contribution characteristics, i.e. under assumptions of **stationarity**). This is done by computing the **net present value** of all cash flows associated with the purchase and operation of the equipment, including the salvage value at its disposal for each possible replacement age. Each of these net present values is then converted into its **equivalent annuity**. The best replacement age option is the one that maximizes the equivalent annuity of net profits.

The second phase determines the marginal discounted net profit for operating the old equipment for another period (including the loss in salvage value). If that value is lower than the equivalent annuity for the optimal replacement age found previously, the old equipment is replaced, otherwise it is kept for another period and the computations are repeated using the latest data both for the old and the proposed new equipment.

If revenues are the same regardless of the option chosen, the computations are based on discounted costs and salvage values alone and the optimal replacement age is the one that minimizes the equivalent annuity of costs.

In reality, neither the performance characteristics nor the costs for any new equipment will remain unchanged over time. The approach makes the simplifying assumption that even if changes occur in absolute terms, performance, costs and revenues will remain the same in relative terms. Owing to the difficulty of obtaining reliable data on future equipment and its performance, replacement models are no more than rough approximations. Often the decision to replace equipment is influenced by other than simply cost considerations, such as greater flexibility, reliability, output quality, public image, etc. *H.G. Daellenbach*

Reference

Daellenbach, H.G. (2001) *Systems Thinking and Decision Making*, Christchurch, REA, ch. 10.

Requisite variety

The law on requisite variety, put forward by one of the foremost **cybernetics** and systems theorists, W.R. Ashby, in the late 1950s, states the almost obvious, namely that if you want to control the behaviour of a **system** your control inputs must have a variety at least equal to the variety of the system. Simply expressed, variety in this context is the number of possible **states the system** can assume. Insufficient variety in the control can lead to system failure or undesirable consequences. For example, if a light signal on a rail line must indicate five different states the next rail segment can be in (such as 'line segment free', 'train ahead on line segment moving in same direction', 'train approaching in opposite direction', 'repair crew working on line', 'track damage ahead, slow down'), then the light signal must also be able to signal five messages. If the signal has two lights (red/green), it can only signal three messages (R,G,RG), assuming that both lights off is not acceptable since it cannot be distinguished from a power failure. A signal with three lights can give seven messages, hence has enough variety. *R. Flood*

Reference

Beer, S. (1985) *Diagnosing the System of Organisations*, Chichester: Wiley.

Rich pictures

Rich picture diagrams, or rich pictures for short, are cartoon-like representations of a **problem situation**. They were popularized by Peter Checkland as part of his **soft**

systems methodology. The representations used are very simple stick-like figures, clouds, blobs and boxes, some slogan-type writing, and arrows depicting connections or time sequences. The talent needed is not a good ability to draw, but a bit of imagination. Do not use computer clipart! The result is stilted and contrived. The rich picture below summarizes my dilemma of whether or not to go to work by bicycle.

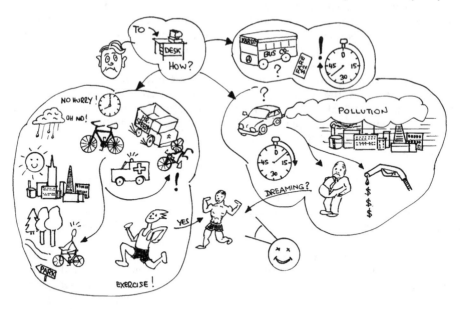

Figure: **Rich picture of my dilemma for going to work by bicycle**
(*Source*: H.G. Daellenbach (2001), p. 63)

Although the prime concern may only be with some particular aspect of a problem situation, it pays to assemble as wide a picture as is reasonably possible, i.e. show the problem in its full context, including other potentially related issues, even if they are not of immediate concern. However, you will still need to strike a sensible balance between the desire for completeness and for parsimony and find an appropriate level of resolution for depicting details. For instance, you may draw a book, inscribed 'rules', rather than showing these rules in full detail. It simply serves as a reminder. Slogans (coming out of some person's head) are often highly effective summaries of details.

The rich picture should include (1) elements of structure, i.e. things that are relatively stable, such as aspects of the organizational divisions, information systems, physical aspects, laws and regulations, policies, long-term goals, (2) elements of processes, i.e. the dynamic aspects, activities, information flows, how long-term goals translate into decisions, the decision rules used and where, and (3) relationships between structure and processes or between processes, i.e. how structure constrains or facilitates activities, how activities affect each other. It should not only include the 'hard' facts, but also 'soft' facts, the organizational climate, such as critical aspects of the informal organization, sources of tension, conflicts or disputes, differing **world views**. Unless the climate is understood, essential aspects of these world views may be missed.

You need to remind yourself that a rich picture is not a **system** description. The term system implies that any interconnectedness is organized and not coincidental. By assuming such organized interconnections you may impose a structure on the situation which may not be present or, if present, focuses your attention in a given direction, rather than encouraging you to keep a completely open mind.

Rich pictures, like **mind maps**, are ideal for communicating between **stakeholders** and analysts, for eliciting and clarifying critical aspects, for identifying data sources, highlighting conflicts or opportunities, and fostering a shared understanding of the problem situation. *H.G. Daellenbach*

References

Daellenbach, H.G. (2001) *Systems Thinking and Decision Making*, Christchurch: REA, pp. 61–68.
Naughton, J. (1984) *Soft Systems Analysis: An Introductory Guide*, Technology Course T301, Block 4, Milton Keynes: Open University Press. Has a workbook with many examples.

Risk analysis

Risk analysis, developed by D.B. Hertz around 1959, is an application of **Monte-Carlo simulation**, i.e. sampling from a complex probabilistic phenomenon, to evaluate multistage or sequential decision problems under **uncertainty**. Investment decisions, such as research and development projects, launch of new products, mergers and take-overs, where initial outlays, operating costs, and returns over time are dependent **random variables**, conditional upon exogenous events, are prime examples. Evaluation of public emergency and disaster policies is another important application.

Conventional **cost–benefit analysis** and investment evaluation substitute expected values or conservative best estimates of the various random cash flows associated with a project. Evaluation of several **scenarios** may be used to capture some aspects of the interdependencies of future events. The answer is always a single number for each scenario – a cost/benefit ratio or a **net present value**, that gives no indication as to its variability and the degree of risk involved in the project. Risk analysis remedies this weakness by producing probability distributions of selected outcomes.

The process is as follows: The **planning horizon** is divided into discrete periods, e.g. years. Each random aspect or event in each period is described by a probability distribution, which may be dependent on other random events. These distributions may be continuous or discrete. They are based on past experience or reflect the subjective best guesses of the decision maker(s) and experts. Often they may be rather crude discrete approximations, such as a high, medium and low value for a given aspect (e.g. the market share captured by the product). A possible sequence of (conditional) events is simulated by generating an observation from each random event in the correct conditional logical and temporal order. This is done with the help of **random numbers** or random variates. For investment projects, the resulting cash flows are usually expressed as present values. Each such simulation yields one possible outcome or sample point for the project for each performance measure of interest, e.g. a net present value of investment. Using different random numbers, this process is repeated hundreds or thousands of times and the sample points collected are summarized in the form of frequency histograms and percentile profiles, which reveal the variability and hence the risks involved.

Several interactive software packages and add-ons to **spreadsheets** facilitate and speed up risk analysis, such as @RISK, or *Insight.xla*. They give access to built-in functions for generating random variates on most theoretical probability distributions and user-specified empirical distributions and correlations. They allow input of conditional sequences of events diagrammatically in the form of event trees or **influence diagrams**, have facilities for easy **sensitivity analysis** (crucial for risk analysis, given the subjectivity of many inputs) and output generators. *H.G. Daellenbach*

References

Evans, J.R. and Olson, D.L. (1998) *Introduction to Simulation and Risk Analysis*, Upper Saddle River, NJ: Prentice-Hall.

Savage, S.L. (1998) *Insight.xla – Business Analysis Tools for Microsoft Excel*, Belmont, CA: Druxbury.
Vose, D. (1996) *Quantitative risk analysis: A guide to Monte Carlo simulation modelling*, Chichester: Wiley.

Robust, model and solution robustness

Robustness in MS/OR refers to how well a model or a solution to a model copes with changes in its environment. The better it copes, the more robust it is.

A solution to a model is robust if it is relatively insensitive to reasonable changes or fluctuations in crucial input data, particularly cost factors and efficiency coefficients. Insensitive in this respect means minor shifts or changes in inputs do not invalidate the solution, which still remains near-optimal. Furthermore, the loss in benefits incurred for not altering the solution to reflect the changes in inputs is relatively modest. For example, the well-known **economic order quantity (EOQ) model** yields robust solutions. An increase in demand or an increase in setup costs of 50 per cent leads to a loss in true benefits of only 2.1 per cent if the old solution continues to be used. Similarly, if the optimal initial decision in a multistage decision problem is not affected by reasonable changes in the probabilities or benefits of future events, it is a robust decision. (See **sensitivity analysis**.)

The solutions of some models lack robustness, in the sense that a minor change causes a drastic change, at least within certain ranges of inputs. The examples are some **queueing models**, where the average waiting time per arrival increases exponentially in the ratio of average interarrival time over average service time. Hence, as that ratio goes beyond about 0.6, the waiting time starts to increase dramatically.

A model is said to be robust if it is flexible. Flexibility here means that it gives valid answers under varied and changing conditions, i.e. the form of its optimal policies does not change (although the optimal value of decision variables may change) and the form of the model itself is not invalidated, or it can be easily adapted to minor structural changes in the environment or the transformation process or activity it models. Again, the EOQ model copes with even drastic changes in the environment, such as a ten-fold increase in demand or a cost factor, the introduction of quantity discounts, or restriction of the EOQ to multiples of some basic unit (e.g. a pallet).

Solution and model robustness are highly desirable properties and increase the confidence that the decision maker can have in the results. Whenever possible, analysts should try to built robustness into the decision process or the model from the start.

Robustness analysis

Robustness analysis is a way of supporting decision making when there is radical **uncertainty** about the future. It addresses the seeming paradox: how can we be rational in taking decisions today if the most important fact that we know about future conditions is that they are unknowable? It resolves the paradox by assessing initial decisions in terms of the attractive future options which they keep open. The approach operates broadly as follows:

1. Identify all alternative initial commitments to be considered.
2. Generate a representative set of possible future conditions that could eventuate in the longer term (5 to 20 years), e.g. through the use of **scenario analysis**.
3. Construct a set of relevant possible future configurations of the system, e.g. by designing systems to fit the different futures in (2).
4. Establish the compatibility of each commitment–configuration pair (= 1 if the configuration is a plausible step towards that configuration, and 0 otherwise).
5. Evaluate the performance of each configuration in each future as either acceptable or not.

Using this information, it is then possible to calculate how many acceptable future options (configurations) are kept open by a particular initial commitment, in a given 'future'. The ratio of this number to the total number of acceptable options in that future is called the *robustness* of that initial decision. This ratio evidently must be between zero ('this commitment is compatible with no acceptable future options') and one ('this commitment is compatible with all the acceptable future options'). The higher the robustness, the more useful flexibility is kept open by that initial commitment under that future. Since keeping future options open is attractive, an initial commitment with high robustness scores across the range of futures is preferred.

The approach, it can be seen, is both sequential and bifocal. It calls for decision making to be carried out not through the construction of master plans for the next *n* years, but step by step. And the criterion it offers is based on the principle of bifocal lenses, combining a focus on the (near) first step with a look ahead at its useful compatibility at the (far) **planning horizon**.

Robustness contrasts sharply with other strategic planning approaches which attempt to tame future uncertainty with elaborate forecasts and then designs an optimal configuration for that imagined future. In contrast, robustness analysis can be carried out either as a backroom analytic activity or, thanks to its low-tech nature, accessibility and openness to subjective inputs, also in a participative workshop environment with those responsible for taking decisions.

Robustness analysis has been applied to decisions about the locations of breweries (United States), the expansion of a chemical plant (Israel), the siting of hospitals (Australia), regional planning of health services (Canada), and for the educational choices of teenagers (UK). *J. Rosenhead*

References
Rosenhead, J. (2001) 'Robustness analysis: keeping your options open', in J. Rosenhead and J. Mingers (eds) *Rational Analysis for a Problematic World Revisited*, Chichester: Wiley.
Best, G., Parston, G. and Rosenhead, J. (1986) 'Robustness in practice: the regional planning of health services', *JORS*, 37: 463–478.

Rolling planning horizon, see planning horizon

Satisficing

Satisficing is a word coined in the late 1940s by the 1978 Nobel-prize-winning economist, Herbert Simon. It means searching for a good or satisfactory solution that meets the most important objective to an acceptable degree rather than trying to find the **optimal solution**. He put forward the principle of *bounded rationality* as an alternative to *economic rationality*, i.e. the human mind's inability to deal simultaneously with many aspects and the brain's limited computational power. Experiments in the 1960s by the psychologist G.A. Miller confirm that most people can only cope with about five to nine different uni-dimensional pieces of information at the same time (referred to as 'the magical number 7 plus or minus 2'). Once this threshold has been exceeded, people tend to disregard certain aspects and reformulate information into bigger and less detailed chunks.

Bounded rationality means that in complex situations the search for and evaluation of decision choices is restricted to only a few. Hence, faced with complexity, human behaviour is prevented from being perfectly rational in an economic sense. It leads to satisficing behaviour. The search behaviour is also affected by the **feedback loop** formed between intensity of search, satisfaction level and aspiration level (the greater the difference between current satisfaction and aspiration, the more intense the search; the easier it is to find good solutions, the higher the aspiration level, etc.).

Satisficing is often used in MS/OR as the only feasible strategy in the face of high complexity, particularly **computational complexity**, i.e. when the number of possible choices is astronomical and/or each evaluation is computationally expensive, or the space of **feasible solutions** is not well behaved in the sense that it contains many **local optima**. In such situations, the analyst may be forced (and even be content) to apply satisficing solution methods, usually based on **heuristics**. These are often efficient and fast in terms of their computational requirements and tend to give good, sometimes even close-to-optimal solutions.

Reference
Miller, G.A. (1963) 'The magical number seven, plus or minus two: Some limits on our capacity for processing information', *Psychological Review*, 63(2):81–97.

Scenario development and planning

Creating a successful organization is as much about understanding the environment as it is about understanding the organization. The dilemma facing decision makers is that the further one looks into the future, the fewer the predictable elements and the greater the **uncertainty** in the environment. Scenario development is one tool for coping with this uncertainty.

A scenario is a story about the future of the social, political, economic and technological environment. It is based on assumptions about underlying drivers of change. Although the number of stories that could be written about the future is infinite, more than three or four scenarios becomes too unwieldy as an aid to strategic thinking. They should cover the range of significantly different possibilities, while still being plausible.

When developing scenarios, it is most important to externalize all assumptions that underlie each scenario. Furthermore, scenarios should not include plans and the market's response to plans.

There is no one best way to develop scenarios. A general outline is as follows:

- Define the issues or variables to be explored in terms of their time frame.
- Identify the **stakeholders**, the power of those that are involved and the effects on those that are affected.
- Identify the basic trends and inherent uncertainties that will affect the variables of interest. The environment should not be seen as a cluster of independent variables, but rather as a network of interdependent relationships.
- Start out with two scenarios: one for a pessimistic future and one for an optimistic future. Making contrasting assumptions about trends and uncertainties will generate alternative scenarios.
- Assess all assumptions underlying each scenario. Check the internal consistency and plausibility of both the assumptions and the scenarios and eliminate aspects that are not credible or even impossible.
- Explore how the key stakeholders would behave in each. This may lead to further adjustments.

It may be useful to explore quantitative scenarios via graphs or quantitative models, such as **system dynamics**.

A set of scenarios provides a backdrop against which an organization can review its strategies:

- Do the strategies enable the organization to do well whichever future it finds itself in, i.e. are they **robust**?
- Is the organization prepared to take the risks associated with pursuing a strategy that is not robust, but fits extremely well for the future thought most likely?
- Which is the most desirable future and what can the organization do to help bring it about?
- Which is the most pessimistic future and how can the organization prevent it from happening?
- What environmental monitoring will help to keep track of the changes in the environment?
- Do these scenarios suggest any completely new strategies for the organization?

As well as assisting strategy evaluation, the process of scenario development has other benefits. During a scenario process participants share their insights and assumptions about the future. Mental models of how the organization's environment works are shared and enhanced. Lateral thinking is encouraged. A scenario process allows the future of the environment to be explored, while at the same time removing the stress associated with attempts to predict the unpredictable.

D. Campbell-Hunt

References

Fahey, L. and Randall, R. (1998) *Learning from the Future*, Chichester: Wiley.
Van der Heijden, K. (1996) *Scenarios – the Art of Strategic Conversation*, Chichester: Wiley.

Scheduling, machine scheduling, job scheduling

The term *scheduling* is used in many areas of MS/OR to mean different things, such as production scheduling, project scheduling, network scheduling, and resource allocation and scheduling. Here the term refers to job and machine scheduling, where machine capacities are the resources to be allocated to tasks or jobs. Each machine is assumed to be able to perform only one task or activity at a time, and a job consists of one or more tasks or activities, referred to as *operations*. A schedule is an allocation of various time intervals on one or more machines to each job, so that for each job the operations are executed in the correct order and all jobs are completed.

A machine environment is typically specified by the number $m \geq 1$ of machines and the type of *machine shop*:

- *Parallel shop:* Each job consists of a single operation that can be processed on any one of m machines. These machines may be identical or different in capacity or speed.
- *Open shop:* Each job consists of a set of n operations and each operation requires a given processing time on a given machine. The order in which the operations are executed is part of the scheduling decision.
- *Flow shop:* All jobs follow the same sequence of operations that have to be executed on different machines in a fixed order.
- *Job shop:* Each job has its own ordered chain of operations that have to be executed on different machines in its own fixed order.

Typical job characteristics are as follows:

- Whether a job may have processing priority that allows it to jump the queue or even pre-empt another job currently being processed on a given machine, i.e. the current job's processing is interrupted and resumed again later.
- Whether there is a precedence relation between jobs, which specifies that such jobs must be processed in a given sequence, i.e. the predecessor must be completed before the successor can start.
- Whether some or all jobs have release dates before which a job cannot be started.
- Whether some or all jobs have due dates, d_j, when all operations on a job need to be completed.

Common performance criteria used are: Minimize

- *makespan*, i.e. the time interval to complete all jobs;
- the sum of lateness or tardiness of all jobs, where tardiness = $\max[0, C_j - d_j]$ and C_j is the job completion time;
- maximum tardiness of any job;
- the number of jobs completed late;
- the sum of all penalties associated with late jobs;

In most instances, the size of the problem and its dynamic nature, with new jobs coming in daily, preclude the use of optimization techniques. Heuristic methods and simulation of various processing rules for each operation or job are the most successful approaches for finding good schedules. Some of the common rules used are: first-come/first-served; shortest operation processing time first; earliest due date d_j first; or lowest amount of slack first, where slack is the difference between the due date d_j for a job and the current date, less total remaining processing time, and others. *B. Chen*

References

Baker, K.R. (1995) *Elements of Sequencing and Scheduling*, Hanover: Baker Press.

Chen, B., Potts, C.N. and Woeginger, G.J. (1998) 'A review of machine scheduling: Complexity, algorithms and approximability', in D.Z. Du and P. Pardalos (eds), *Handbook of Combinatorial Optimization* (Vol. 3), Dordrecht: Kluwer, pp. 21–169.

Graves, S.C., Rinnooy Kan, A.H.G. and Zipkin, P.H. (eds) (1993) *Handbooks in Operations Research and Management Science (vol. 4): Logistics of Production and Inventory*, Amsterdam: North-Holland.

Scheduling, see applications of MS/OR to crew scheduling

Scientific method and the MS/OR process

It is often claimed that MS/OR uses the scientific method. It is therefore interesting to compare the two critically. The scientific method originated in the natural sciences. It consists of five major steps:

- Based on observations, logical reasoning, and analogies (or plain hunches), the scientist forms hypotheses about a phenomenon, such as 'X is true'.
- The scientist then designs experiments to observe certain things under strictly controlled conditions. This design spells out in detail what has to be done and how, the sampling plan, and so on.
- The experiment is conducted according to the design and the results observed and recorded.
- The results are analysed, which may involve extensive **statistical analysis**. This will either confirm or refute the hypotheses or lead to an inconclusive answer.
- Based on this, the scientist may form theories about the phenomenon.

The process is such that any other scientist can replicate it under the same conditions. Its aim is to advance scientific knowledge. The scientist supposedly remains at all times an 'objective' outside observer who cannot affect the results.

Although there may be some weak parallels, the general process of MS/OR is a far cry from this. First, the aim of MS/OR practice is normally not to advance knowledge, except maybe in the case of **action research**. It is intervention for problem solving, aimed at achieving some objective, often of a pecuniary nature, or improving the **effectiveness** or **efficiency** of something. Each project is executed in the real, often turbulent world and is relevant for specific conditions at a given point in time. Rarely will it be possible to replicate it under the same conditions.

However, most importantly, the MS/OR practitioner and the **stakeholders** of the problem are not objective, but they all bring their own **world views**, values and goals to the project. The choices as to which particular issue to investigate, the selection of the **boundary** of the **system** and what is controllable and what is not, the level of resolution for modelling the system, the solution method, all involve a large dose of **subjective** judgement and rightly so. Finally, by the very fact that the analyst intervenes, he or she is not an objective outsider, but an active participant who may even compete for funds from the same pool as the other stakeholders.

So claiming that MS/OR uses the scientific method is misleading. All we can do is to make sure that we follow a scientific spirit of conduct, such that what we do, why we do it and how is transparent, that we state all assumptions made, **validate** and **verify** our models, disclose their weaknesses as well as strengths, and provide complete documentation.   *H.G. Daellenbach*

References

Daellenbach, H.G. (2001) *Systems and Decision Making*, Christchurch: REA, ch. 5 and section 12.2.

Midgley, G. (2000) *System Intervention: Philosophy, Methodology, and Practice*, Dordrecht: Kluwer Academic, pp. 179–186.

Search heuristics

Search heuristics incrementally change an initial solution or set of solutions to a problem with the aim of finding a good quality solution. In this respect search heuristics are similar to solution improvement heuristics. The latter start from an initial solution, examine a set of changes to that solution, move to the best solution found and repeat the process until no further improvement is found. They have a major problem in that they may get trapped in a **local optimum**. Search heuristics avoid this by allowing non-improving moves which enable the search to move out of a local optimum, thereby increasing the likelihood that the heuristic will find a **global optimum**. Search heuristics only stop searching when a particular *stopping criterion* is met, such as a set amount of computer time has elapsed or a certain number of iterations have been executed. Three popular search heuristics are **genetic algorithms**, **simulated annealing** and **tabu search**.

An essential element of any search heuristic is how the search space is defined. For example, for a **knapsack problem**, i.e. the problem of deciding which items should be packed into a knapsack from a set of n possible items, the search space could be defined as being any combination of items. Each combination can be represented by a vector or list of n elements. A zero in the ith position indicates item i is not taken, a one that it is taken. Some possible combinations of ones and zeros do not yield feasible solutions because they exceed the space or weight limit of the knapsack. A search heuristic could either avoid producing infeasible candidates or penalize them in the objective function to discourage their construction. Alternatively, the knapsack problem search space could be redefined so that infeasible candidates never get generated. For example, in the knapsack problem all items could be ranked in order of priority. When the search considers a particular element of this search space (i.e. a ranked order of items) the ranking is converted into a solution to the problem by filling the knapsack in the order of each item's rank until the knapsack is full. No candidate can ever produce an infeasible solution, but different rankings may end up packing the same items.

Related to the search space definition is the method for moving from one point in the search space to another, often called the *neighbourhood scheme*. This can involve manipulating one or more points to produce one or more new points. Using the knapsack example above, changing a single element in the 0/1 list from 0 to 1 or vice versa or interchanging the rank order of two items are moves to a neighbourhood point.

R. James

Reference

Reeves, C.R. (ed.) (1995) *Modern Heuristic Techniques for Combinatorial Problems*, London: McGraw-Hill.

Search theory

Search theory is the set of mathematical techniques, probability theory, and communication and positioning technology used for locating an object in a defined space. For example, it could be a boat missing at sea, an aircraft or person lost in a wilderness area, or a boat or aircraft sunk in the sea. The object could be stationary or moving. It could even try to evade being located, such as an enemy submarine. In fact, search theory has its beginning in World War II for locating submarines and is still extensively used for military purposes. In recent years, computer, communication and satellite positioning technology have greatly increased search capability and search theory is now used for search and rescue missions and commercial applications, such as exploration for mineral resources, search in commercial fisheries, and salvage projects, as well as medicine.

In the simplest case for a stationary object, the optimal search procedure is based on **Bayesian decision analysis** concepts: The search area is divided into cells – not an easy problem in itself. A probability is assigned to each cell that the object is there. These probabilities may be based on prior information about the likely location and previous movement of the object. They could be subjective educated guesses. The search starts with the cell that has the highest probability. If the object is not found, the prior probabilities of the cells are updated using *Bayes' Theorem*, and the next cell searched is again the one that has the highest updated probability. In essence, the search strategy is constantly revised on the basis of the search results.

Search of each cell itself poses serious problems, since the searcher passing over or through the cell has only a narrow sweep width. In rugged territory, the sweep bands may not be straight and hence miss out parts of the cell. Similarly, a high cost or valuable search time lost for moving from one cell to another may work against the optimal sequence.

For evading objects or targets, **game theoretic** approaches may be useful. In complex searches, such as when the object moves and the search area changes continuously (e.g. sea surface due to current and wind action), or where there is great uncertainty about the object's movement prior to its being reported as lost, computer **simulation** techniques are used to build up a probability distribution for its location.

Some major successful applications of search theory include location of gold bullion ships, the lost submarine *USS Scorpion*, and US Coast Guard search and rescue missions.   *J. Strümpfer and H.G. Daellenbach*

References

Washburn, A.R. (1981) *Search and Detection*, Operations Research Society of America.

Washburn, A.R. (1987) 'Search theory', in L.C. Thomas (ed.) *Golden Developments in Operational Research*, Oxford: Pergamon Press.

Secretary problem

The secretary problem is concerned with sequential choice. Assume a decision maker is presented with a sequence of decision alternatives, one at a time, without any knowledge of those which are still to follow. After inspecting an alternative, he or she may choose or reject it. Rejected alternatives become unavailable – the decision maker cannot return to them again. Once a choice has been made, the process stops – no further alternatives will be presented. The decision maker's objective is to choose the best of all alternatives available for presentation.

Because the sequence in which the alternatives are presented is assumed to be random, it is not possible to devise a strategy which guarantees that the optimal alternative will be selected. Therefore, the best strategy has the objective of maximizing the probability that the best alternative will be chosen. An argument from probability shows that the best rule is to inspect about 37 per cent of the items, and then select the first item which is better than any inspected so far. The probability of successfully choosing the best is also approximately 37 per cent.

There are various modifications to the basic problem. For example, if one knows the probability distribution of the quality of the items, and wishes to maximize the expected quality, then the problem becomes one of **dynamic programming**. One may not know the number of items in advance. Or one may be content with choosing one of the best two or three items. Biologists have observed that certain birds select their mates by rules which approximate to the behaviour assumed in this problem.

The name of the problem comes from the parallel with appointing a secretary, under conditions where nothing is known about the capability of any applicant until the interview, and where a decision must be made at once. In a fast-moving housing market, where offers may remain open only for a short period, a house buyer may be faced with a situation that reflects similar characteristics.   *D.K. Smith*

Reference

Ferguson, T.S. (1989) 'Who solved the secretary problem?', *Statistical Science*, 4(3):282–296.

Self-regulation of natural systems, see feedback loops

Sensitivity analysis

Sensitivity analysis is the systematic exploration of how the optimal solution responds to changes in its inputs. For instance, how do the optimal value of the objective function and the decision variables of a **linear program** change in response to a change in the right-hand side of a constraint? Such analysis is usually performed separately for each input, e.g. cost of activities, demand levels to be satisfied, resource constraints, assuming that all other inputs remain unchanged. Sensitivity analysis is particularly crucial for (often) arbitrary policy constraints, such as quality standard or budget allocations, that are of a 'soft' nature. The decision makers should know the 'cost' of such policy choices. Sensitivity analysis for resource constraints is equivalent to finding the **shadow price** of the resource, i.e. the rate of change in the optimal value of the objective function for a unit change in the resource availability. Linear programming software automatically produces extensive sensitivity analysis with respect to the right-hand side of the constraints and the objective function coefficients.

Sensitivity analysis, in this sense, compares optimal solutions for different inputs. There is another form of sensitivity analysis, called *error sensitivity analysis*. It is in the form of hypothetical 'what-if' questions: 'what is the loss in true benefits or increase in true cost if a solution is used which is based on an input that is in error by x per cent?' That solution is obviously only pseudo-optimal. Error analysis compares the true benefit or cost for the pseudo-optimal values of the decision variables, based on the 'correct' inputs, with the optimal value of the objective function for the correct inputs. For example, for the **economic order quantity (EOQ) model**, error sensitivity analysis gives the following surprisingly simple expression for the percentage error:

$$\left[\frac{k+1}{2\sqrt{k}} - 1 \right] \times 100\%$$

where k is the error factor, i.e. [input value used] $= k \times$ [correct input value]. For instance, if the true demand is 2 760 and the incorrect value used is 50 per cent higher or 4 140, then $k = 1.5$. The formula then states that the increase in cost is 2.1 per cent. The figure below shows the percentage increase in true costs of the EOQ model as a function of the error factor.

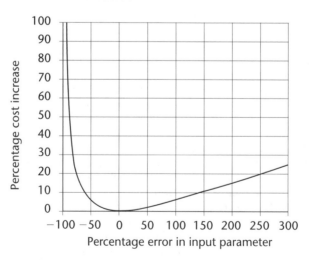

Figure: **Error sensitivity of EOQ model**
(*Source*: H.G. Daellenbach, 2001, p. 192)

If a solution or a model is relatively insensitive to changes or errors in the crucial input factors, it is termed **robust** – a highly desirable property. Sensitivity analysis is also an important input into evaluating the **validity** of a model and for determining how accurate the inputs need to be to capture most of the benefits and for setting critical ranges on the inputs that should trigger an update of the solution using the new updated input values.   *H.G. Daellenbach*

Reference
Daellenbach, H.G. (2001) *Systems Thinking and Decision Making*, Christchurch: REA, pp. 186–194.

Sequencing, Johnson's algorithm

Sequencing deals with **scheduling** jobs for processing through machines. One of the simplest situations is where m jobs have to be processed first by machine A and then by machine B and the objective is to minimize *makespan*, i.e. the time between starting the first job on machine A to the time the last job has finished on machine B. In 1954, S.M. Johnson published the following surprisingly simple **algorithm** which will find the optimum sequence:

Step 0: Set $k = 1$ and $j = 0$. All jobs form the initial list of unprocessed jobs.
Step 1: In the list of unprocessed jobs, find the job with the smallest processing time on either machine. Break ties arbitrarily.
Step 2: If that smallest time is on machine A, schedule it as the kth job and increase k to $k+1$. If the smallest time is on machine B, schedule it as the $(m-j)$th job and increase j to $j+1$.
Step 3: Remove that job from the list. If the list is empty, stop; otherwise return to Step 1.

The table below shows an example for seven jobs. Job 5 has the lowest time of 20. Since it is on machine A, $k = 1$, the first job to be processed, and k is increased to 2. Job 7 with 25 has the smallest time of those remaining on the list. Since it is on machine B it is scheduled last, and j is increased to 1, and so on.

Machine	Job	1	2	3	4	5	6	7
A	Process	30	50	60	60	20	120	70
B	time	60	60	50	25	40	100	80
Sequence		2	3	6	7	1	5	4

The optimal processing sequence of jobs on both machines is 5–1–2–7–6–3–4. The total elapse time to complete all jobs on both machine is 465 minutes.

Unfortunately, the algorithm cannot be extended to three or more machines (except for a trivial case) and still find the optimal makespan sequence. In fact, scheduling jobs through machines, particularly if each job may have a different order of machines, is computationally one of the most difficult problems in MS/OR. Usually **heuristic methods** are used to get good schedules. (See also **scheduling**.)
                                                                H.G. Daellenbach

Reference
Johnson, S.M. (1954) 'Optimal two- and three-stage production schedule with setup times included', *Naval Logistics Quarterly*, March.

Seven tools of quality management

Since the early 1960s, seven tools have formed the basis for solving quality problems. The strength of the tools comes from their use within a systematic method, where the output from one tool is used as input for another, and so together become a key part of a robust problem-solving process. The tools are:

1. *Pareto charts:* They separate the 'vital few' causes from the 'useful many' and provide a visual representation of the rank order of causes contributing to a quality problem. Their interpretation is based on the 'Pareto principle' or 80–20 Rule where 80 per cent of all problems are attributable to 20 per cent of the causes (the vital few).
2. *Cause-and-effect diagrams* (also called *Ishikawa* or *fishbone diagrams*): They classify potential causes of a particular effect or problem. Branches (bones) off a central spine are labelled for major categories of factors/causes usually grouped as: methods, materials, measurements, machines, people and environment. Note that 'people' refers to characteristics of groups, not to individuals. Figure 1 gives an example.

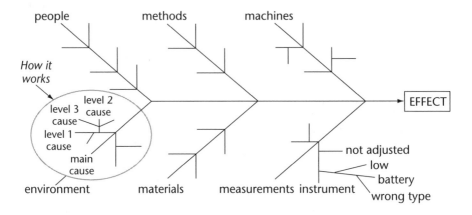

Figure 1: **Ishikawa or fishbone diagram**

3. *Check sheets:* a simple method for collecting and recording data in a clear, easily filled in, easily read form. Check sheets are used to answer the question 'How many times are certain events happening?'
4. *Histograms:* graphs of frequency distributions of numeric data, revealing a distribution's skewness, i.e. the possible lack of symmetry of the length and shape of the tails, the range or spread of the data, and the central tendency of the data, such as the mode, 50 percentile, or the average.
5. *Scatter plots:* They show if a plot of paired measurements on two variables reveals any visual relationship between them. The shape of the scatter may point to the existence of a positive or negative correlation, or a lack of correlation between the variables, as demonstrated in Figure 2.
6. *Graphs and control charts:* These labels cover a variety of tools, including run charts (or time plots) used to display data in the sequence in which they are collected over time. Control charts are used to plot the average and range of sample data over time to determine if the process is under control (see **statistical quality control**).
7. *Stratification:* It divides data into meaningful subsets or strata. Often stratification is replaced by another tool such as flowcharts or *brainstorming*. Brainstorming is a team-based method for generating ideas about characteristics of a problem and potential causes as the starting point for many other basic tools. Flowcharting maps a process flow and provides a common reference for teams when exploring a process. *D.J. Houston*

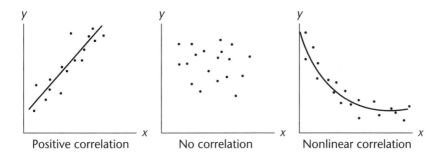

Figure 2: **Scatter plots**

References
Brassard, M. and Ritter, D. (1994) *The Memory Jogger II*, Methuenm MA: *GOAL/QPC*.
Ishikawa, K. (1982) *Guide to Quality Control*, Tokyo: Asian Productivity Organization.

Shadow prices, Lagrange multipliers, dual prices

Shadow prices are associated with constrained **optimization**. They measure the rate of change in the optimal value of the objective function as the amount of a resource is changed by one unit. For example, assume that a product, produced in batches, needs to be stored under refrigeration. The optimal unconstrained production batch which minimizes the sum of the annual production setup and stock holding costs, derived from the **economic order quantity model**, is 20 m³. The current refrigeration space available – our resource – restricts the batch to at most 8 m³. If the space available is increased marginally by a small amount (i.e. the resource constraint is relaxed), the total annual cost decreases at a rate of about $410/m³. This is the shadow price of the resource constraint 'refrigeration space' at the current level of 8 m³. The figure below shows how this shadow price decreases as the volume of refrigeration space increases. Once the resource increases beyond 20 m³, the shadow price drops to zero.

The shadow price measures the value of acquiring additional resources. Three points are important for the correct interpretation of this concept. First, the shadow price is a *rate of change* valid for a particular amount of the resource available. As that amount changes, so may the shadow price. Second, it refers to the rate of change in the objective function at the **optimal solution** for a given resource availability, not simply any arbitrary **feasible solution**. Third, once the resource constraint does not restrict the optimal solution anymore, the shadow price drops to zero.

In the above example the shadow price – recall it is a rate of change – changes continuously. The reason for this is that the objective function is nonlinear. Hence, increasing the space from 8 to 9 m³ does not decrease the minimum cost by $410, since at 8.1 m³ the shadow has already fallen to $398, and around 9 m³ it is down to $306. The actual cost decrease is only $356.

For **nonlinear programming problems**, the shadow price is also referred to as a *Lagrange multiplier*. It serves as a penalty to prevent the optimal solution from violating a constraint.

The concept of shadow price can be extended to any type of constraint, not just resource constraints. It is particularly important in **linear programming**, where it indicates how the optimal value of the objective function responds to changes in the right-hand side of a given constraint under the condition that all other inputs remain unchanged. As a constraint is relaxed, the shadow price decreases in discrete steps (i.e. at values of the right-hand side that result in a change of basis for the **basic**

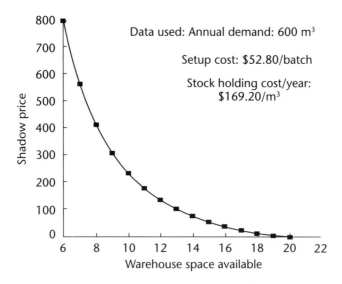

Figure: **Shadow price for resource constraint**
(*Source*: H.G. Daellenbach, 2001, p. 51)

feasible solution), but remains constant in between such changes. This is due to all relationships being linear. To draw the correct inference, it is important to verify the range for which a given shadow price is valid.

The shadow prices are equal to the optimal values of the dual variables, and therefore are also called the *dual prices*. H.G. *Daellenbach*

References
Daellenbach, H.G. and McNickle, D.C. (2001) *Systems Thinking and OR/MS Methods*, Christchurch: REA, pp. 49–55.
Daellenbach, H.G., George, J.A. and McNickle, D.C. (1983) *Introduction to Operations Research Techniques*, 2nd edn, Boston, MA: Allyn & Bacon, ch. 19. Good coverage of shadow prices in general and Lagrange multipliers.
Hillier, F.S. and Lieberman, G. (2001) *Introduction to Operations Research*, 7th edn, New York: McGraw-Hill, ch. 6. Covers dual prices in LP.

Shewhart's PDCA cycle, see **statistical quality control**

Silver–Meal heuristic

The Silver–Meal heuristic finds a close-to-optimal, if not the optimal sequence of production batches for the case where the known demand varies in a deterministic pattern from period to period. As for the **dynamic economic order quantity model**, each batch covers the demand for its own period and possibly one or more consecutive periods. The objective is to minimize the average cost per period which consists of setup costs and holding costs on stock carried forward to later periods. Using **incremental analysis**, the **algorithm** compares the average cost per period for successively longer and longer trial intervals. It continues adding additional periods as long as the average cost decreases and stops when it starts to increase.

The table below shows the calculations for four weeks. The first trial is a batch that covers the demand of 800 for week 1 only. Since all units produced are used up in week 1, the only cost incurred is the setup cost of $960. This is also the total cost as well as the average cost for this first trial replenishment. The second trial covers the demand for weeks 1 and 2. This adds the cost of carrying 220 units to period 2, at $0.40/unit per period, resulting in a total cost of $1048, or an average per period of $524. Since the average decreases, we continue. The third trial also covers period 3, adding the cost of carrying the week 3 demand for two periods. The average decreases again. The average increases at the fourth trial. Hence trial 3 has the lowest average, calling for a batch of size 1 470.

Week	Demand	Trial batch	Incremental cost ($)	Total cost ($)	Average cost ($)	Best batch
1	800	800	960	960	960	1470
2	220	1020	88	1048	524	
3	450	1470	360	1408	469	lowest
4	1200	2670	1440	2848	712	

The algorithm now starts again, with period 4 being the first period, and so on. However, we are really only interested in the first period batch, which is the only one that can be implemented right away. A new best batch is determined at the beginning of period 4 with updated demand data.   *H.G. Daellenbach*

Reference
Silver, E.A., Pyke, D. and Peterson, R. (1998) *Inventory Management and Production Planning and Scheduling*, New York: Wiley, ch. 6.

Simplex method, reduced costs

The simplex method, developed in 1947 by the US mathematician, G. Dantzig, is one of the methods for finding the **optimal solution** to a **linear programming** problem. The **algorithm** assumes that all constraints are *equalities*. This can easily be achieved by introducing **slack and surplus variables** to inequality constraints. Assume there are a total of n variables and m constraints.

The algorithm works with **basic feasible solutions**. A basic feasible solution has $n-m$ variables, called *nonbasic variables*, set equal to zero, allowing the remaining m variables, called *basic variables*, to assume positive values. Each basic solution corresponds to a corner point or *vertex* of the **feasible** region. Two vertices are *adjacent* if they differ by only one basic variable. This is depicted in the figure below which corresponds to the same problem as the one in the entry on **linear programming**. It has a total of $n = 6$ variables (2 decision variables and 4 slack variables) and $m = 4$ constraints.

There are four basic feasible solutions, i.e. A, B, C and D. At A, x_1 and x_2 are nonbasic at zero and the four slack variables are basic, while at vertex D, x_2 replaces the slack variable for the shelving constraint as the basic variable.

The simplex algorithm starts with an initial basic feasible solution, such as the origin, vertex A. It then explores if moving to an adjacent vertex (i.e. replacing a basic variable by a nonbasic variable) will improve the value of the objective function. In the graph this is either vertex B or D. This is where the *reduced costs* come into play. The reduced costs of basic variables are always equal to zero, while those of nonbasic variables measure the change in the objective function if the variable is set equal to 1. The algorithm selects as the next candidate to become a basic variable the one that gives the largest improvement in the objective function per unit increase of the variable. In our example, this is x_1. As x_1 increases from zero, some current basic variables will change in value, some increasing, others decreasing. The one that goes to zero first is replaced by the candidate. When x_1 reaches 27.6, the slack variable for the shelving constraint becomes 0 and is replaced by x_1 as the new basic variable. The new basic feasible solution corresponds to vertex D.

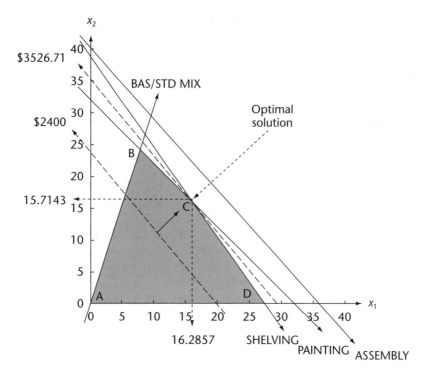

Figure: **Graphical representation of basic feasible solutions**

This iterative process of introducing a new nonbasic variable to replace basic variables continues until all reduced costs signal that there are no adjacent vertices with better values for the objective function. The algorithm has found the **optimal solution**. This occurs at vertex C in our example. (See also **Big-M method**.)

At the optimal solution, the reduced costs show by how much the objective function will deteriorate if the corresponding nonbasic variable is given a unit value. If a nonbasic variable has a reduced cost of zero, this means that there are **alternative optimal solutions**.

H.G. Daellenbach

Reference

Hillier, F.S. and Lieberman, G.J. (2001) *Introduction to Operations Research*, 7th edn, New York: McGraw-Hill, ch. 4.

Simulated annealing

Simulated annealing (SA) is a class of **search heuristic** originally proposed in the early 1980s. However, the mathematical models on which the search is based date back to 1953. The technique was inspired by the way atoms move in a metal as heat is applied and then cooled, and how different rates and durations of heating and cooling cause the metal to have different structural properties and energy levels when the metal comes back to its normal state. The laws of thermodynamics state that during this type of heating process the solid will go from a higher energy level to a lower energy level automatically, while it will go from a lower energy level to a higher energy level with a probability proportional to $e^{-\delta/kt}$, where δ is the change in energy level, t is the temperature of the system, $t \geq 0$, and k is a constant.

If this analogy is applied to optimization problems, the state of the metal becomes a point in the search space; the level of energy of the metal is the value of the objective function and the temperature is a control parameter that determines by how much the current point in the search space may deteriorate in a single move. An SA algorithm simply chooses a random neighbour from the current starting point and tests if the new solution makes an improvement on the current solution. If it does, the algorithm makes the move; if it does not, it will accept the move with a probability of $e^{-\delta/kt}$. The algorithm may then, depending on user-defined parameters and the problem formulation, reduce the temperature t and repeat the process. As long as the temperature is greater than zero, $e^{-\delta/kt} > 0$, and therefore there is a positive probability that the search will move away from its current point, even if all neighbours are worse, in contrast to *improvement heuristics* which would remain stuck there. Since the point reached may be a **local optimum**, this is an important feature that allows SA to continue searching for the **global optimum**.

The three key parameters that control an SA search are the initial temperature, the method used for decreasing the temperature and the stopping criterion for the search. These three elements combine into what is referred to as the *cooling schedule*. The quality of the solution obtained can be strongly dependent on the choice of these parameters.

One unique feature of SA is that it has been proven that it will always find the optimal solution to a problem – the only hitch is that it requires a very slow cooling schedule that would require more computational effort than completely enumerating every possible point in the search space.   R. James

References

Eglese, R.W. (1990) 'Simulated annealing: a tool for operational research', *EJOR*, 46:271–281.
Reeves, C.R. (ed.) (1995) *Modern Heuristic Techniques for Combinatorial Problems*, London: McGraw-Hill.

Simulation

Webster's Ninth Collegiate Dictionary defines simulation as 'the imitative representation of the functioning of one system or process by means of the functioning of another'. By this definition, a film is a simulation. In OR/MS, simulation is used to study complex commercial, industrial, technical or physical systems, operations or phenomena, such as the operation of an emergency clinic, a factory, an entire firm, or a multistage chemical process, with a view to understanding and controlling its **system**.

A computer model containing the logic of the system's operations is run to record in exact chronological order of *simulated time* (i.e. imitated time) each and every change in the **state of the system** that would have occurred had the actual system been operated in equivalent real time. Simulated time is incremented either in fixed discrete time units or more commonly by variable increments coinciding with occurrence of the next event. Each simulation run covers a sufficiently long time interval so as to observe a large number of each possible event type. For example, simulation of the minute-by-minute operation of an emergency clinic would cover several hundred weeks. Information and statistics are collected on various performance measures, such as averages and measures of variation of the waiting time of patients, and the idle time of doctors and nurses. Running the program for different policy choices or control inputs, e.g. different staffing levels, different priority processing rules, it is possible to determine safe staffing levels at various times of the day and days of the week.

If the behaviour of a system involves **stochastic** aspects, such as random exogenous inputs (e.g. arrival times of patients, type of injury, etc.) and random events inside the system (e.g. treatment times, occurrence of haemorrhages, etc.), they are generated using **random numbers** from which *random variates* mimicking observations from specified probability distribution are produced.

Simulation provides the operations researcher with a laboratory technique for systems experimentation similar to the experimental methods available in the natural sciences. There are several types of simulations: *Discrete event simulation* assumes that the state of the system changes by fixed amounts at arbitrary points in time (i.e. time is continuous) or at fixed points in time. For *continuous-state simulation* and **system dynamics** the state of the system is continuous and changes occur continuously at various (possibly changing) rates. **Monte-Carlo simulation** refers to generating a large number of individual sample points or observations from a (possibly complex) probability distribution or **stochastic process**.

While it is possible to write simulation programs in general-purpose computer languages, it is more efficient to use simulation packages. These automatically update the state of the system, advance simulated time, and provide facilities for random variate generation and performance evaluation. Some allow the model to be entered interactively by simply drawing its logic on a graphics screen or entering the **activity cycles** of all types of entities.   *H.G. Daellenbach*

References

Daellenbach, H.G. and McNickle, D.C. (2001) *Systems Thinking and OR/MS Methods*, Christchurch: REA, ch. 7.
Pidd, M. (1992) *Computer Simulation in Management Science*, 3rd edn., Chichester: Wiley & Sons.

Single-, double-, and triple-loop learning

Single-loop learning has been defined by some authors (notably, Argyris and Schön, 1985) as learning by detecting errors in a system, but without altering (or placing in question) the underlying values of the people involved in a **system**. The process amounts to monitoring some performance outcomes, comparing these with what is desired, and initiating necessary corrective action as determined by existing policy frameworks. Double-loop learning, in contrast, amounts to (re)examining the underlying values and frameworks in terms of which organizational behaviour is considered meaningful.

Single-loop learning can be considered as appropriate for getting jobs done by concentrating on developing the most efficient way of achieving outcomes. Double-loop learning, however, raises questions concerning what the purposes are that make specific jobs relevant in the organizational context. Raising 'what' questions may involve challenging the mental models that people are used to upholding. Argyris and Schön suggest that double-loop learning requires an attitude in which people are less defensive (than in single-loop learning) about protecting assumptions that they may have held dear in the past.

The concept of triple-loop learning draws on (and extends) these considerations (Flood and Romm, 1996). Triple-loop learning implies that people can ask questions of different types as they conduct themselves in their organizational and social settings. One type of question that they might ask is: 'Are we doing things right?' For example, when teaching in an educational institution, we may consider whether we are adopting the right methods to organize our teaching activities. Other questions we may wish to pose are: 'Are we doing the right things in the first place?' 'What are we trying to achieve here?' 'Are we trying to get students to understand the teachers' thinking?' 'Or are we, say, trying to set up contexts of mutual inquiry on the part of both the teachers and the students?' Posing such questions implies (re)considering the values guiding our activities.

Yet another line of questioning that may be undertaken relates to the issue of who is defining what is 'right'. The question can be asked whether conceptions of rightness are being buttressed (unfairly) by the power of 'mightiness'. For instance, for an educational institution, we may ask whether definitions of appropriate teaching are being devised by people in positions of authority in the organizational setting. And should this in itself afford these definitions legitimacy? And why should this be so? (See also **knowledge-power**.)

Triple-loop learning as a concept points to a consciousness that is able to bring together 'how', 'what' and 'why' questions into one overall process that involves continual looping between them. In terms of such a consciousness, people can then, ideally, act more responsibly in managing the insoluble issues faced in their organizational and social existence.   *N.R.A. Romm*

References

Argyris, C. and Schön, D. (1985) *Strategy, Change and Defensive Routines*, Cambridge, MA: Ballinger.

Flood, R.L. and Romm, N.R.A. (1996) *Diversity Management: Triple Loop Learning*, Chichester: Wiley.

Senge, P.M. (1990) *The Fifth Discipline: The Art and Practice of the Learning Organisation*, New York: Currency Doubleday.

Swieringa, J. and Wierdsma, A. (1992) *Becoming a Learning Organisation: Beyond the Learning Curve*, Wokingham: Addison-Wesley.

Slack and surplus variables

A slack variable measures the difference between the right-hand side (RHS) and the left-hand side (LHS) of less-than-or-equal type constraints (such as in a **linear program**). If the constraint refers to the usage of a scarce resource, such as the amount of funds or the time available, or the capacity of a piece of equipment, the slack variable represents the amount of the resource left unused by a given solution.

A surplus variable measures by how much the LHS of a larger-than-or-equal type constraint exceeds the RHS value. For example, if the RHS refers to the minimum required content of a given nutrient in an animal feed-mix formula, then the value of the surplus variable indicates by how much this minimum requirement has been exceeded by the ingredients used in the feed mix.

Since the **simplex method** for solving linear programming problems works with equations, slack and surplus variables need to be introduced to convert all inequality constraints to equalities. Their objective function coefficients are set equal to zero. Linear programming software performs these operations automatically.

Social systems sciences, see **interactive planning**

Socio-ecological systems

Socio-ecology, also known as *open systems theory* (OST), has developed over the past fifty years from work with real people in their natural settings of organizations and communities. Socio-ecology takes people to transact with their social environment, i.e. they are affected by it and change it and their own behaviour to meet their purposes.

Socio-ecology views organizations as structured by one of two fundamental design principles. The first design principle, DP1, is called *redundancy of parts*. There are more parts (people) than the organization needs for its productive capacity, i.e. some have only supervisory roles. The responsibility for coordination and control is not located where the work, learning or planning is taking place, but with supervisors. The second design principle, DP2, is called *redundancy of functions*. More functions and skills are built into each person than the person can apply at any one time. Responsibility for coordination and control is located where the work, learning or planning is done. The basic module of a DP2 structure is a self-managing group of any form, multi-skilled, specialist or temporary project team. DP2 structures provide the conditions for the development and maintenance of intrinsic motivation and effective communication, without which projects and organizations may flounder.

Socio-ecological concepts and principles are used to design unique processes and events for the successful planning of desired outcomes. Socio-ecology has developed two major methods, the *search conference* for planning, and the *participative design workshop* for structural design and redesign. Used in combination, they are known as the *two-stage model of active adaptation*. The participants are those with the knowledge and responsibility to implement the desired outcomes.

Since 1959, the search conference has been viewed as a reliable method for strategic planning. It claims to be still the only method that fully explores the global social field of values and expectations that is so influential in determining future customers' needs and what people will expect of their organizations. Knowing how to monitor changes in this field keeps planners in touch with, and anticipating, social trends.

There are two forms of the participative design workshop, one for structural design, the other for redesign. Within that for redesign, those who work in the organization use their detailed process, technical and other organizational knowledge, together with DP2, to fundamentally restructure for responsible, motivated work: analysing the effect of the organization on people's productive activity, skill development and distribution; redesigning the organization from DP1 to DP2; setting measurable group goals, training requirements and career plans. The aim is a highly productive, flexible and adaptable organization.

The workshop for designing from scratch is used to bring into being project teams which cover all skills required, including those for implementation, and task forces, new organizations and/or plants, based on similar previous experiences.   *M. Emery*

References

Ackoff, R.L. and Emery, F.E. (1972) *On Purposeful Systems*, London: Tavistock.

Emery, F.E. (1998) *Futures We're In*, revised and updated, Southern Cross University Press.

Emery, M. (ed.) (1993) *Participative Design for Participative Democracy*, Canberra: Centre for Continuing Education, Australian National University.

Emery, M. (1999) *Searching: The theory and practice of making cultural change*, Amsterdam: John Benjamin Publishers.

Trist, E. and Murray, H. (1990, 1993, 1997) *The Social Engagement of Social Science: A Tavistock Anthology*, 3 volumes, Philadelphia: University of Pennsylvania Press.

Soft operational research (Soft OR)

Soft OR emerged in the 1970s as a response to the inability of traditional OR to deal with complex, ill-structured **problem situations**. The protagonists of soft OR refer to traditional OR and engineering **systems analysis** as 'hard' methodologies. They claim that these are suitable only if

- the problem has been clearly defined, the **objectives** of the decision maker are known and there exist **criteria** to ascertain when they have been achieved, the alternative courses of action are specified, either as a list of options or sets of decision variables, the constraints on the decision choices are known, and all input data needed are available;
- the problem is relatively well structured, meaning that the relationships between the variables are tractable, they can be expressed in quantitative form, and the computational effort for determining the **optimal solution** is economically feasible;
- the problem can be sufficiently well insulated from its **wider system of interest**;
- the problem is of a technical nature, largely devoid of human aspects; and
- the decision maker can enforce **implementation** of the solution.

They claim that, although such problems exist, the majority of the real-life problem situations in business, industry and government violate many, if not most of these assumptions, and that hard OR is thus largely dealing with either simple or technical issues.

Soft OR approaches address complex problem situations, which are messy, ill-structured, ill-defined, not independent of people; in other words, where different stakeholders with different **world views** have different, possibly conflicting perceptions about the problem situation and major issues, and where there may be no agreement about the appropriate objectives, or even the set of possible actions. They are characterized by

- structuring the problem situation, rather than by problem solving;
- facilitating dialogue between the various stakeholders with the aim of achieving a greater degree of shared perceptions of the problem situation, rather than providing a decision aid to the decision maker;
- 'what' questions, more than by 'how' questions, i.e. 'what is the nature of the issue?', 'what are appropriate objectives?' given the various world views of the stakeholders; 'what is the appropriate definition of the system for the issue considered?', 'which changes are systemically desirable and culturally feasible?', and only then 'how are these changes best brought about?'
- eliciting the resolution of the problem from the stakeholders themselves, rather than from the analyst; and
- changing the role of the 'problem solver' to one of becoming a facilitator and resource person who relies on the technical subject expertise of the stakeholders.

Note that 'how' questions, i.e. which means are the best for achieving the desired objectives, must ultimately also be addressed by soft systems methodologies. But they are often an anti-climax, almost obvious, rather than being centre-stage as in most OR/MS projects.

Methods include **hypergame analysis, metagame analysis, interactive** management, **robustness analysis, soft systems methodology, strategic assumption surfacing and testing, strategic choice approach, strategic options development and analysis, drama theory**, and the most recent addition, the **theory of constraints.**

D.P. Dash and H.G. Daellenbach

References
Journal of the Operational Research Society (1993) 44(6) Special Issue.
Rosenhead, J. and Mingers, J. (eds) (2001) *Rational Analysis for a Problematic World Revisited,* Chichester: Wiley.

Soft systems methodology or SSM

SSM is a **soft operational research** or **problem structuring methodology** for dealing with messy problem situations, where there is no clear definition of the problem among the people concerned or the **objectives** to be achieved. It was developed by Peter Checkland in the early 1970s through a rigorous programme of **action research**, and rests on solid theoretical premises and **systems concepts**. Checkland sees problem solving within a management context as a never-ending learning process. This is clearly reflected in his methodology which has itself evolved from its original 'methodology-driven' mode 1 to its 'situation-driven' mode 2.

Mode 1 is a seven-stage learning system, as depicted in the figure below, that starts in the real world, moves into the world of systems thinking, and returns to the real world.

At stages 1 and 2 the problem solver, together with the various stakeholders, seeks to build the richest picture of the **problem situation**, i.e. the problematic issues within their context. Stage 3 involves structuring the issue selected for detailed study

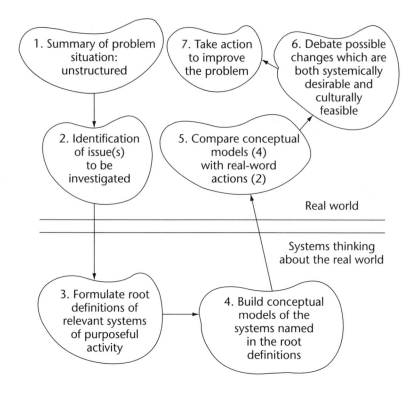

Figure: **Flow diagram of Checkland's soft systems methodology**

in systems terms, i.e. setting the **boundary** (or environment) of a *notional system* (as if it existed in the real world), defining its activities or transformation process, the 'owners' of the system who have decision power over it, their 'world view' (which gives meaning to the notional system), the 'actors' who do the activities, and the 'customers' who suffer or benefit of the consequences – easily memorized by the CATWOE mnemonic. Checkland recommends that several different definitions of notional systems are explored in parallel, each encapsulated in a so-called *root definition*. Stage 4 derives for each root definition a *conceptual model* which shows the logical relationships between all activities needed for the transformation defined in the root definition, including monitoring and control of activities and performance assessment. At stage 5 the conceptual model is brought back into the real world and compared with what is currently happening. The aim is to prepare an agenda for discussion topics, such as 'what activities are missing or problematic (when viewed from the perspective of the notional system)?' The emphasis is on 'what should be done' rather than on 'how to do it'. This agenda is used at stage 6 to generate a debate with a view to gaining greater insights and from which changes that are **systemically** desirable and culturally feasible may emerge. Regardless of whether any changes are implemented or not (stage 7), each complete cycle of this process transforms the original problem situation into a new one that may become the trigger to engage in a new learning cycle. With sufficient training, managers can apply this methodology without the need for a facilitator.

While mode 1 is driven by the seven-step methodology, in mode 2 the decision makers and managers are assumed to have internalized this learning process to a degree that it becomes a natural thinking process, which they can invoke informally or formally, if the need arises, to reflect on and make sense of a problematic situation they face. *H. Lópes-Garay*

References

Checkland, P.B. (1981) *Systems Thinking, Systems Practice*, Chichester: Wiley. A must for any serious systems thinker.

Checkland, P.B. (1990) 'Soft system methodology' and 'An application of soft system methodology', in J. Rosenhead (ed.) *Rational Analysis for a Problematic World*, Chichester: Wiley, pp. 71–120. A good starting point.

Checkland, P.B. and Scholes, J. (1990) *Soft Systems Methodology in Action*, Chichester: Wiley.

Soft systems thinking, see interpretive systems theory

Solution space, see feasible solutions

Spreadsheets

Spreadsheets are $m \times n$ tables of cells, where m is the number of columns and n the number of rows. Each cell is referred to by its column and row index (e.g. C7, denoting the third column, labelled C, and row 7). Cell entries can be text, numbers, mathematical or logical formulas, expressions and functions which reference entries in other cells. For example, the formula 'SUM(C7:K7)' in cell L7 computes the sum of the nine numerical contents in the cell range C7 to K7 and stores it in cell L7, while 'MAX(C7–C8,0) inserts the difference of cells C7 and C8 provided it is positive or zero otherwise. The reference cells can in turn be functions of other cells. Whenever an entry in a cell is changed, all other cells that refer to that cell entry will normally be updated automatically. Hence the cell contents are always up to date. One of the prime strengths of spreadsheets is the ease with which **sensitivity** and **scenario analysis** can be performed.

Spreadsheet software, such as Microsoft Excel, Lotus 1–2–3 or Quattro Pro, provides a complete set of algebraic, trigonometric, logical, statistical and financial functions. They offer the full range of manipulation of content and formatting available in word processors, such as copying, editing, sorting, saving, linking or interfacing to other data sets or spreadsheets, as well as macro facilities. They also have facilities to represent the numerical contents of sets of rows and/or columns in various graphical forms. Some have limited statistical analysis or **optimization** capabilities.

Spreadsheet add-ons, such as @RISK, SIM.*xla* for **risk analysis** and *Excel Solver* for **linear** and **nonlinear programming**, have been developed that facilitate the formulation, use, and solution of MS/OR-type problems, thereby enhancing the analysis and problem-solving capabilities of spreadsheets.

Spreadsheets have become the indispensable everyday tool for any type of analysis from business to technical. On the surface, spreadsheets are easy to write. Unfortunately, studies have shown that 1 to 4 per cent of cells have errors. So spreadsheets are best for small applications, where formulas can be checked easily. Here are some guidelines for writing good spreadsheets:

- Make your spreadsheets read from left to right and top to bottom. Formulas should reference cells above and to the left, not to the right and below.
- Be concise and try to keep the spreadsheet on one screen, since it is easier to navigate.
- All input data (constants, costs, coefficients, etc.) should not be entered in cell formulas, but instead entered in cells which are referenced by formulas in other cells.
- Carefully verify the correctness of all cell formulas and references. Numerical formulas must be checked by hand. *J.F. Raffensperger*

Stakeholder

Most MS/OR projects involve people who have vested interests in the issue analysed or are affected by its outcomes. They are the stakeholders of the problem. Several classifications are used in the systems literature. One looks at their different roles from the point of view of the **narrow system of interest**:

- The *problem owners* who have decision power over the choice of actions and can stop the project: Most often they are also the actual decision makers, but may have delegated some decision making to others within set guidelines. They are the ones who 'own' the narrow system of interest, and control essential resources in the **wider system of interest**.
- The *problem users* who use the results of the project: They have no independent decision power beyond executing the actions approved (which may be in the form of decision rules). Their cooperation is, though, essential for the decisions to be implemented correctly and to a high degree.
- The *problem customers* who benefit from the results or are the victims of the consequences.
- The *problem analysts* or solvers who analyse the problem and develop solutions for approval by the problem owners.

Although described here as different individuals with different functions, in fact their roles may overlap. The same person can, and often does, assume several or even all four roles.

W. Ulrich, one of the originators of **critical systems thinking**, divides the stakeholders into two mutually exclusive groups (see **critical systems heuristics**):

1. Those involved in the design, who are
    - the clients of the system design, i.e. those whose problem it addresses; for public projects they are usually different from the actual decision makers;
    - the decision makers, i.e. those who have control of the choice of action;
    - the planners or analysts, i.e. those who provide the expertise for the design.
2. Those affected by the design, but not involved, i.e. the silent victims or beneficiaries of the results. This latter group may include future generations, as well as species other than human or even the environment.

Obviously, the clients and decision makers are also affected by the results, but they are also involved, in contrast to the last group. Correct identification of the various stakeholders is important for a number of reasons. Failing to identify properly all stakeholders may lead to solving the wrong problem for the wrong persons. The **world views** of the problem owners and decision makers will usually be the ones that shape the objectives of the project. The capabilities of the problem users will affect what are appropriate decision rules. The choice of system **boundary** determines who (or what) are the silent victims and beneficiaries, and what, if anything, should be done about them. *H.G. Daellenbach*

References

Daellenbach, H.G. (2001) *Systems Thinking and Decision Making*, Christchurch: REA, pp. 58–61.
Urlich, W. (1994) *Critical Heuristics of Social Planning*, Chichester: Wiley, pp. 244–264.

Stakeholder analysis

The origin of the term *stakeholder* in the management literature can be traced back to 1963, when the word appeared in an international memorandum at the Stanford Research Institute. Stakeholders are any group or individuals who can affect or are affected by the achievement of the firm's objectives (Freeman, 1984). A systematic analysis of stakeholders is an essential aspect of most MS/OR projects.

The management literature offers several approaches for analysing stakeholders. Freeman's framework consists of three levels of stakeholder analysis – rational, process and transactional. Mitchell *et al.* (1997) develop a stakeholder typology model to understand the dynamics of stakeholders. Based on the approaches developed by Freeman, Mitchell and others, Elias, Cavana and Jackson (2001) propose an eight-step procedure:

- *Rational-level stakeholder analysis:* (1) Develop a stakeholder map of the project. This should include the major groups of stakeholders affected by the project or organization. (2) Prepare a chart of all the specific stakeholders by expanding the stakeholder map. (3) Identify the stakes of the stakeholders. (4) Prepare a two-dimensional grid. The first dimension categorizes the stakeholders by stake and the second dimension by power. 'Stake' can be classified into equity, economic or influencers, and 'power' can be formal (or voting), economic or political.
- *Process-level stakeholder analysis:* (5) Conduct a process-level stakeholder analysis. Here, the emphasis is on understanding and evaluating how management implicitly or explicitly manages its relationships with its stakeholders.
- *Transactional-level stakeholder analysis:* (6) Conduct a transactional-level stakeholder analysis. At this level, understand the set of transactions or bargains between management and its stakeholders and deduce whether these negotiations fit with the stakeholder map and the organizational processes for dealing with stakeholders. Successful transactions with stakeholders are built on understanding the legitimacy of the stakeholder and having processes to routinely surface their concerns.
- *Stakeholder management capability index:* (7) Determine the stakeholder management capability of a project or organization. For this, first judge whether management understands its stakeholder map or not. Then, rate its organizational processes and transactions for dealing with its stakeholders. Thus, stakeholder management capability of a project/organization can be defined as its understanding or conceptual map of its stakeholders, the processes for dealing with these stakeholders and the transactions which it uses to carry out the achievement of its purpose with its stakeholders.
- *Stakeholder dynamics:* (8) In the final stage analyse the dynamics of stakeholders, for example by using the model developed by Mitchell *et al.* They propose that classes of stakeholders be identified by the possession or attributed possession of one or more of three relationship attributes: power, legitimacy and urgency. The salience or prominence of stakeholders will change as their power, legitimacy and urgency change. Thus it is possible to analyse the dynamics of stakeholders by using systems or MS/OR methodologies to trace the changes in these attributes.

A.E. Elias and R.Y. Cavana

References
Elias, A.A., Cavana, R.Y. and Jackson, L.S. (2001) 'Stakeholder analysis to enrich the systems thinking and modelling methodology', *Proceedings of the 19th International Conference of the Systems Dynamics Society,* Atlanta, Georgia, USA (Paper in CD-Rom).
Freeman, R.E. (1984) *Strategic Management: A Stakeholder Approach,* Boston: Pitman Publishing.
Mitchell, R., Agle, B. and Wood, D. (1997) 'Towards a theory of stakeholder identification and salience: defining the principle of who and what really counts', *Academy of Management Review,* 22(4):853–886.

State variables, state of system

A *state variable* is the numeric value associated with an **attribute** of a system component, relationship or activity. For example, in an irrigation scheme fed from a reservoir, the water level of the reservoir is the attribute of the system component 'reservoir', the rate of outflow is the attribute for the activity 'outflow'. Each of these is captured by a state variable that indicates the numeric value of the corresponding attribute as of a given point in time. Categorical attributes can also be expressed as numeric state variables, by associating a numeric value with each form or condition of a given attribute. For instance, a machine can be idle, working, down in need of attention or down under repair. The corresponding state variable assumes values 0 for idle, 1 for working, 2 and 3 for the other two conditions. A system may have thousands of state variables.

In addition to state variables that are directly associated with system components, relationships and activities, systems modellers tend to introduce artificial state variables that measure the cumulative aspects of the performance of the system, such as the cumulative total waiting time of all arrivals in a **queueing system** or the cumulative total idle time of all servers. This is particularly so for **simulation** modelling. Such artificial state variables simplify the analysis of system performance. In fact, the analyst will be more interested in these state variables than the details of all other state variables.

The *state of the system* at any given point in time is given by values assumed by the totality of all state variables. Whenever a state variable changes its value, there is a change in the state of the system. The behaviour of the system is therefore completely defined by the changes in the state of the system. In the majority of cases, the analyst is rarely interested in the details of these state changes, but only in the aggregate measures accumulated in the artificial state variables. Knowing the total waiting time of all customers in the queue in front of a given server, and knowing the number of customers that arrived at that server, the average waiting time per customer – one of the measures of performance of the system – can easily be computed.

The *state space* is the multidimensional range of potential values that the state of the system may assume, i.e. the combined set of potential values of all state variables.

Stationarity

Stationarity refers to the nature over time of a phenomenon or the uncontrollable inputs into a **system**, such as the nature of the demand for a good or service, or the nature of the failure rate for a piece of equipment over a given time period. If the behaviour of the aspect in question is known or predictable with certainty, stationarity implies that the aspect remains constant over the entire **planning horizon** (i.e. the period explicitly considered for analysis). For example, the demand for a product occurs at a known constant rate, or the market price of a product remains fixed at a known level over the planning horizon.

If a phenomenon or the uncontrollable input into a system is a **random variable**, then stationarity implies that the underlying probability distribution remains unchanged over the planning horizon. For example, the daily demand for a product may have a normal distribution with a mean of μ and a standard deviation of σ, both of which remain constant over the period of analysis of, say, one year.

In reality, the environment of most systems or the nature of economic phenomena are rarely stationary, but undergo constant changes, i.e. are dynamic. If these changes are slow, such as for natural phenomena, then stationarity may serve as a useful modelling approximation, simplifying the analysis considerably. In fact, the assumption of stationarity is often the only practicable way to derive a tractable analytic **mathematical model** for a system. Abandoning this assumption may result in a model that is computationally beyond our current computer technology.

If the purpose of the analysis is the exploration of the dynamic behaviour of a non-stationary phenomenon or system over time, **simulation** may be the most practical technique, since it allows the modelling of the dynamic nature of a phenomenon or of inputs from the system's environment.

Stationarity should not be confused with the concept of **steady-state**, or equilibrium of a system. The latter concepts refer to the long-run average behaviour of a system.

Statistical analysis, descriptive statistics, statistical inference

Statistics is the science of collecting, analysing, interpreting and presenting data, numerical and attribute, for the purpose of describing distinctive features of some phenomenon (*descriptive statistics*) and drawing inferences on characteristic aspects of it (*statistical inference*).

Descriptive statistics deals with organizing, classifying, summarizing and presenting data. It highlights particular features hidden in the mass of numbers and attributes by reducing them to a form where they can be more easily comprehended. Of particular interest are measures of *central tendency*, such as the mean, median or mode, and *measures of dispersion* or variation, such as the variance or standard deviation, as well as various percentiles or even a frequency distribution.

Statistical inference draws conclusions from a sample of data about the statistical population from which the data were drawn. The aim is to seek confirmation or refutation of *hypotheses* (or conjectures) about the population. Statistical inference is based on the premise that data collection follows valid principles of *random sampling*, e.g. any element in the population surveyed has an equal chance of being selected into the sample, in contrast to judgemental sampling where the elements are chosen according to other criteria, such as perceived representativeness. Inferences drawn are subject to two types of errors. A *type I error* occurs when it is concluded that a hypothesis is not true when, in fact, it is true. The frequency of committing that error is controlled by the *level of significance*, usually set to a small percentage, such as 5 or 1 per cent. The more serious the consequences of this error, the smaller the level of significance. The *type II error* occurs when the hypothesis is accepted as being true when, in fact, it is not. Its frequency of occurrence is also related to the level of significance. The risk of both errors can be reduced by increasing the sample size.

Statistical inferences requires designing a plan. This consists of

- defining the population and the variables or attributes to be sampled;
- formulating hypotheses about the population of interest and specifying the level of significance;
- selecting a sampling plan, i.e. the process of sampling the population, including the size of the sample;
- collecting the sample data according to the specified sampling plan;
- processing the data to a form required for testing the hypotheses;
- testing the hypotheses which leads either to their acceptance or rejection.

This process is often identified with what is known as the **scientific method**. The current world would be almost unthinkable without statistical analysis. It pervades all walks of life, from government to commerce and industry, from fundamental research in the physical and biological sciences to applied industrial testing, from economics and social science to market research and political polls. Some form of statistical analysis is an integral aspect of most MS/OR projects.   *S. Stray*

Reference
Smith, G. (1998) *Introduction to Statistical Reasoning*, Boston, MA: McGraw-Hill, or any text on statistics.

Statistical bootstrapping

The term 'bootstrapping' is used by statisticians quite differently to the way it is used in **judgemental bootstrapping**. Statistical bootstrapping refers to a group of techniques that involve *resampling* the observed data, so as to generate a series of samples that have similar properties to the observed sample. This enables the analyst to assess the properties of results derived from the observed data. The name is derived from the saying 'lifting oneself by one's own bootstraps'.

Suppose that we collect a sample of n observations, $x_1, x_2, \ldots x_n$. It may be reasonable to assume that the data are taken from a normal distribution with mean μ and standard deviation σ. The unknown parameter μ is often the quantity of prime interest, but will have to be estimated from the data. An intuitively sensible way to do this is to use the *sample mean*, $\bar{x} = \Sigma x_i / n$, as an estimate. Statisticians write $\hat{\mu} = \bar{x}$, where the 'hat' over μ indicates that we have an estimate of μ. The accuracy of this estimate is given by the *standard error*, $\sigma/\sqrt{n}$. In other words, the larger the sample size n, the smaller will be the standard error and hence the more precise will be the estimate.

For a normally distributed population, a 95 per cent *confidence interval* for μ is given by $\bar{x} \pm 1.96(\sigma/\sqrt{n})$. Suppose, instead, that we notice that the distribution is skewed or that there are one or two outliers (observations that are a long way from the rest of the data). Then results using normal theory no longer apply. We could try if another distribution is appropriate, such as the gamma distribution, but this can lead to theoretical and practical difficulties. A completely different approach is provided by bootstrapping, which does not require so many assumptions about the model underlying the data.

Suppose that we have collected a sample of size n that contains one clear outlier. Then it would be imprudent to assume normality. Instead we could construct a series of *bootstrap samples*, of the same size n, by sampling the observed data with replacement. Hence a new bootstrap sample might not include the outlier at all or might include it twice. What we are doing is taking samples of size n from the observed empirical distribution where each observation is chosen with probability $1/n$. The sample means of the bootstrap samples generate the empirical distribution of the sample mean from which we can assess the likely variability in $\bar{x}$ and hence find a 95 per cent bootstrap confidence interval for μ. This interval need not be symmetric about the (original) sample mean, $\bar{x}$. The key idea is that, if we do not know the population distribution, then the distribution of values in the sample is the best guide to what would happen in the population. Thus, to assess what would happen if the population was resampled, a sensible strategy is to to resample the observed data.

When applied to *time series data*, extra care is needed because the observations are ordered through time and the order matters. Suppose that we want to get a prediction interval for a future forecast, when the prediction error distribution is suspected of being non-normal. Then it would be invalid to construct bootstrap samples by taking random samples from the observed time series, as they would not reflect the natural ordering of the data. Instead, one way to proceed is to fit a time series model to the given data, compute the residuals from the model, and then take a series of samples, size n, from the residuals. If we have fitted a 'good' model, then the residuals should be approximately independent, so that their order does not matter. These bootstrap samples of residuals can be used to generate a bootstrap time series sample, conditional on some assumed starting values for the series. Values of the bootstrap time series are generated sequentially using the assumed model, by adding on a random residual. Notice that we do need to assume that we know a (parametric) model for the time series to do this, but that we do not need to assume that the error terms are normally distributed.

Although computationally intensive, these techniques are being increasingly used as appropriate software becomes more readily available, particularly whenever there are doubts as to what parametric model assumptions can reasonably be made about a particular set of data.   *C. Chatfield*

References

Chatfield, C. (2001) *Time-Series Forecasting*, Boca Raton: Chapman & Hall. Section 7.5.6 gives a brief up-to-date review of the use in time series analysis and forecasting, and further references.

Efron, B. and Gong, G. (1983) 'A leisurely look at the bootstrap, the jackknife and cross-validation,' *American Statistician*, 37:36–48. Nice, clear introduction, commendably brief.

Efron, B. and Tibshirani, R.J. (1993) *An Introduction to the Bootstrap*, New York: Chapman & Hall. Standard statistical monograph on the topic; many references; rather advanced.

Manly, B.F.J. (1997) *Randomization, Bootstrap and Monte Carlo Methods in Biology*, 2nd edn, London: Chapman & Hall. Introductory text, albeit aimed at Biology students. Chapter 3 is of particular interest.

Statistical quality control

The earliest approaches to quality control in manufacturing relied on end-of-line inspection of completed products. It was expensive, prone to errors, and failed to contribute to process or product improvement. In the 1920s Walter Shewhart demonstrated that variation in the production process leads to variation and therefore product defects and quality problems. His theories of the causes of variation, his invention of *control charts*, the application of statistical quality control and the Shewhart 'Plan, Do, Study, Act' cycle (Figure 1) became the principal means of the management of quality.

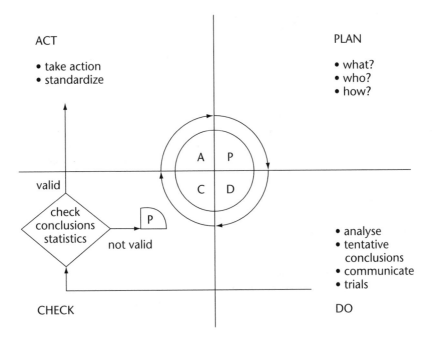

Figure 1: **Shewhart's 'Plan, Do, Study, Act' cycle**

Current statistical quality control

- focuses on the process–product relationship and the detection and control of quality problems;
- involves testing samples of product at critical points in the production process for statistical compliance with specifications, and
- provides a basis for systematic improvement.

Shewhart recognized that variation is a fact of industrial life and that management must distinguish between controlled and uncontrolled variation, the latter signalling problems. He identified two types of causes of variation:

- common causes which are a normal part of the system, e.g. the quality of raw materials, the characteristics of machines, and which contribute to controlled variation; and
- special or assignable causes of uncontrolled variation which are not a normal part of the system (e.g. operator error, changes to methods, changes in materials) and which can be traced and assigned to a particular circumstance.

Control charts are used to distinguish common causes from special causes of variation and to measure the state of statistical control of a process. Once a process is in a state of statistical control, improvement can follow.

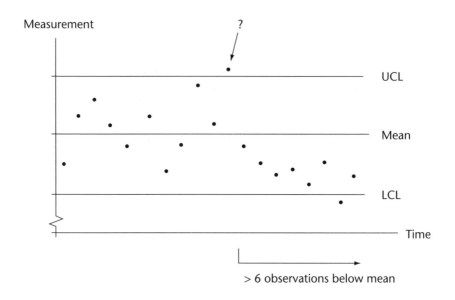

Figure 2: **Control chart**

Referring to Figure 2, a control chart consists of:

1. A time plot of the data. The actual points plotted are usually averages of samples taken over time, rather than individual measurements. Shewhart discovered that a plot of averages from any stable system followed a normal distribution. Hence, no matter what kind of data, the same theory and interpretations can be used.
2. A centre line which represents the mean of the distribution of the data.
3. Upper and lower control limits (UCL and LCL) giving the range of values that would normally be expected from controlled variations under stable conditions only.

Control charts can be created to analyse both variable and attribute data. Other considerations in selecting the appropriate type of control chart are the sample size and whether or not the sample size remains constant. Control charts are carefully checked for patterns indicating a possible lack of control, such as two or more points on the same side outside the control limits, six or more points in a row increasing or decreasing, eight consecutive points on one side of the centre line, all indicating possible non-random variations. *D.J. Houston*

References

Juran, J.M. and Godfrey, A.B. (1999) *Juran's Quality Control Handbook*, 5th ed., New York: McGraw-Hill, ch. 41.

Leitnaker, M.G., Sanders, R.D. and Hild, C. (1996) *The Power of Statistical Thinking: Improving industrial processes*, Reading: Addison-Wesley, chs 3, 4, 5 and 7.

Steady-state, system equilibrium, equifinality

The steady-state or the state of equilibrium of a **stochastic** system refers to the long-run behaviour an **open system** may gravitate towards, if it exists. This state of equilibrium is independent of the initial starting state. Most biological and social systems, if they remain undisturbed, tend to gravitate towards an equilibrium. Von Bertalanffy, the father of **general system theory**, called this property of open systems *equifinality*.

Steady-state is an unfortunate misnomer, since the concept refers to an average stable long-run behaviour of the **state of the system** – not a specific state, defined by a given set of values for all state variables – the temporary values assumed by all **attributes** that completely characterize the system at a given point in time. A system is in 'steady-state' if the long-run averages of these attributes, such as the average waiting time of customers in a **queueing system**, remain more or less constant. An alternative definition of the steady-state is that the probability of finding the system in a given state at a time in the far future, relatively speaking, is the same as at any other time in that far future, and that these steady-state probabilities are independent of the initial state.

Systems with positive **feedback loops** may be unstable in the sense that they tend to veer off in a given direction, with their averages increasing or decreasing constantly. Other systems undergo regular cyclic fluctuations (e.g. weather pattern over each year). No stochastic system ever reaches its 'steady-state', but only approaches it asymptotically, with random disturbances throwing it off course from time to time. After a disturbance, the system may tend towards the same equilibrium or a new equilibrium if the disturbance was big enough. Obviously, a system whose inputs are not **stationary** never approaches an equilibrium.

Steepest ascent methods, see gradient methods

Stella, VENSIM, *ithink*, system dynamics diagrams

Stella (systems thinking experiential learning laboratory with animation) is a user-friendly software package for simulating **system dynamics** models. In discrete-event **simulation**, system behaviour is described by the activities of the entities. In system dynamics the behaviour is described by the change in stock or level variables and the flows between different stock variables.

Stella is best demonstrated by a simple example, reproduced in the multi-part figure below.

The parents of a baby born in 2000 wish to secure their child's future college education, which will begin in 2018 (Eq. 4 in part (a)) and will last four years. They start saving $5 000 a year (Eq. 2 and 7) in a special account that pays 11.25 per cent interest per year (Eq. 9), compounded annually. Inflation over the next 21 years is expected to average 6 per cent annually (Eq. 8). This increases annual college expenses to around $40 000 (Eq. 6), for years 2018 through 2021.

Rather than having to enter all these equations directly, Stella allows the user to submit the system interactions in the form of a *system dynamics diagram*, as shown in part (b), which uses its own specific diagrammatic shapes and notation. While the user is creating the diagram, the software automatically generates the equations required for the simulation. All the user has to add are the various numeric constants.

(a) Level Variable:
     College Fund(t) = a special deposit account
     College Fund(2000) = 0 {$}   (1)
     Rate Variables:
     deposits = deposit amount - STEP(deposit amount, 2018) {$/year}   (2)
     gain = interest * College Fund {$/year}   (3)
     withdrawals = 0 + STEP(college expenses, 2018) {$/year}   (4)
     loss = inflation * College Fund {$/year}   (5)
     Auxiliary Variables & Parameters:
     college expences = 40,000 {$/year}   (6)
     deposit amount = 5000 {$/year}   (7)
     inflation = 0.06 {dimensionless}   (8)
     interest = 0.1125 {dimensionless}   (9)

(b)

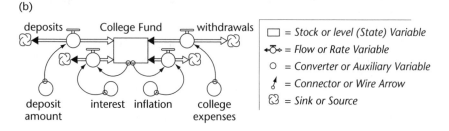

(c)

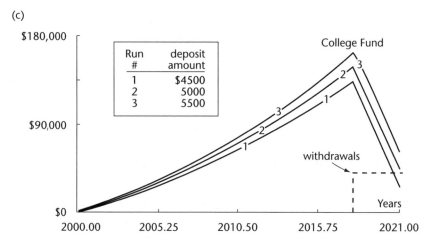

Figure: **Stella input and result windows**

Part (c) shows the results of the simulation for three different annual rates of deposits. The exponential growth patterns from 2000 to 2018 are produced by the **feedback loop** of monetary gains caused by the net annual growth of 5.25 per cent (equal to the interest rate minus the inflation rate) and the yearly fixed deposits. In 2018, Stella steps deposits down to zero (Eq. 2) and starts the withdrawal stream by stepping up withdrawals from zero to match the $40,000 annual college expenses (Eq. 4). In 2021, after financing the last year in college, a small balance is left for a modest graduation gift. The sharkfin-shaped pattern of the funds balance over time is typical of other personal finance schemes, such as retirement funds.

Special built-in functions, such as steps, ramps, statistical distributions, etc., are part of all system dynamics software. In Stella's **sensitivity analysis** window, users can vary parameters incrementally, arbitrarily, or according to a statistical distribution function.

System dynamics software facilitates the process of exploring dynamic processes with a considerable degree of realism and thereby stimulates intuition and creativity. Other software packages with similar features are *ithink*, Vensim and Powersim.

N.C. Georgantzas

References

Morecroft, J.D.W. (1992) 'Executive knowledge, models and learning', *EJOR*, 59(1):9–27.

Richmond, B. *et al.* (2000) *Stella: The power to understand!*, Hanover, NH: High Performance systems, Inc.

Sterman, J.D. (2000) *Business Dynamics: Systems thinking and modeling for a complex world*, Boston, MA: Irwin/McGraw-Hill.

Web sites

ithink and Stella: http://www.hps-inc.com/
Powersim: http://www.powersim.com/
Vensim: http:/www.vensim.com/

Stochastic, see deterministic, random variables

Stochastic dynamic programming, see dynamic programming

Stochastic mathematical programming approaches

Almost all decision making involves some degree of **uncertainty**. What will happen in the future is uncertain, such as the demand for a product over the coming few months. A production decision will have to be made before it is known. Even the current state of things may involve uncertainty, such as the accuracy of measurements made.

If the degree of uncertainty is minor, it may be acceptable to ignore it and decide or plan on the basis of averages or some other conservative estimates. However, if uncertainty is substantial, it needs to be modelled explicitly. Several approaches have been taken for dealing with uncertainty in **mathematical programming** models. Their main differences come from when and how information can be used and how uncertainties are reacted to. Probability distributions are required for all random data. Power generation by a hydro-station will be used as an example to demonstrate the concept. For simplicity, the uncertainty is assumed to be confined to the right-hand side of constraints, such as resource availability or requirements – in the example, to the amount of water available and the demand for power. The objective is to minimize total costs.

Stochastic programming with recourse or **recourse programming** schedules a so-called recourse action or decision for the case where a constraint is violated. In the hydro-station example, lack of water or high demand is countered by making up the generation shortfall through purchases of power at set prices from another electric power company for a small set of possible outcomes, each occurring with a given probability.

Chance-constraint programming does not provide for recourse decisions, but instead allows the constraints to be violated with a specified probability. In the hydro-station example, planned generation may have to cover the possible demand with a probability of, say, 0.9, specified as an (arbitrary) input, rather than found as part of the optimization.

Both of these two approaches have found rather limited practical use.

Stochastic dynamic programming – a version of **dynamic programming** – allows for the most realistic modelling of probabilistic events. However, its computational demand prevents it catering for more than two constraints. Uncertainty can be approximated in fair detail by discrete probability distributions. In the hydro-station example, both the water inflows into the hydro-reservoir and the power demand would be expressed as discrete distributions for each period. The amount of water in the reservoir is the resource constraint, which is tracked over the planning horizon for all possible values it may assume.

There are also real-time online approaches that assume that no information is known about the probability distributions of the random variables. Uncertainty is only reacted to when it has been resolved, i.e. the event outcome is known. These methods concentrate on worst-case results and hence lead to highly conservative decisions. *S. Dye*

References
Kall, P. and Wallace, S.W. (1994) *Stochastic Programming*, Chichester: Wiley.

Wets, R. and Ziemba, W. (1999) *Stochastic Programming. State of the art 1998. Annals of Operations Research*, vol. 85, JC Baltzer Science. Extensive bibliography on stochastic programming in the preface.

Stochastic process

A stochastic process or random process is a process which happens over some index set (frequently related to time), and whose behaviour is driven by some kind or kinds of probability distributions. The name, which comes from a Greek word which can, perhaps significantly, be translated as 'proceeding by guesswork', covers a very wide range of possibilities – too wide, in fact, to be of much use for deriving useful properties in MS/OR models. Almost invariably we have to specify more properties of the distributions to get tractable results.

For example, we might assume for modelling purposes that the successive times at which calls arrive at a telephone exchange come from a series of probability distributions. So if X_n is the time at which the nth call arrives we assume there is some probability distribution F_n such that $P(X_n \leq t) = F_n(t)$. The sequence of random variables $\{X_1, X_2, ...\}$ is a stochastic process. Here the index set is the numbers 1, 2, ... This is an example of a *continuous-valued process* over a discrete index set. Alternatively, our stochastic process of interest could be $\{Y_t, t \geq 0\}$, the number of phone calls in progress at time t. Now we have a *discrete-valued process* (the number of calls in progress) on a continuous index set, time. Thus on the same physical process or situation we can define different stochastic processes.

To return to the $\{X_1, X_2, ...\}$ times-of-arrival process, if we allow the sequence of probability distributions $\{F_n, n = 1, 2, ...\}$ to be quite general we will end up with an infinite number of probability distributions to analyse, and it is unlikely that we will be able to get any useful formulas to help us model our telephone exchange. In practical MS/OR modelling we will frequently also assume that the times between successive calls are independent and come from a particular negative exponential distribution. Under these assumptions the stochastic process is also called a *Poisson process*, and is much easier to deal with.

A **Markov chain** is an example of another stochastic process where making assumptions about the probability distributions leads to tractable results. *D.C. McNickle*

References
General books on Stochastic Processes tend to be highly mathematical. Look for specific processes in MS/OR texts.

Ross, S.M. (1997) *Introduction to probability models*, 6th edn, San Diego: Academic Press.

Strategic assumption surfacing and testing (SAST)

SAST is a **problem structuring method** that evolved in the 1970s and owes much to C. West Churchman's dialectic approach of submitting opposing positions to detailed scrutiny by playing devil's advocate. Its main area of use is for strategic planning and controversial policy decisions that may challenge the status quo, particularly when there is polarization between groups of entrenched protagonists. The core aspect of SAST is the testing and questioning of assumptions that underlie strategies, and by extension the strategies themselves.

The method consists of four phases:

1. *Group formation:* A wide cross-section of individuals with vested interest in the policies are put into groups, such that the views within each group are similar and close, while the views between groups are highly diverse. Each group should have at least one other group intent on challenging its position.
2. *Assumption surfacing:* Each group is asked to identify the **stakeholders**, i.e. the people involved and affected by their preferred strategy, and the assumptions they have made about them. For example, if a preferred strategy is to implement quality management, one group of stakeholders will be the employees, and an assumption might be that they have the skills and competencies to cope with implementation of quality, which may or may not be valid. Each group then positions their assumptions on an *assumption-rating chart,* as shown in the figure below. The idea is to find out which ones are most important and how certain the group is that their assumptions are correct. An important but uncertain assumption identifies a weakness in the strategy. These are the ones falling into the south-east quadrant.

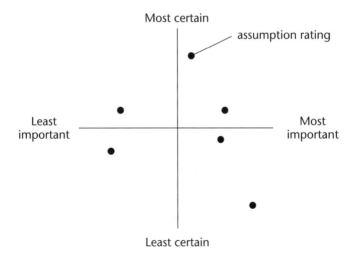

Figure: **Assumption rating chart**

3. *Dialectic debate:* Once the groups have completed an assumption-rating chart, an outsider facilitates an *adversarial* debate between the pairs of opposing groups, where each group critically evaluates the assumption-rating chart of the other. The aim is to identify weaknesses in the other group's assumptions, thus causing them to rethink their approach. Weaknesses might include lack of awareness of some potential stakeholders, failure to recognize assumptions made about them, and flaws in assumptions.
4. *Synthesis:* Following adversarial debate, each group rethinks their strategy, then meets with the other group with the aim of synthesizing a new approach agreeable to both sides. This might be possible following the learning process that SAST encourages. *R.L. Flood*

References
Jackson, M.C. (2000) *Systems Approaches to Management*, New York: Kluwer Academic, pp. 226–232. A succinct but informative treatment.
Mason, R.O. and Mitroff, I.I. (1981) *Challenging Strategic Planning Assumptions*, New York: Wiley.

Strategic choice approach

The Strategic Choice Approach (SCA) is a **problem structuring method** for managing intensive workshop sessions for groups of people who may represent different organizational or community interests, under the guidance of a skilled facilitator or facilitation team. It is a flexible and interactive vehicle for planning under conditions of (a) pressures for early decisions and (b) a range of sources of **uncertainty** – those of a technical nature relating to the environment, those of a political nature, involving value judgements, and those stemming from the interrelatedness of decision areas, all of which need to be managed. The approach helps people to make sustained progress together towards agreed decisions and also towards clearer appreciation of each other's perspectives. The principal output of an SCA workshop is agreement on the content of a carefully balanced progress package. This takes the form of a grid in which commitments to early action in some areas of decision are accompanied by a set of agreed exploratory steps – investigations, policy consultations or negotiations – designed to address key areas of uncertainty. In this way, more confident foundations are built for further decisions on appropriate future time horizons.

As a framework for communication, the facilitators introduce a range of mainly graphical methods, enabling the group to work flexibly among four complementary modes of decision making:

- *Shaping:* Formulating and structuring the interconnected areas of concern, referred to as decision areas to be addressed; exploring the intensity of their linkages and selecting a subset for immediate attention.
- *Designing:* Listing the choice options for each decision area and generating a range of feasible combinations of options or strategies.
- *Comparing:* Predicting the effects of various strategies and comparing them over multiple criteria, while drawing out key areas of uncertainty, particularly those related to value judgements. The aim is to identify one or more preferred strategies.
- *Choosing:* Incrementally constructing successive progress packages.

These four modes are used iteratively, switching from one to another (and back), in response to progress made or to impasses encountered. During a workshop, a cumulative record of progress is built up on flip charts ranged around the walls of the room. An interactive computer program known as STRAD – short for Strategic Adviser – is now available to complement this low-technology approach. This has proved especially helpful for small informal groups to work together around a desktop computer.

The roots of the strategic choice approach are to be found in two applied research projects conducted by mixed teams of operational research scientists and social scientists from the Tavistock Institute in London. One project (1966) was concerned with issues of communication in the building industry, while the other (1977) was concerned with the processes of policy development and planning in city government. The approach has since been refined and applied in many countries in both commercial and public policy settings. It has proved especially productive in helping people to work constructively together across inter-organizational boundaries.

J. Friend

References
Friend, J. (1995) 'Supporting developmental decision processes: the evolution of an OR approach', *International Transactions in Operational Research*, 2: 225–32.

Friend, J. and Hickling, A. (1997) *Planning under Pressure*, 2nd edn, Oxford: Butterworth-Heinemann.

Strategic option development and analysis (SODA)

Strategic option development and analysis is a **problem structuring method** to deal with complex decision situations that involve several **stakeholders** with potentially different perceptions of the problem, different values and goals, and possibly different vested interests. It was largely developed by Colin Eden in the late 1970s. It has psychological/sociological roots with four perspectives:

- the individual who uses language to express concepts, captured by **cognitive maps**, based on Kelly's 1955 theory from clinical psychology – SODA's main theoretical anchor,
- the nature of the organization (politics, coalitions, negotiation)
- the nature of consulting practice (facilitation, negotiation), and
- technology or technique (quantitative analysis, computer software Decision Explorer, previously known as COPE).

SODA's aim is to bring about a degree of consensus on a complex issue or set of issues and a commitment to action among a group of 2–10 people, actively and substantively involved in the situation. It consists of four main steps:

Step 1: Construct a cognitive map for each stakeholder involved, based on one or two extended individual interviews with each.

Step 2: Discuss each map with its owner, clarify ambiguities and contradictions. Work through the map in order to explore the **hierarchical** nature of both goals and action options. Through this process, each member should have a good understanding of his or her own views and perception of the situation.

Step 3: Merge the maps into a *strategic map*, by combining identical or similar constructs, retaining the richer construct wording. Possible conflicts between individual views are revealed. The aim of the strategic map is to develop *emerging themes* and *core concepts*. Emerging themes or issues are usually seen in the form of clusters of constructs with many links between them and fewer links with clusters outside the group. Core concepts are those that have either many arrows leading into it or many leading away. The Decision Explorer software facilitates the merging of maps and the identification of clusters.

Step 4: A facilitator leads a SODA workshop with all participants where the merged map is explored, reconciled, and individual views are 'appreciated'. Then one or more related clusters are chosen for deeper analysis with the aim of getting a shared view, if not on goals then at least on options. The workshop ends with a commitment to action. If major differences in views and goals exist between individuals, a facilitator skilled in **negotiation** is needed to bring about a sufficient degree of consensus for a commitment to action.

Colin Eden and his associates at the University of Strathclyde have applied SODA successfully to numerous projects as part of their consulting practice. They have developed Decision Explorer for drawing, analysing, and merging of maps.   *J. Holt*

References

Eden, C., Jones, S. and Simms, D. (1983) *Messing About in Problems*, Oxford: Pergamon Press.

Eden, C. and Simpson, P. (1989) 'SODA and cognitive mapping in practice', in J. Rosenhead (ed.) *Rational Analysis in a Problematic World*, Chichester: Wiley.

Eden, C. (1989) 'Using cognitive mapping for strategic options development and analysis (SODA)', in J. Rosenhead (ed.) *Rational Analysis in a Problematic World*, Chichester: Wiley.

Web site
http://www.banxia.com/dexplore/debiblio.html

Subjectivity, see objectivity

Suboptimization, see optimization

Subsystem, see systems concepts, hierarchy of systems

SUMT, see barrier methods

Sunk costs, see costs, type of

Supply chain management, see logistics, supply chain management

Surplus variables, see slack and surplus variables

Sustainability

Increasing concern during the 1980s about the environmental impact of economic development and growth, and the recognition that environmental pollution threatens prosperity and even the survival of species, was generated particularly after the release of *Our Common Future* (United Nations World Commission on Environment and Development). This book is seen as responsible for the increasingly widespread use of the terms *sustainable development* and *sustainability* as well as fuelling international debate in this area.

No universally agreed definition of sustainability exists. Broadly, it means the ability of countries and economies to develop such that they meet the present generation's needs without jeopardizing future generations' abilities to meet their own needs. Some people interpret this to mean that each generation should pass on to future generations an undiminished stock of natural resources. This definition fails to recognize that each generation passes on other valuable assets to future generations, such as an increased body of knowledge and technological developments. Therefore, others argue that it is possible for reductions in some resources to be more than offset by increases in other resources. Consequently, sustainability occurs if each generation passes on stocks of net resources per capita that are at least equal to the per capita stocks it inherited. This includes natural and environmental resources, as well as knowledge, technology, and physical and human capital.

Two aspects are central to the debate on sustainability. First, there is the recognition that planet Earth cannot sustain indefinitely the consumption of resources, both renewable and non-renewable ones, and the degradation of its environment, at the currently projected rates. Secondly, the debate rightly focuses on the needs of the world's poor and their current and future development, as well as serious questions of distributional equity between developed, developing, and underdeveloped countries. Therefore, at least in the intermediate future, it involves consideration of strategies for helping underdeveloped regions reach legitimate goals. This implies continued change and development rather than consolidation to some equilibrium based on the current situation.

The issue of sustainability is important for MS/OR, since much of the planning for it will involve MS/OR modelling on various levels, some dealing with local concerns about the exploitation and use of local resources or the reduction in local environmental degradation, others addressing regional or global issues, such as regional development, sustainable harvesting of the oceans and control of greenhouse gases. Analysts addressing such problems bear a social and moral responsibility to work for sustainability.   *S. Stray*

Reference

Pearce, D.W., Markandya, A. and Barbier, E.B. (1989*) Blueprint for a Green Economy*, London: Earthscan.

Symbolic model

Symbolic models represent the relationships between entities or concepts by means of symbols. The daily newspapers, professional magazines, TV news and most technical books display all sorts of symbolic models: graphs of how one variable varies as a function of another variable, diagrams of how various concepts are connected via cause-and-effect relationships, precedence diagrams between concepts, tasks or actions, flow diagrams that show how material or information is processed or depict the process to get to a decision, decision trees, statistical charts of all sorts, organization charts showing the chains of command, geographical maps, and so on.

Mathematical models are also symbolic models. They use variables, coefficients, functions, inequalities – all symbols – to represent a set of relationships.

While mathematical models form the core of 'hard' MS/OR, non-mathematical symbolic models are extensively used in **soft operational research** and **problem structuring methods**.

Synchronous manufacturing, see Drum–Buffer–Ropey, OPT

System analysis

This terms is used to refer to two completely different activities. In computer science, systems analysis is the design of computer hardware and software systems. In MS/OR it describes the investigation into complex issues, usually in the **public policy** field. It has its origin in The RAND Corporation where it was initially used for strategic military decision making, but was later popularized, largely by H.J. Miser and E.S. Quade, as a methodology for any type of strategic management and policy decisions. Its general format has been adopted by traditional 'hard' MS/OR. The flow chart below depicts the process. Although shown as a sequence of steps, Miser and Quade stress that in fact it is an iterative process that may require several partial or full passes through it before a project is completed. Wherever possible, there is recourse to quantitative tools and techniques, particularly for the models used to evaluate the performance of alternative courses of action. The tools used are mainly **statistical analysis**, **cost–benefit analysis**, econometric modelling, **simulation**, **system dynamics**, and **gaming**, as well as **mathematical programming**.

Its proponents see system analysis largely as a craft that is acquired through experience and the norms of teachers and colleagues. However, it strives to follow the principles of the **scientific method**, taking an engineering perspective. This is reflected by the label 'research' given to the middle phase. It implies that the analyst is viewed as an 'objective' outsider, capable of selecting the appropriate boundaries of the system and evaluating the consequences of decision choices free of any **subjectivity**. The human aspect is largely absent or captured by supposedly neutral numeric measures, such as the number of people affected, the increase in economic welfare, lives saved, and so on. This has led to considerable criticism.

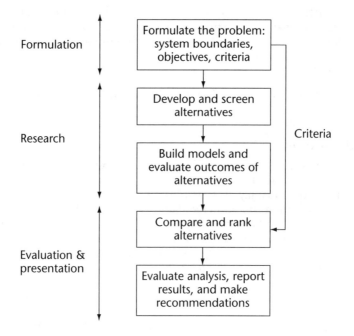

Figure: **System analysis process**
(*Source*: adapted from Miser, 1995, p. 217)

It has been and still is the basis on which many public policy decisions are made. Often it is difficult to say whether the application leans more towards traditional MS/OR or system analysis. Example applications are: San Diego clean air project; storm flood protection in the Netherlands; water management projects; blood bank operation; fire protection services; energy strategy analysis; and so on.

H.G. Daellenbach

References
Miser, H.J. (ed.) (1995) *Handbook of System Analysis: Cases*, New York: Wiley.
Miser, H.J. and Quade, E.S. (eds) (1985) *Handbook of System Analysis: Overview of Uses, Procedures, Applications, and Practice*, New York: North-Holland.
Miser, H.J. and Quade, E.S. (eds) (1988) *Handbook of System Analysis: Craft Issues and Procedural Choices*, New York: Elsevier Science.

System dynamics

System dynamics is a methodology for studying complex dynamic system behaviour from a holistic perspective. As a methodology, it is based on three prime ideas:

- its emphasis on the way the **system** structure determines system behaviour, referred to as the endogenous view;
- its explicit recognition of **feedback loops**;
- its use of computer **simulation** to explore system behaviour.

A system dynamics model expresses system behaviour in the form of **differential equations** that relate the time trajectory of system variables, called *stocks* or *levels*, and *rates-of-flow*. Stocks usually express amounts of various resources, inventories, and cumulative measures of performance. Rates-of-flow show how stocks change over time. Rates may in turn be functions of one or several stocks, leading to feedback loops. The changes may be lagged, either in the form of a fixed time delay or a gradual or exponential change. Feedback loops may be positive or negative. A positive feedback loop has a destabilizing effect, i.e. certain crucial stocks grow or shrink without bound – usually an undesirable property. A negative feedback has a stabilizing effect, i.e. system behaviour tends to gravitate towards an *equilibrium* or a **steady-state**. The dynamic aspect of system behaviour is captured in the form of time series of key system variables.

For example, the process of filling and emptying the toilet cistern is a dynamic system with a negative feedback loop. When the toilet is flushed, the cistern plug opens, allowing the water level (= stock variable) to drop fast. Once close to empty, the plug closes, preventing further water to escape. The lowering of the water level in turn opens the water inflow valve via the level of the ball cock, allowing water to fill the cistern. As the level rises, it causes the valve to close gradually, reducing the inflow (= the rate-of-flow variable) to zero.

Causal-loop diagrams and **influence diagrams** are effective means to show the dynamic cause-and-effect relationships between various system variables and the resulting feedback loops. Some system dynamics software allows the input to be in the form of causal-loop diagrams (see **Stella**).

A number of typical causal structures or archetypes have been identified, such as the 'limits-of-growth' archetype and the 'shifting of the burden' archetype. The first describes how a growth process can be curbed or even crash as the result of its interactions with a stabilizing process. The second describes how overdependence on short-term solutions at the expense of addressing the more fundamental problem may have side-effects that may be even more undesirable than the original problem.

System dynamics is the brainchild of Jay Forrester at MIT, who in the late 1950s used his background in servomechanism to analyze industrial processes, culminating in the now classical text *Industrial Dynamics* in 1961. He subsequently broadened his research to study urban development and global resource issues. It is without doubt one of the most used modelling approaches to study the dynamic behaviour of a wide range of economic, industrial, social, health, environmental and even political issues. The field is supported by an active international society, an international journal, and efficient computer software, such as *ithink*, Stella, Powersim, Vensim, and Forrester's own Dynamo. *J. Barton*

References

Forrester, J. (1961) *Industrial Dynamics*, Cambridge, MA: The MIT Press.
Richardson, G.P. (ed.) (1996) *Modelling for Management* (2 vols), Aldershot: Dartmouth Publ. Co.
Sterman, J.D. (2000) *Business Dynamics*, Boston: McGraw-Hill Irwin.
Wolstenholme, E.W. (1990) *System Enquiry*, Chichester: Wiley.

Web sites

www.systemdynamics.org/
www.strategydynamics.com

System dynamics diagrams, see **Stella** for an example

Systemic, systematic

These two concepts are often confused, even by dictionaries. People use systematic when in fact they mean systemic, or worse, they believe that if something is systematic it is a **system**. Systematic means 'methodical in procedure or plan', 'marked by thoroughness and regularity'. A classification system is (hopefully) systematic if it facilitates retrieving entries. An effective plan or efficient procedure to perform a task must be systematic, showing each step in its sequence, including fallback positions. A search for something is more likely to be successful if it is systematic, i.e. thorough and covering all search areas. However, the relationships between system components are not systematic, they are systemic. Systemic means 'pertaining to systems', or viewing things and relationships in the system in terms of their role towards contributing to the objectives or purpose of the system, i.e. seeing things in terms of **systems thinking** rather than **reductionist** or **cause-and-effect thinking**. Obviously, a systematic procedure could be part of a system transformation process or a system activity could be methodical.

Systems concepts

A system involves the following concepts and properties:

1. It is an *organized* assembly of *components*. 'Organized' means that there exist special relationships between the components. For example, viewing a school as a system for educating students, its components include some physical entities, such as pupils, teachers, administrators, facilities and other resources, and abstract things, such as the curriculum, processes and rules. They are organized in order to educate students. A chaotic aggregate, such as a pile of stones, is not a system.
2. Each component contributes toward the behaviour of the system and is affected by it. No component has an independent effect. For example, whether or by how much a gifted teacher lifts the educational performance depends on the response of the students and his or her colleagues. System behaviour is changed if any component is removed. For example, without an educational curriculum a school would become a social interaction system.
3. A system exhibits **emergent properties** that none of its components have individually, i.e. that emerge from the relationships and interaction between the components. For example, an emergent property of a school is pupils' absorption and mastery of knowledge in a chosen subject. However, social conditioning is also an emergent property. All need the interactions of the various components to 'emerge'. Producing such desirable emergent properties is usually the purpose of **human activity systems.**
4. Subgroups of system's components may themselves have properties 1, 2 and 3, i.e. they form *subsystems*. For example, a class may be viewed as a subsystem of the school.
5. A system has an outside, called its *environment*. Certain parts or aspects are viewed as being inside the system, others are viewed as belonging to the environment, i.e. the system has a **boundary**. For example, the school's environment is the community in which the school resides and its social and educational climate, the educational regulations, and the resources provided by the government and the community.
6. A system does something, i.e. it *transforms inputs* from the environment into *outputs* to the environment.

There are two ways of interpreting systems. The 'out-there' interpretation sees systems as actually existing in the real world. The system is an objective entity. For example, an estuary is seen as an existing ecological system. The natural and physical sciences, as well as **general systems theory**, take this stand.

The second approach views systems as (**subjective**) human **conceptualizations** – 'the inside-us' view. They do not exist *per se*, but only in the mind of an observer who finds it useful to view something as a system for a given purpose. This is the view taken in the management sciences. For example, the principal of a school may want to see his or her school as a system for educating pupils.

This **interpretive systems** view implies that the choice of a system's boundary is arbitrary. Different observers may draw different boundaries, depending on the purpose of defining the system, and the available resources (time, funds, training of analyst, computational power, etc.). However, even for the 'out-there' view, the boundary selection is arbitrary too – what one observer sees as a component of the system, another may place into the environment. Boundary choice should be submitted to critical evaluation (see **boundary critique**).

The environment may be subdivided into a part that actively interacts with the system and, in particular, has control over its major inputs, called the **wider system of interest**, and the rest which provides a further background. The original system now becomes the **narrow system of interest**. _H.G. Daellenbach_

References

Daellenbach, H.G. (2001) *Systems Thinking and Decision Making*, Christchurch: REA, ch. 3.
Flood, R.L. and Carson, E.R. (1988) *Dealing with Complexity*, New York: Plenum Press, chs 1–2.

Systems thinking

Traditional Western thinking is largely based on two major ideas. The first is *reductionism* – the belief that everything can be reduced, broken down or disassembled into ultimately simple indivisible parts. Explaining each part is assumed sufficient to understand and explain the behaviour of the whole. In terms of decision making this implies that the best overall course of action is made up of the coordinated best individual actions for each part. The second idea is that all phenomena, events and activities are explainable by *cause-and-effect* relationships, where effects become the cause of other effects, ends become the means for other ends.

However, everyday experience shows that this type of linear thinking fails to explain and predict adequately the behaviour of many complex phenomena or events, both natural and human-created. They exhibit types of behaviours and properties that none of the individual components have by themselves and that cannot be explained in terms of 'A causes B'. They are called **emergent properties** and are the result of mutual interactions and synergies. To harness the desirable ones is usually the reason for 'creating' **human activity systems**. Even if the individual components can be separated out and their one-on-one relationships identified, this may not explain the emergence of such new properties.

Therefore, **reductionist** and **cause-and-effect thinking** need to be complemented with a third type of thinking, *systems thinking*. Rather than building an understanding of the system and its behaviour from the properties of its basic components and the causal links between them – from the bottom up, so to speak, systems thinking takes a top-down approach. It explains the behaviour of components or groups of components in terms of their **systemic** role in the transformation process of the system.

An important aspect of systems thinking is the search for appropriate **boundaries** to the system, i.e. what is considered as a part of the system, what is seen as its environment, and what is seen as irrelevant. Boundary choices always involve some degree of arbitrariness and need to be challenged and justified by way of **boundary critique**.

When evaluating system performance, rather than putting the main emphasis on the **efficiency** of system activities, i.e. how well resources are used for each activity, its main focus is on the **effectiveness** of the activities, i.e. how well the activities achieve the system's overall mission, **objectives** or goals. Part of systems thinking deals with setting these objectives and goals, such that they are congruent with the objectives and goals of the **wider system of interest**. Little is gained by making the **narrow system of interest** highly efficient, only to lose any gains made by a worse performance of the wider system. Furthermore, some outputs of the system may affect other parties that have no say in the operation of the system. Such effects may be highly detrimental. Examples of this are various forms of pollution. Ethical considerations dictate that such effects cannot be dismissed as irrelevant by those in control of the system. *H.G. Daellenbach*

References

Daellenbach, H.G. (2001) *Systems Thinking and Decision Making*, Christchurch: REA, chs 2–5.

Jackson, M.C. (2000) *Systems Approaches to Management*, New York: Kluwer Academic.

Tabu search

Tabu search (TS) is a **search heuristic** first proposed by Glover in 1986. It uses several types of memory to stop the search from becoming trapped in **local optima**. TS starts from a single point in the search space and attempts to improve on that point by examining a set of possible changes to that solution, which are referred to as *candidate moves*. The set of all candidate moves is defined by a procedure for incrementally changing solutions, called the *neighbourhood scheme*. For example, in a **knapsack problem**, i.e. the problem of deciding which items should be packed into a knapsack from a set of n possible items, the neighbourhood scheme could change one item from being taken in the knapsack to not being taken or vice versa. The candidate list may include all possible solutions defined by this neighbourhood scheme or it may reduce the number of neighbours considered in order to speed up the search. The objective function value of each new neighbour is calculated and if the move is the best candidate move found to date then the short-term memory, called a *tabu list*, is checked to see if the move is allowed. If the move is not allowed, i.e. it is a *tabu*, then the *aspiration criterion* is checked. This is a rule that overrides the tabu status of a move. It could be in the form of 'allow the move if it produces a solution better than the best solution found so far'. If the move is allowed, or it is tabu, but allowed by the aspiration criterion, the move becomes the best candidate move, and the search continues evaluating the other candidates. Once all candidate moves have been assessed, the search executes the best candidate move it has found so far. A *tabu restriction* is added to the tabu list, preventing the move from being reversed in a subsequent iteration. The short-term memory will 'forget' this restriction after a given number of iterations. If the search is trapped in a local optimum, this will make it possible for the search to escape again. The number of iterations a move is remembered, called the *tabu tenure*, and the type of tabu restrictions help control how the search will be guided through the search space. They also control the level of search intensification, i.e. how carefully the search examines the current region, and search diversification, i.e. how quickly the search moves to other areas of the search space.

Long-term memory stores information as the search progresses, such as the number of times a problem element has been set to a particular value. *Intermediate memory* is used to generate new starting points, exploiting good combinations that have been found during the previous search, hence intensifying the search. Alternatively, long-term memory is used like intermediate-term memory, but allows combinations that have not been tried during the search, hence diversifying the search.

R. James

References
Glover, F. and Laguna, M. (1997) *Tabu Search*, Boston, MA: Kluwer Academic Publishers.
Reeves, C.R. (ed.) (1995) *Modern Heuristic Techniques for Combinatorial Problems*, McGraw-Hill.

Taguchi methods for quality control

Taguchi methods, named after their inventor, refer to techniques of quality engineering that focus on two basic ideas:

- Quality should be measured by the deviation from a specific target value, rather than by conformance to tolerance limits.
- Product design and its manufacturing processes must be optimized to achieve best performance and **robustness** against uncontrollable variations.

Traditional methods define quality in terms of upper and lower tolerances between which an item is acceptable. Taguchi's response to quality differs greatly from this goalpost philosophy. According to Taguchi, performance begins to deteriorate as soon as the design parameters deviate from their target values. He places a great deal of emphasis on minimizing variation as a means for quality improvement and for bringing the mean of the process to the design-mandated target.

Taguchi defines quality as 'the (minimum) loss imparted to society from the time the product is shipped'. This includes the traditional quality-associated costs, such as scrap and rework, but adds the costs stemming from customer dissatisfaction and environmental costs. Traditional measures may therefore underestimate the costs of poor quality substantially.

Taguchi also introduced the concept of a loss function for quantifying the cost of product or process variation. According to him, the loss increases approximately proportionally to the square of the deviation of the performance characteristic from its target value, as depicted in the figure below. This shows that a reduction in variability about the target leads to a decreased loss and a subsequent increase in quality. The loss is at its minimum when the product or service is at the target value.

Taguchi makes a distinction between *online* and *off-line* quality control. He argues that most existing quality control methods use online control by focusing on the

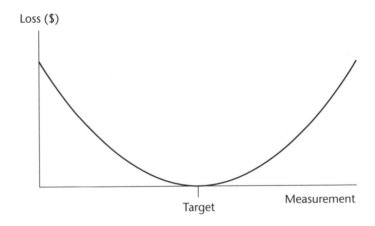

Figure: **Taguchi loss function**

actual production operation, when greater attention to product and manufacturing process design, i.e. off-line quality control, offers a more cost-effective and more robust approach to high quality.

Robust product and production process design is achieved if manufacturing performance is insensitive to hard-to-control disturbances, called noise factors. Robust design improves product quality, manufacturability and reliability, usually at lower cost. Improving quality at the design stage pushes quality control back from the online manufacturing process to the design and development stage where it belongs, resulting in fewer defective products being produced. *P. Venkateswarlu*

References

Tsui, K. (1992) 'An overview of Taguchi method and newly developed statistical methods for robust design', *IIE Transactions*, 24(5) (November).

Taguchi, G. and Clausing, D. (1990) 'Robust quality', *Harvard Business Review*, 68(1) (Jan–Feb).

Ross, P. (1988) *Taguchi techniques for Quality Engineering: Loss Function, Orthogonal Experiments, Parameter and Tolerance Design*, New York: McGraw-Hill.

Bendell, A., Disney, J. and Pidmore, W.A. (eds) (1989) *Taguchi Methods: Application in World Industry*, Bedford: IFS Publications.

Team building

In the current business environment, most of the work of organizations takes place in teams. Teams, therefore, constitute the building blocks of organizations. A team is a collection of individuals, typically between two and twelve. Teams consisting of more than twelve individuals are likely to be unwieldy and ineffective. Teams can be temporary or permanent. They may have an appointed leader or may be self-managing with different individuals in the team taking on the leadership role. There are three common types of teams: problem-solving teams, cross-functional teams and self-managing teams.

A team may consist of individuals with diverse backgrounds but they all share a common goal. In order to work effectively, individuals must have clearly defined roles. A team must have appropriate procedures and processes to achieve its goal. Finally, the interpersonal relationships between team members must be healthy. Team members should be tolerant of differences, mutually respectful, flexible, open and honest. These four aspects – shared goals, clearly defined roles, appropriate procedures and good interpersonal relations – are characteristic of high-performance teams. Team building consists of assembling a team with these four characteristics. It requires hard work, persistence and patience. A sense of humour helps! Teams not only achieve organizational goals but also provide individuals with a sense of identity. They can enhance an individual's self-esteem and contribute to his or her personal growth.

There are some common concerns, which individuals face when they join teams. The most common worries are:

- membership or inclusion worry (Who am I in this team?);
- influence or control worry (Will people listen to me? How much influence do I have in this team?); and
- worry of getting along (Will people like me? Will I get along with others?).

These issues or concerns may not be spoken about, but they are common to most individuals.

High mutual trust among members is a basic feature of high-performance teams. Research on trust has identified three important dimensions of trust:

- Integrity: Truthfulness and honesty.
- Ability: Competence, knowledge and skills in a specific area.
- Benevolence: Wanting to do good to others.

Manager, team leaders and team members can build trust among themselves by following simple behaviours:

- Be fair and open.
- Share their feelings.
- Be consistent in their speech and actions.
- Demonstrate competence in their area of work.
- Demonstrate ability to work for others' interests as well as their own.
- Maintain confidences.

There are ten ingredients for a successful team:

- Clear team purpose and goals, understood and shared.
- An action plan to achieve its goals.
- Clearly defined roles, duties and responsibilities for each member.
- Clear and direct communication.
- Behaviours that make discussions and meetings effective.
- Well-defined decision procedures.
- Balanced participation by all members.
- Established ground rules of acceptable behaviour.
- Awareness of the group process by all members.
- Problem solving and decision making based on data and facts rather than opinions.

V. Nilakant

References

Dyer, W. (1971) *Team Building: Issues and Alternatives*, Reading, MA: Addison-Wesley.

Katzenbach, J.R. (ed.) (1998) *The Work of Teams*, Boston: Harvard Business School Press.

Robbins, S.P., Millett, B., Cacioppe, R. and Walters-Marsh, T. (1998) *Organizational Behaviour*, 2nd edn, Sydney: Prentice Hall.

Mayer, R.C., Davis, J.H. and Schoorman, F.D. (1995) 'An integrative model of organizational trust', *Academy of Management Review*, 20(3):709–734.

Scholtes, P. and Associates (1990) *The Team Handbook*, Madison, WI: Joiner Associates.

Theory of constraints or TOC

The theory of constraints is a systems-based problem solving/**problem structuring methodology** based on the work of Eliyahu Goldratt. TOC first emerged in the early 1980s as the manufacturing scheduling method, **synchronous manufacturing/OPT**. Since then, the theory underlying this scheduling method has been expanded into a thinking process that can be applied to policy, strategic and tactical decision making. Underpinning the approach is the assumption that in any system there are usually a small number of key constraints or core problems that lower the system's performance. Goldratt asserts that the best strategy is to focus on how to use or change these key constraints so that the entire system can perform better. If the constraint is a physical one, then the TOC five focusing steps and/or the synchronous manufacturing/OPT approach can be used. Where the constraint is non-physical, or where accelerated change is desired, we use the *thinking processes*.

The TOC thinking processes (TPs) guide the process of change, by providing the tools to help decision makers diagnose what needs changing, what to change to, and how to cause the change. The TPs comprise five tools, which can be used in sequence or on their own. The tools use one of two types of logical analysis, cause-and-effect analysis using *sufficiency logic* (i.e. 'if A then B') or *necessity logic* (i.e. 'to have X, we must have Y').

Using these types of logic statements, decision makers can build up trees, firstly describing the current situation – a *Current Reality Tree* (CRT). To build the CRT one generally works from the symptoms of a problem towards the likely causes, until one has isolated the core problem which is currently leading to most of the symptoms. Eliminate this core problem and many of the symptoms that arise from it are removed. But this is unlikely to be easy because the core problem is usually due to some dilemma or conflict. The *Evaporating Cloud* (EC), the first of the necessity logic tools, is designed to 'evaporate' the conflict, by surfacing the reasons for the conflict, and ways to resolve it. The *Future Reality Tree* (FRT) can then be built using the same logic as the CRT, but now depicting the future arising from having removed the core problem. This represents the grand strategy – the vision. Any obstacles standing in the way of implementation are mapped using a *PreRequisite Tree* (PRT) (what are the prerequisites for the change to be successful?) Finally the *Transition Tree* (TT) maps the detailed tactical steps.

Although devised to be used as a suite, particularly for significant problems, the tools are very effective individually for one-off problems: e.g. the EC for resolving conflicts or dilemmas; the PRT for planning how to achieve an ambitious target; and the *Negative Branch Reservation* (a subset of the FRT) for giving constructive feedback.

The methodology has been used for a wide range of applications and organizations, in addition to the manufacturing context. TOC facilitates teamwork and employee involvement in devising and enacting changes. *V. Mabin*

References

Dettmer, H.W. (1998) *Breaking the Constraints to World-Class Performance*, Milwaukee,WI: ASQC Quality Press.

Goldratt, E.M. (1994) *It's Not Luck,* Great Barrington, MA: North River Press.

Mabin, V.J. and Balderstone, S.J. (2000) *The World of the Theory of Constraints: A Review of the International Literature,* APICS Series on Constraints Management, Boca Raton, FL: St Lucie Press.

Scheinkopf, L. (1999) *Thinking for a Change: Putting the TOC Thinking Processes to Use,* APICS Series on Constraints Management, Boca Raton, FL: St Lucie Press.

Timetabling

Timetabling problems arise whenever a set of events have to be scheduled into a set of fixed time slots so that no one is required to attend two events at the same time. The most obvious occurrences are in education (school timetabling, course timetabling and examination scheduling), but similar problems arise in other areas such as sports, conference and personnel scheduling.

Although the basic problem is that of finding a feasible timetable subject to physical constraints, such as the number of rooms or seats available, most practical applications also require a good quality timetable. The definition of 'good' will vary, but typical considerations are spreading lessons in the same subject throughout the week, meeting requests for double or triple slots, and organizing sports events so as to provide a fair schedule for all competitors.

The problem of finding a clash-free timetable can be modelled as a well-known problem in **graph theory** – that of colouring a graph in k colours so that adjacent vertices are different colours. Each event is represented by a vertex, and two vertices are adjacent if the events they represent cannot take place simultaneously. Each colour in the solution represents a different time slot. Exact or **heuristic** approaches to this colouring problem have been used to produce feasible timetables in practical situations.

However, when the secondary objectives relating to timetable quality are taken into account, the problem becomes more difficult and heuristic approaches are typically applied. The simplest of these are single-pass greedy construction methods that build up the timetable one event at a time. Events are chosen in order according to some measure of the perceived difficulty in scheduling that event. This ordering may be determined from the start (e.g. largest event or events with the most potential clashes first), or may be determined dynamically as the heuristic progresses. Each event is then allocated to the best slot with respect to the various objectives.

With an appropriate choice of ordering rules such approaches can be very effective, but there are many examples of problems for which they are unable to yield satisfactory solutions. In this case more sophisticated approaches are available. These can be broadly partitioned into three classes. The first simply extend the above by allowing some backtracking (i.e. undoing one or more of the recent decisions and allocating those events elsewhere). The second uses a technique from **artificial intelligence** known as constraint satisfaction. The final approach is to use meta-heuristic techniques such as **simulated annealing, tabu search** and **genetic algorithms,** all of which have proved successful in solving practical problems. However, no single approach has been found to out-perform the others across the broad spectrum of problem characteristics. *K.A. Dowsland*

References

Carter, M.W. and Laporte, G. (1996) *Practice and Theory of Automated Timetabling. First International Conference. Selected Papers*, pp. 2–31.

Schaerf, A. (1999) 'Survey of automated timetabling', *Artificial Intelligence Review*, 13(2):87–127.

Total quality management

The term *total quality management* (TQM) was first used in the mid-1980s to introduce the Japanese quality philosophy of *total quality control* or company-wide quality control to American industry, while avoiding the negative connotations of 'control'. Portrayed as a systematic approach to organizational improvement, TQM was promoted as the answer to the crisis of productivity and competitiveness faced by American industry at the time. From manufacturing, TQM spread rapidly to service industries, healthcare and education, where *continuous quality improvement* became the preferred label.

There are no generally agreed definitions of what TQM is or what it does. The internationalization of the term was accompanied by a proliferation of definitions and growing criticism of TQM's failure to deliver its promised benefits. Disagreement continues over the precise nature of TQM and the benefits that it provides. In some cases it is promoted as a new philosophy of management, while elsewhere it is seen as an approach to organizational change. It has even been presented as a specific technique for continuous incremental improvement of organizational processes. The broadest definitions encompass all of these concepts, i.e. TQM is a philosophy of management for organizational transformation and learning built around continuous improvement of organizational processes.

Despite the lack of agreed definition, there is general agreement on the underlying concepts of:

- focusing on the outside customer
- understanding and managing systems
- understanding and using data
- understanding people
- mastering improvement
- providing direction and focus for the organization.

In common usage the key concepts are often reduced to customer focus, total participation and continuous improvement. There also is a recurring emphasis on management leadership.

The origins of TQM can be traced back through the works of W. Edwards Deming,

Kaouro Ishikawa, Joseph Juran, Philip Crosby and Armand Feigenbaum, all key figures in the quality movement in Japan and the USA, to the **statistical quality control** techniques developed by Walter Shewhart in the 1920s and 1930s. Interestingly, several of these key figures rejected the term TQM as meaningless and a distortion of their ideas and theories.

Statistical process control, through the use of *control charts*, can be seen as the core control and improvement method of TQM. The range of TQM tools has expanded to include tools for product and process improvement, such as the **seven basic tools of quality management**, and techniques for management planning and control. There is, however, considerable debate about the relationship between various tools and the underpinning principles of quality management, the appropriateness of including some of them within the TQM toolkit, and the problem of equating TQM directly with the use of particular techniques and tools.   *D.J. Houston*

References

Bounds, G., Yorks, L., Adams, M. and Ranney, G. (1994) *Beyond Total Quality Management: Towards a new paradigm*, New York: McGraw-Hill.

Dale, B.G. (1994) *Managing Quality*, 2nd edn, New York: Prentice Hall.

Garvin, D. (1988) *Managing Quality: The strategic and competitive edge*, New York: Free Press.

Total systems intervention (TSI), critical system practice

Total systems intervention or critical system practice is an ambitious undertaking, originated by R. Flood and M. Jackson in the early 1990s, of creating a *meta-methodology* for guiding the practice of systems intervention. They see TSI as firmly based on the tenets of **critical systems thinking** (CST), i.e. critical and social awareness of the paradigm underlying each systems intervention methodology (including its strengths and weaknesses), striving for human well-being and emancipation, and complementarism of the various methodologies.

TSI endeavours to operationalize CST for use in practical interventions. Following Jackson's account (2000) it consists of three phases:

Creativity: The decision makers are encouraged to identify various issues faced by the organization and highlight which ones are currently crucial and dominant and which ones are dependent on others. TSI suggest that this is done using metaphors or analogies of systems, such as viewing the organization as
- a machine, with coordinated division of labour, a clear chain of command, like a bureaucracy
- an organism, consisting of systemically interrelated parts, whose needs must be met for the system to survive
- a brain, such as an information processing and learning system
- a culture, such as a social system of shared values and beliefs
- a political system, consisting of coalitions that seek and use power
- a coercive system, held together through enforced consensus.

Choice: Having chosen one or a group of interconnected issues, a system-based methodology or a group of methodologies is identified as suitable for both the issue(s) and the organizational metaphor. One is chosen as the dominant methodology, with others as support whenever appropriate.

Implementation: Application of the chosen dominant methodology to the issue(s) to bring about change, using the supporting methodologies as and when appropriate.

Flood and Jackson emphasize that, like most MS/OR methods, both 'hard' and 'soft', these three phases are used iteratively, with continual linkages and references back and forth between them. It may be useful to explore switching or interchanging dominant and supportive methodologies, as more is learned about the issue and its organizational and human context.

TSI places high demands on the analyst-facilitator: in-depth theoretical and practical knowledge about a whole range of methodologies and practical expertise in their use, from functionalist approaches (e.g. 'hard' MS/OR, **system dynamics**, the **viable systems model**), to **interpretive systems** approaches (e.g. **interactive planning**, **strategic assumption surfacing and testing**, and the **soft systems methodology**). Only then will he or she be able to select the most appropriate combination and correctly apply them. This is clearly its major practical weakness. Being familiar and able to view organizational issues through various metaphors also puts high demands on the decision makers. TSI is still evolving and may yet emerge as a 'viable' meta-methodology.

H.G. Daellenbach

References

Flood, R.L. and Jackson, M.C. (1991) *Creative Problem Solving: Total Systems Intervention,* Chichester: Wiley.

Jackson, M.C. (2000) *Systems Approaches to Management,* New York: Kluwer/Plenum, chs. 10 and 11.

Transportation problem

Transportation of goods from several sources to several destinations is a widespread commercial problem. The transportation problem is a special type of **linear program** (LP) for determining a transportation schedule that minimizes total transportation costs.

The figure below shows the typical **network** structure for a small transportation problem with $m = 3$ sources and $n = 5$ destinations. Each link represents a possible decision choice for shipping x_{ij} units from source i to destination j at a cost per unit of c_{ij}.

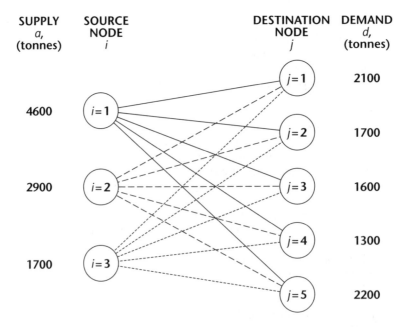

Figure: **Network for transportation problem**

This results in the following LP:

$$\text{Minimize } \sum_{i=1}^{m} \sum_{j=1}^{n} c_{ij} x_{ij},$$

subject to

$$\sum_{j=1}^{n} x_{ij} \le a_i, \quad \text{for } i = 1, ..., m \qquad \text{(Supply constraints)}$$

$$\sum_{i=1}^{m} x_{ij} \ge d_j, \quad \text{for } j = 1, ..., n \qquad \text{(Demand constraints)}$$

$$\text{and } x_y \ge 0, \quad \text{for all } i \text{ and } j.$$

The supply constraints say that no source can ship away in total more than what is available; the demand constraints say that each destination must receive total amounts at least equal to its requirements. There is the explicit assumption that total availabilities are at least equal to total demands.

Although no integer restrictions are imposed on the decision variables, all optimal x_{ij} will assume integer values provided all a_i's and all d_j's are integers. The problem can be solved with a highly efficient special version of the **simplex method**, called the *stepping-stone algorithm*.

Variations of the basic transportation problem may involve one or more of the following situations. If total availabilities are less than total demands, it makes more sense to allocate goods to destinations so as to maximize the total profit that can be achieved. The inequalities of the two sets of constraints are then reversed. There may be minimum and/or maximum (capacity) requirements for some or all links. The problem may not involve transportation of goods over space, but over time, such as production in regular time and/or overtime in period i to meet demand in period j, with no back-ordering allowed, and the cost of 'shipping' one unit of goods consisting of all variable production costs, including overtime if appropriate, and the cost of carrying stocks from period i to period j. This is known as the *regular time/overtime production scheduling problem*. Another variation is the **transshipment problem**, where goods may first be shipped to intermediate nodes, where they may be reallocated and then transshipped to the final destinations.

H.G. Daellenbach

Reference

Hillier, F.S. and Lieberman, G. (2001) *Introduction to Operations Research*, 7th edn, New York: McGraw-Hill, ch. 8.

Transshipment problem

The transportation of goods in a supply chain or a distribution system usually involves supply points (sources), demand points (destinations) and transshipment points. Supply points (e.g. factories) can provide up to a known maximum level of goods; demand points (e.g. customer locations) have a known minimum require-ment for goods; transshipment points (e.g. warehouses and depots) allow for the transfer of goods, but may also act as a net source or destination. The possible rout-ings that the goods may take are represented by a **network** of arcs and nodes (the points), as shown in the figure below, where S1 and S2 are the supply points, D1 and D2 the demand points, and T1 a transshipment point. It is assumed that the capacity for flow of goods along each arc is unlimited and that the total supply bal-ances with the total demand, which in the example below both amount to 50. The objective is to find a set of feasible shipping routes which minimizes the total distribution costs.

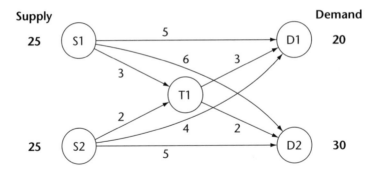

Figure: **Transshipment problem network**

A problem which contains only supply and demand points is known as a **transportation problem**. If, more realistically, the problem includes at least one transshipment node, it is called a *transshipment problem*. Any transshipment problem can be converted into a transportation problem by calculating the cheapest per-unit flow costs between each source–destination pair, by travelling either directly or via a transshipment node, where the latter option may include a handling cost. For instance, in our example it is cheaper to move goods from S1 to D2 via T1 (at a total per-unit cost of 3+2=5) rather than directly (at a per-unit cost of 6). For the case of shipping from S1 to D1, however, the direct cost is 5, whereas the cost via T1 is 3+3=6, so routing via the transshipment node is more expensive.

An alternative method is to treat each transshipment point as both a demand and a source point, with a demand and an availability equal to the total supply. We also allow each transshipment point to supply itself at a zero cost. (If the *stepping-stone algorithm* is used to solve the problem, all inadmissible shipments from transshipment points back to supply points are heavily penalized, so that they will never be optimal.) If the transshipment points have handling capacity constraints, then these are used as their corresponding demand and supply. These tricks will guarantee that the algorithm will find the true optimal solution.

Not surprisingly from its description, the transshipment problem is also a special case of a minimum cost **network flow** problem, in which the capacity of each arc is unlimited, which can be solved by specially efficient versions of the **simplex method**. *J.W. Giffin*

Reference
Hillier, F.S. and Lieberman, G.J. (2001) *Introduction to Operations Research*, 7th edn, New York: McGraw-Hill, ch. 7.

Travelling salesperson problem

The travelling salesperson problem (TSP) is undoubtedly the most illustrious member of all **combinatorial optimization** problems, with notoriety not only amongst mathematical circles, but also in the popular scientific press. It deals with the problem of finding the minimum distance (or cost or time) itinerary or *tour* to visit each of *n* cities once and return to the starting place.

Although this problem may seem like a puzzle, it has many important applications for pick-up and delivery services and its extension to the **vehicle routing problem** of scheduling deliveries by a fleet of vehicles, connecting components on a computer board, and finding a minimum time sequence for jobs to be processed on a machine, to name just a few.

Although all these problems can, in principle, be formulated as **integer programs**, the problem size and computational time explode as the number of cities increases. Hence, a variety of **heuristic methods** have been developed to find good, if not optimal solutions to fairly large problems of hundreds of cities in a reasonable amount of computer time.

To demonstrate the nature of some heuristics, consider the nine-city problem shown in the figure below, where city 1 is the starting point. The *nearest neighbour heuristic* sequentially builds a complete tour by always selecting the closest unconnected neighbour at each iteration, as shown on the left-hand-side diagram. City 2 is the closest to 1, city 4 the closest to 2, etc. Note that the tour intersects itself – not a desirable feature, since it means extra distance. Interchanging the cities on the pair of links 7–8 and 6–9 to 7–6 and 8–9 gives a shorter tour.

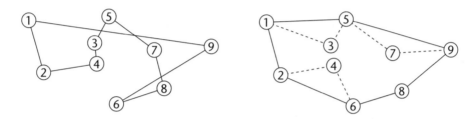

Figure: **Different tours for the TSP**

The diagram on the right-hand side demonstrates the highly successful Norback and Love heuristic. It first forms a partial tour from the *convex hull* that encloses all cities. This is given by the set of solid lines, i.e., 1–2–6–8–9–5–1. It then inserts the remaining cities one by one, each time inserting the city with the largest angle between two consecutive connected cities. In the example, 7 forms the largest angle between any two consecutive cities. It is inserted between 9 and 5 (the broken lines). The next is 4, and finally 3. The resulting tour is 1–2–4–6–8–9–7–5–3–1, which in this instance is the optimal or shortest tour.

The *cheapest-insertion heuristic* also starts with selecting the closest city first. At each subsequent iteration, it inserts into the partial tour that unconnected city which adds the minimum increase in total distance, until a complete tour has been formed.

Once a complete tour has been formed, most heuristics apply a second phase, such as an *interchange heuristic*. Pairs of links are interchanged. If that reduces the length of the tour, the interchange is retained. This is explored for all pairs of links.

A.D. Tsoularis

References

Lawler, E.L. *et al.* (eds) (1985) *The Travelling Salesman Problem*, Chichester: Wiley.

Norback, J.P. and Love, R.G. (1977) 'Geometric approaches to solving the travelling salesman problem', *Management Science*, July, 1208–1223.

Reinelt, G. (1994) *The Travelling Salesman: Computational Solutions for TSP Applications*, Berlin: Springer Verlag. An advanced text.

Winston, W.L. (1999) *Operations Research: Applications and Algorithms*, 3rd edn, Belmont: Duxbury. Easy treatment of both optimization and heuristic approaches.

Triangulation

Triangulation is the use of a multiple of theories and research methods to improve internal validity of qualitative research in a single research study. Methods triangulation refers to both different methods of research and different types of methods or procedures for data collection. For example, a project dealing with the development of an appropriate information system for a national police force could investigate the problem using different systems theories in parallel or in sequence, such as an **information systems** theory approach that studies the technological specifications, a **cybernetic** view via a **viable systems model** of the system functions (task execution, coordination, control, development and policy), and a **soft systems methodology** that focuses on the human constraints of the project. Together they provide a much richer understanding of the complexity of the problem than one approach alone. Similarly, different data collection procedures used in parallel, such as interviews, questionnaires and focus groups, complement each other and provide data validation.

Data triangulation refers to the use of multiple data sources using a single qualitative method. Multiple interviews are an example of the use of multiple data sources using a single method. Multiple observations would be another example of data triangulation. Another important part of data triangulation may involve collecting data at different times, at different places, and from different people.

Most importantly, triangulation overcomes the potential or actual weaknesses and limitations of a single method or theory or single data source and, therefore, provides a better and more reliable and valid basis for drawing conclusions or making recommendations. *G. Alexander*

References

Marshall, C. and Rossman, G. (1999) *Designing Qualitative Research*, 3rd edn, Thousand Oaks, CA: SAGE Publications.

Miles, M. and Huberman, M. (1994) *Qualitative Data Analysis: An expanded sourcebook*, 2nd edn, Thousand Oaks, CA: SAGE Publications.

Triple-loop learning, see single-loop learning

Type I and Type II errors, see statistical analysis

Uncertainty

Uncertainty refers to incomplete knowledge about something – an event, a phenomenon, a process, a future outcome, or something in the past that has not been revealed fully. The uncertainty could be about the **attribute(s)** of something, e.g. its gender, or its numeric value, or its timing. The degree of uncertainty may vary from knowing almost nothing, e.g. we only have a partial list of potential outcomes, to knowing almost everything, e.g. we may know that its numeric value lies inside a narrow band.

Uncertainty is an everyday occurrence and we have many words related to the concept: chance, likelihood, probability, risk, hazard, random, stochastic, odds, to name just a few. What are the causes of uncertainty?

- *Ignorance:* The thing is not understood in sufficient detail. If we completely understood the processes that cause earthquakes, we could predict the timing and severity of the next earthquake occurring in San Francisco.
- *Incomplete information:* It is technically possible to get complete information, but for various reasons, such as limited funds or time, we only collect information about a sample of the thing. For instance, TV ratings for various programmes are based on a small fraction of all viewers. Hence they are inaccurate. It is again ignorance, but by design.
- *Inability to predict the moves of other parties involved:* The final outcome of a process may be the joint result of actions taken by competing parties – economic competitors, opposing teams. Their actions or moves may be taken independently of each other or in response to or anticipation of the competitors' moves.
- *Measurement errors:* They may be due to mistakes made by observers or inaccuracies in the measuring instruments.

Probability statements are an approach to quantify the degree of uncertainty. Such statements may be 'objective', i.e. either based on past observations, such as the observed frequency of a given outcome, e.g. past sales of newspapers, or derived by deductive reasoning from the nature of the process, e.g. of rolling two dice. They may be '**subjective**', expressing a person's strength of belief that an outcome will eventuate.

Various methods of prediction or **forecasting** are attempts to reduce the degree of uncertainty. Most are based on the interpolation or extrapolation of past data; others are based on expert judgement.

The **decision theory** literature distinguishes between decision making under complete uncertainty and decision making under risk. The former assumes that we only know the possible outcomes, but have no information about the likelihood of each, while the latter assumes knowledge about the probability of each outcome. The two lead to completely different decision-making approaches and **criteria**.

Behavioural research (Tversky and Kahneman, 1974) indicates that we are subject to various biases when faced with uncertainty, such as wishful thinking, judging based on stereotyping or on how easy it is to imagine something.

H.G. Daellenbach

References

Daellenbach, H.G. (2001) *Systems Thinking and Decision Making*, Christchurch: REA, ch. 5.

Tversky, A. and Kahneman, D. (1974) 'Judgment under uncertainty: heuristics and biases', *Science*, 185:1124–31 (also reprinted in Daellenbach above).

Upper bounding in LP

Whether a **linear programming** (LP) problem is being solved by the **simplex algorithm** or by an **interior point algorithm,** a good estimate of the amount of computational time required to find the optimal solution(s) is proportional to the cube of the number of constraints in the problem. For efficiency, it therefore makes sense to try to reduce the number of constraints that must be explicitly included in the problem, and the *upper bounding* and *generalized upper bounding* techniques are two commonly used approaches. Consider the following set of constraints for an LP:

$$2x_1 + 4x_2 + x_3 \le 25$$

$$2x_1 + x_2 - x_3 \le 20$$

$$x_1 \le 4, \; -1 \le x_2 \le 5, \; 1 \le x_3 \le 4$$

$$\text{and } x_1 \ge 0.$$

The first two constraints in this LP are 'normal' constraints, each including more than one decision variable; the remaining ones each specify upper and lower bounds on one variable. Note that all LP solution algorithms assume that the decision variables are nonnegative. Hence, the lower bounds for x_2 and x_3 have to be put into this form by variable substitutions, i.e. $x_4 = x_2 + 1$ and $x_5 = x_3 - 1$, which changes the original constraint set to one with simple upper bounds:

$$2x_1 + 4x_4 + x_5 \le 28$$

$$2x_1 + x_4 - x_5 \le 22$$

$$x_1 \le 4, \; x_4 \le 6, \; x_5 \le 3$$

$$\text{and all variables} \ge 0.$$

Since the simplex algorithm works with **basic feasible solutions** (which correspond to corner points of the **feasible region**, defined mathematically as consisting of a *basic variable* for each constraint – selected from the combined set of decision, **slack and surplus variables** – with the remaining *nonbasic variables* set equal to zero), five basic variables are needed if each of these constraints is considered explicitly. This number can be reduced to the number of normal constraints by extending the definition of a nonbasic variable to allow it to be set to the value of its upper bound (as an alternative to the conventional lower bound of 0, as used by the simplex algorithm). In the above example, the initial basic feasible solution has x_1, x_4 and x_5 as the nonbasic variables and s_1 and s_2, the slack variables for the normal constraints, as the two basic variables needed. Valid initial basic feasible solution values now include $(x_1 = 0, x_4 = 0, x_5 = 0, s_1 = 28, s_2 = 22)$, $(4, 0, 3, 17, 17)$ and $(0, 6, 3, 1, 19)$.

Note that the combination (4, 6, 0, −4, 8) is infeasible. Therefore, checks for infeasibility are necessary. The uniqueness of any basic feasible solution description has also been lost. Furthermore, 'bookkeeping' steps have to be added to the simplex algorithm, to keep track of which variables are at their upper bounds and prevent variables exceeding their upper bounds or going below their lower bounds. However, the reduced storage requirements of the upper bounding procedure far outweighs the added computational cost.

J.W. Giffin

Reference

Winston, W.L. (1994) *Operations Research: Applications and Algorithms*, 3rd edn, Belmont CA: Duxbury Press, ch. 10.

Utility functions, reference lotteries, certainty equivalent

In many decision situations, the intrinsic worth of outcomes corresponds to the monetary outcomes, particularly when they fall well within the decision maker's normal range of experience. Decision criteria based on expected monetary outcomes are then appropriate. However, strategic and unique decisions may not only involve huge potential gains and losses but also high **uncertainty**. The decision maker may not be willing to 'play the average' and risk financial disaster. Car, house, medical and life insurances are proof most of us behave this way. In 1944, Von Neumann, a mathematician, and Morgenstern, an economist, proposed an index designed to quantify the personal subjective worth of risky outcome. They called it *utility* – a rather unfortunate choice, given the discredited usage made of this term by 19th-century economists.

A utility function expresses an individual's valuation of risky outcomes on an arbitrarily numerical point scale, such as from 0 to 1. The endpoints reflect the worth of the most desirable outcome and the least desirable outcomes, with all other outcomes in between. Figure 1 depicts three shapes of utility functions. The same person may be *risk averse* on some ventures, e.g. for insurance cover, but *risk seeking* on others, e.g. occasional gambling. A *risk neutral* person does not need a utility function.

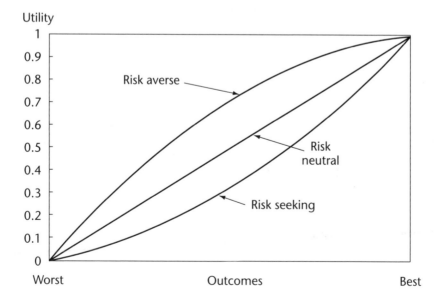

Figure 1: **Three basic shapes of utility functions**

The approximate shape of a utility function can be obtained by a simple procedure. Consider the hypothetical example in Figure 2, with the worst outcome of −1.6 and a best outcome of 4.9 million dollars, assigned a utility of 0 and 1, respectively. We present the decision maker with the following choices: (1) a so-called 50–50 *reference lottery* involving the worst and the best outcomes, and (2) a certain outcome of X. Trial and error will pinpoint the value for X for which the decision maker is indifferent between the two options. Assume that X = 0.45. This is referred to as the *certainty equivalent* to the reference lottery. Since the decision maker is indifferent between 0.45 and the gamble, both must have the same utility. But the gamble has an expected utility of [0 × 0.5 + 1 × 0.5] = 0.5. So 0.45 has a utility of 0.5. This process is repeated a number of times, each time selecting values for the 50–50 gamble for which we have already assessed the utility. Usually five or nine points provide a sufficiently good approximation. Each decision choice is then ranked in terms of the expected utility of outcomes. Unfortunately, research has raised serious doubts about the reliability of this method.

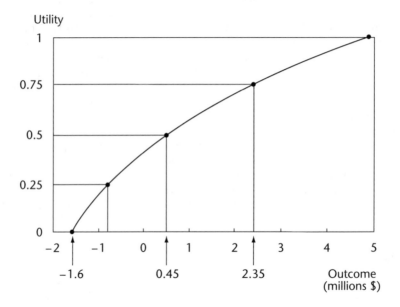

Figure 2: **Five-point assessment procedure**

The utilities obtained in this manner reflect both the intrinsic worth of the outcome and the attitude towards risk for a given individual and a given situation. It may not be valid to extend it to other situations. They are not valid for riskless situations. It is also not possible to claim that because the utility of a given outcome is twice as high as that of another, the first outcome is twice as preferred.   H.G. Daellenbach

References

Daellenbach, H.G. (2001) *Systems Thinking and OR/MS Methods*, Christchurch: REA, pp. 246–255. Elementary discussion.
de Neufville, R. (1990) *Applied Systems Analysis*, New York: McGraw-Hill, chs 18 and 19.

Validation, verification

Model validation has two facets: checking *internal validity*, particularly for **mathematical models**, and establishing *external validity*. Both are crucial aspects of MS/OR modelling.

Internal validity or *verification* checks if all mathematical expressions correctly represent the assumed relationships and are logically consistent, e.g. that the dimension of the left-hand and right-hand sides of an equation are the same. Verification requires numerically checking the results by hand or by detailed computer tracing for a sufficiently wide range of inputs. All data must be checked for correctness in form and format. Beginners often assume that because something is done on a computer it is automatically correct, when in fact this is far from true. It is estimated that between 1 and 4 per cent of all entries in spreadsheets contain errors.

External validity establishes if the model is a sufficiently valid representation of reality. This is far more difficult to establish than internal validity. What is or is not a close enough approximation is largely a question of judgement. The purpose of building the model and the intended use of its solution are highly relevant aspects to consider.

All **stakeholders** should realize that it is not possible to prove a model to be externally valid. It is only possible to show it to be wrong. External validity can often be assumed if the model accurately mimics reality. If a model cannot be pitted against a past system (e.g. because none exists or data are not available), it should be submitted to the detailed scrutiny of people thoroughly familiar with the situation. They must judge if the results seem plausible and as expected.

Validation also implies performing **sensitivity analysis**. The analyst needs to ascertain the responses of the model to changes in inputs – are they as expected and if not, why? Complex **systems** often exhibit **counterintuitive behaviour**. A model's validity is put into question unless convincing explanations can be found for such behaviour. If it is due to factors not included in the model, the model may need to be partially reformulated.

Validation and verification is often viewed as a task performed once the modelling phase is essentially completed. This is the wrong approach. Not only is it much more costly to revisit the model at that time, but various details of the model may not be fresh in the mind of the analyst anymore, and it may be difficult or impractical to get access to other key people involved in the original analysis. Furthermore, if errors are discovered at that late stage, a lot of work may have to be redone. Verification and validation of the model should be ongoing and continual. *H.G. Daellenbach*

References
Landry, M., Malouin, J.-L. and Oral, M. (1983) 'Model validation in operations research', *EJOR*, (Nov) 14(3):207–20.
Landry, M. and Oral, M. (1993) Special Issue: Model Validation, *EJOR*, (April) 66(2).

Variable cost, see costs, type of

Vehicle routing and scheduling

Vehicle routing and scheduling (VRS) involves finding a set of routes or itineraries, originating and terminating at a central depot, using a fleet of vehicles with various capacities, to transport loads to a given set of customer locations, each load using a given volume or weight of a vehicle's capacity. Normally a delivery to a customer cannot be split between vehicles, but must be made by a single vehicle. The objective may be to minimize the total cost of making the deliveries, consisting mainly of mileage cost, or to minimize the total distance travelled by all vehicles, or minimize the number of vehicles needed to make the deliveries.

Variations of the problem may have more than one originating depot, sharing the delivery vehicles. The problem may involve pick-ups rather than deliveries, or both. In addition to the constraint of vehicle capacity, there may also be restrictions on the maximum travel distance or the maximum travel time (including loading and unloading). For the transport of liquids, such as milk or beer, the total vehicle capacity may consist of several separate compartments with different capacities. Loads to or from different customers cannot share compartments. There may also be the additional complication that unloading or loading of compartments must be done in a given sequence in order to ensure vehicle road stability. The customer may only be able to receive the goods or the service before a latest or after an earliest time or during one or more limited *time windows*. The VRS may not involve transportation of goods, but provision of services, such as repairs, at the customer locations, each requiring a certain length of time.

The 'vehicles' need not be trucks, but could be ships, airplanes, or people.

The notorious **travelling salesperson problem** is a special type of VRS, in which a salesperson must visit once each of a number of cities before returning to the home city, with the objective of minimizing the total distance travelled.

Almost all VRS problems are computationally difficult or so-called **NP-hard**. For any reasonably sized problem good or near-optimal solutions are derived by **heuristic methods**. Most good vehicle routing trips involve so-called *petal routes*, the graph of the overall solution similar to a daisy flower. *B. Chen*

References
Bodin, L., Golden, B.L., Assad, A.A. and Ball, M. (1983) 'Routing and scheduling of vehicles and crews: The state of the art', *Computers and Operations Research*, (10):63–211.
Golden, B.L. and Assad, A.A. (eds) (1988) *Vehicle Routing: Methods and Studies*, Amsterdam: North-Holland.

Venn diagrams

Venn diagrams show the relationships between groups or sets. A set is defined as a collection of objects, which share (or lack) some specified properties. As shown in the figure below, a Venn diagram usually represents the set of all objects as a rectangle, and subsets as circles inside, although the shapes do not matter. The overlap of two (or more) circles is the set of objects which share the same two (or more) properties.

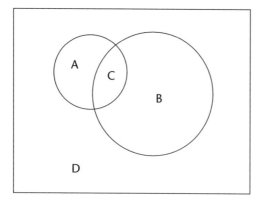

Figure: **Venn diagram**

The figure could, for instance, depict the following situation: The rectangle represents all public share companies. Of these, 100 have been chosen to make up the stock exchange index, such as the British FTSE 100 index. Circle A covers these. The companies in circle B are those whose turnover grew by at least 10 per cent last year. It overlaps circle A. Area C covers all those companies which are included in the FTSE 100 index and also grew by at least 10 per cent last year. Finally, area D covers all those companies which are neither included in the FTSE 100 index nor grew by 10 per cent.

Venn diagrams are valuable for visual illustration and for identifying useful or important subgroups of objects. Mathematically, set, subset and overlap membership relationships can be expressed by *Boolean algebra.*   *D.K. Smith*

Reference
Concise Dictionary of Mathematics, Buckingham: Open University Press.

Verification, see validation

Viable systems models and theory

The viable systems model (VSM) and its underlying theory was developed by one of the world's great systems thinkers, Stafford Beer, in 1959. His project was to discover the laws of system viability, by which he meant the ability of a **system** to maintain an independent existence. Through studying the way in which the nervous system organizes and controls the human body – known to be a viable system – Beer arrived at a set of laws that he claims govern the functioning of all viable systems. Although his **analogy** draws mainly on the example of the human being, he said that he would have reached the same conclusions had he chosen other viable systems, including single biological cells such as the amoeba, whole animal species, or complex social systems, such as enterprises and cultures.

Looking at the structure of the VSM, Beer found that all organisms displaying viability shared five basic properties. They all exhibited, in some shape or form:

- autonomous *operational elements* that directly interface with the external environment, that enact the identity of the system;
- *coordination functions*, that ensure that the operational elements work harmoniously;
- *control activities*, that maintain and allocate resources to the operational elements;

- *intelligence functions*, that consider the system as a whole – its strategic opportunities, threats, and future direction; and finally,
- an *identity function*, that conceives of the purpose or *raison d'être* of the system, its 'soul', and place in the world.

As a viable system, the human body exhibits an efficient interplay between operational elements – such as skin, limbs and vital organs; coordination functions like the nervous system; control, through the pons medulla at the base of the brain allocating resources, like blood; intelligence, characterized by the brain's use of the senses and thought; and identity in what might be described as an individual's sense of self, or soul. A viable human being, like any organism, requires all five functions to be in place and working in unison. Beer later derived the same principles from **cybernetics**.

Beer's main thesis is that the conditions outlined above: the five **systemic** elements and the various communication channels running between them, and between them and the environment, must be present in all viable systems, including organizations. Starting from a model of the human nervous system, Beer ends up with a macro-level picture of what structural and communication arrangements are necessary in order to facilitate ongoing organizational viability.

VSMs have been applied by Beer and others in numerous consulting projects to help design or redesign enterprises and governmental agencies. The most ambitious attempt was the model developed for the Chilean social economy when Allende became president of Chile in 1970.   *J. Brocklesby*

References
Beer, S. (1972) *Brain of the Firm*, London: Allen Lane.
Beer, S. (1979) *Heart of Enterprise*, Chichester: Wiley.
Beer, S. (1985) *Diagnosing the System for Organization*, Chichester: Wiley.
Brocklesby, J. and Cummings, S. (1996) 'Designing a viable organization structure', *International Journal of Strategic Management: Long Range Planning*, 29(1):49–57.
Espejo, R. and Harnden, R. (1989) *The VSM: Interpretations & Applications of S Beer's VSM*, Chichester: Wiley.

V.I.S.A., see **multiattribute value functions**

Waiting lines, see queueing

Weltanschauung, world view

This German word, first used by C.W. Churchman for **systems thinking**, is loosely translated into *world view*. It reflects an individual's personal values, beliefs and biases, as affected by upbringing, cultural and social background, education and experience. It has been compared to coloured glasses that taint what we see, hear or read, and acts as a filter to emphasize some things, and obscure, distort or ignore others. It is used to interpret and give meaning to observations and situations and largely determines the stand we take on issues that involve judgement, uncertainty and values. Few people are fully aware of their own world view, and we usually can only get glimpses of and guess other people's world view as it reveals itself partially through their statements and actions. Disagreements and conflict often arise because of differing world views. Problems are seen differently by different **stakeholders** because people have different world views.

For example, the three partners in a graphic arts design company may each view their firm and its aims and purposes from a different perspective. The first takes a materialistic view and sees it as a means to increase wealth; the second, an idealistic artist, sees it as a means to exercise her and her co-designers' creative artistic drive; and the third, a humanitarian, as a way to provide local employment for talented people.

The world view of stakeholders, particularly the decision makers or client of a project, largely determines what the goals or **objectives** should be. One of the skills MS/OR analysts have to learn is to see the situation through other people's world view. Hopefully, this will make them more aware of their own.

Problem structuring methods are all concerned with externalizing the world views of the various stakeholders, at least partially. This will allow all stakeholders to gain better insight and appreciation into where the other stakeholders are coming from, thereby bringing about some convergence of world views or at least acceptance of other stakeholders' values. An individual's world view cannot be dismissed as wrong, no matter how strongly we object to it. All we can do is to try to make the person aware of its implications for other people and the environment and try to change it. However, world views are resistant to change or in Churchman's words: 'No data can ever fatally destroy a world view.'   *H.G. Daellenbach*

Wider system of interest, see **hierarchy of systems**

Winter's method, see **exponential smoothing**

World-class manufacturing

World-class manufacturing (WCM) is a comprehensive approach and philosophy to the production of goods and services, based on three main principles: **just-in-time** production (JIT), **total quality management** (TQM), and *total people involvement* (TPI). Triggered largely by the rise of Japanese industrial might and the ensuing erosion of the world dominance experienced by US manufacturers in the early 1980s, WCM has arisen from the study of the performance of world-class companies involved in both manufacturing and the service industries located all over the globe. What was discovered in Japanese factories was a unique combination of JIT, TQM and TPI.

The objectives of JIT are to eliminate all forms of waste, shorten production **lead times**, reduce costs, and strive for continuous improvement (Hall, 1993). These objectives are achieved by attention to workplace organization, visibility, *pull* (rather than *push*) production control, timing and flexibility. Workplace organization involves tidiness, simplification, equipment reliability, equipment and **facility location**, and production **scheduling**. Visibility involves communication, **facilities layout**, and limits on work-in-process levels. JIT specifies that each operation is supplied with exactly the quantity of materials it needs, only when the workers in that operation request them, hence the term 'just-in-time'. This reduces work-in-process to a minimum. Timing involves the setting and progressive reduction of cycle times – the time between recurring activities. Flexibility covers product mix flexibility (often improved by setup time reduction), volume flexibility, people flexibility and design flexibility. For JIT to be successful, top product quality is crucial.

TQM is concerned with holistic quality, involving all aspects of an organization, from product conception to after-sales service (Bounds *et al.*, 1994). It involves definitional, motivational and technical aspects. The concept of product quality covers suitability for use, reliability, and consistency in terms of customer value. The motivational aspects of TQM are closely related to TPI in the sense that for the endeavours related to quality to be 'total', there must be genuine acceptance of the ideals by all employees. The field of practical techniques for the improvement of quality is a vast one. They include measuring, monitoring, the **seven tools of quality management**, **Shewhart's PDCA cycle**, **statistical quality control**, failsafe methods, reducing feedback time, standardization, and **quality control circles**.

TPI is the most important factor in achieving WCM (Huge and Anderson, 1988). It is concerned with the genuine and respectful involvement of all employees in organizational improvement. It involves motivation, job enlargement, job enrichment, increased worker responsibility, participation, empowerment, teamwork, constructive performance measurement, personal goal setting and gain sharing. The overall objective of TPI is to harness the initiative, creativity, intelligence and energy of every employee in the organization.

There have been some remarkable successes in companies that have adopted WCM (Schonberger, 1996).   *L.R. Foulds*

References

Bounds, G., Yorks, L., Adams, M. and Ranney, G. (1994) *Beyond Total Quality Management*, New York: McGraw-Hill.

Deming, W.E. (1985) 'The roots of quality control in Japan', *Pacific Basin Quarterly*, Spring/ Summer: 1–4.

Hall, R.W. (1993) *Attaining Manufacturing Excellence*, Chicago: Irwin/APICS, ch. 4.

Huge, E.C. and Anderson, A.D. (1988) *The Spirit of Manufacturing Excellence*, Chicago: Irwin/APICS, ch. 7.

Schonberger, R.J. (1996) *World Class Manufacturing: The Next Decade*, New York: Free Press.

World view, see Weltanschauung

Zero-sum games

Game theory is a scientific approach to the study of human interactions in which the outcomes depend on the individual strategies of two or more persons, who have opposing or sometimes mixed motives.

A two-person zero-sum game is a game between two *players* whose individual *pay-offs* (rewards) based on their respective strategies add up to zero. In other words, a zero-sum game is a game in which what one player gains, the other loses. Consider the following highly simplified situation: Two manufacturers, A and B, manufacture the same product. Company A has a fixed cost of $40,000 per period, while B's fixed cost is $50,000, whether they sell anything or not. Each company has to choose between a high price ($10) or a low price ($5). The current market conditions dictate the following rules:

- A total of 10 000 items can be sold at $10 for a revenue of $100,000.
- A total of 20 000 items can be sold at $5 for a revenue of $100,000.
- If both companies charge the same price, they split the sales evenly between them.
- The company that charges a higher price loses sales completely.

This simple game can be represented in the following payoff matrix, where each entry corresponds to the difference between revenues and costs for A and B, respectively, for a given combination of strategies:

Payoffs for [A; B]		Price charged by company B	
		$5	$10
Price charged by company A	$5	$10,000; $0	$60,000; –$50,000
	$10	–$40,000; $50,000	$10,000; $0

Although a priori the payoffs for B are not the negative of A's and vice versa, subtracting $10,000 from each of A's payoffs produces that property, while not affecting the outcome or the relative payoffs of the game.

Which strategy should A and B choose? Being risk averse, A will reason as follows: Choosing $5 gives a higher payoff than $10 no matter what strategy B chooses.

Hence A's best strategy is $5. Similarly, B's best strategy is also $5, resulting in the outcome shown in the shaded cell above. This result is in accordance with the **minimax criterion**, which states that each company should choose the strategy that minimizes its maximum loss or, equivalently, maximizes its minimum payoff. In this game there is a stable solution; neither player has any incentive to defect from the $5 strategy. Each sticks to one choice – a so-called *pure strategy*.

Some games do not have a set of pure minimax strategies, as e.g. the game shown below:

A \ B	B1	B2	min
A1	10; –10	40; –40	10
A2	60; –60	0; 0	0
min	–60	–40	

Each player now has an incentive to defect from the minimax strategy (shaded cells). If A really plays A1, then B should switch from B2 to B1, in which case A should switch to A2, and so on. Rather than choosing a pure strategy, each player can do better (in terms of expected outcome) by choosing a strategy at random. For instance, A could select A1 with probability 0.6 and A2 with probability 0.4, thereby keeping the opponent guessing. Such a choice is called a *mixed strategy*. John von Neumann, the founder of game theory, proved that every two-person zero-sum game has a minimax solution in mixed strategies. In our case, A should choose A1 with probability 2/3 and A2 with 1/3, while B should select B1 with probability 4/9 and B2 with 5/9. The expected outcome for A is then $26\frac{2}{3}$ and that of B the negative of this.

A.D. Tsoularis

References

Casti, J. (1996) *Five Golden Rules*, New York: Wiley. Delightful book covering popular accounts of five landmark mathematical theories of the 20th century.

Von Neumann, J. and Morgenstern, O. (1944) *Theory of Games and Economic Behaviour*, Princeton, NJ: Princeton University Press. A classic still in print.

Williams, J.D. (1986) *The Compleat Strategyst*, Mineda, NY: Dover Publications. Most readable and entertaining.

Winston, W. (1994) *Operations Research: Applications and Algorithms*, Belmont, CA: Duxbury Press, ch. 15.

Appendix

Primary journals

Administrative Science Quarterly
Annals of Operations Research
Asia-Pacific Journal of Operations Research
Behavioral Science
Computers and Operations Research
Cybernetics and Systems
Decision Sciences
European Journal of Operational Research (EJOR)
Human Relations
INFOR (Canadian Journal of OR & Information Processing)
Interfaces
International Journal of Systems Science
International Transactions in Operational Research
Journal of Applied Systems Analysis
Journal of Behavioral Decision Making
Journal of the Operational Research Society (JORS)
Management Science
Mathematical Programming: Series A and Series B
Mathematics of OR
Naval Research Logistics
Omega
Operations Research
Operations Research Letters
Opsearch
OR Insight
OR/MS Today (Informs membership journal)
OR Newsletter (UK Society)
Queueing Systems
Simulation
Systemic Practice and Action Research
Systemist
System Research and Behavioural Science
Systems Research

Abstracting Journals

International Abstracts in Operations Research

Secondary Journals

Academy of Management Review
Advances in Management Studies

American Journal of Mathematical and Management Sciences
Archival Science
Belgian Journal of Operations Research, Statistics and Computer Science
Cahier du Centre d'Étude de Recherche Opérationelle
Central European Journal for Operations Research and Economics
Computational Optimization and Applications
Constraints
Cybernetics and Systems Analysis
Czechoslovak Journal for Operations Research
Discrete Event Dynamic Systems
Futures
Fuzzy Optimization and Decision Making
General Systems
Group Decisions and Negotiation
Groups and Organization Studies
Health Care Management Science
Human Systems Management
IEEE Transactions in Systems, Man, and Cybernetics
IIE Transactions
Information Systems and Operations Research
INFORMS Journal on Computing
International Journal of Flexible Manufacturing Systems
International Journal of Game Theory
International Journal of General Systems
International Journal of Information and Management Sciences
International Journal of Operations & Production Management
International Journal of Physical Distribution and Materials Management
International Journal of Production Economics
International Journal of Project Management
International Journal of Quality and Reliability Management
Journal of Combinatorial Optimization
Journal on Computing
Journal of Global Optimization
Journal of Heuristics
Journal of Intelligent Information Systems
Journal of Intelligent Manufacturing
Journal of Management Studies
Journal of Multi-criteria Decision Analysis
Journal of Network & Systems Management
Journal of Operations Management
Journal of Productivity Analysis
Journal of Quality Management
Journal of Risk and Uncertainty
Journal of System Engineering
Journal of the Korean OR/MS Society
Journal of the Operations Research Society of Japan
Knowledge and Process Management
Long-Range Planning
Malaysian Journal of Management Science
Manufacturing & Service Operations Management
Mathematical and Computing Modelling
Mathematics and Computers in Simulaton
Monist
Networks

Networks and Spatial Economics
Neural Networks
Neural Processing Letters
Open Systems & Information Dynamics
Operations Research-Spektrum
Operations Research Verfahren
Optimization and Engineering
Organization
Organization Science
Organizational Dynamics
Production and Inventory Management Journal
Production and Operations Management
Production Planning and Control
Quadernas d'Estidistica, Sistemes, Informatica i Investigacio Operativa
Quality and Quantity
Queueing Systems
Revue d'Automatique d'Informatique et de Recherche Opérationelle – Operations Research
SIAM Journal on Algebraic and Discrete Methods
SIAM Journal on Applied Mathematics
Simulation Transactions
Socio-Economic Planning Sciences
Synthese
System Dynamics Review
The International Journal of Production Research
The Journal of Quality and Participation
The Total Quality Management Magazine
Theory and Decision
Total Quality Management
Transactions of the Society for Computer Simulation
Transportation Research A: General, B: Methodology, C, D: Environment, E: Logistics
Transportation Science
Yugoslav Journal of Operations Research
Zeitschrift für Operations Research (Germany)

Journals in Related Disciplines

Advances in Applied Probability
Communications in Statistics: Stochastic Models
Computational & Mathematical Organization Theory
Computers & Industrial Engineering
Computers and Mathematics with Applications
Data Mining and Knowledge Discovery
Decision Support Systems
Engineering Costs and Production Economics
European Journal of Information Systems
Information & Management
Information Systems Frontiers
Information Systems Journal
Information Systems Research
Information Technology and Management
International Journal of Forecasting
Journal of Applied Mathematics and Stochastic Analysis
Journal of Applied Probability
Journal of Forecasting

Journal of Information & Optimization Sciences
Journal of Information Systems
Journal of Management Information Systems
Journal of Management Inquiry
Journal of Marketing Research
Journal of Mathematical Analysis for Business Decisions
Journal of Policy Analysis and Management
Journal of Policy Modeling
Journal of the Royal Statistical Society (Series A)
Marketing Science
Mathematical Social Sciences
Organizational Studies
Research Policy
Review of Quantitative Finance and Accounting
Stochastic Models
Technological Forecasting and Social Change
The Journal of Strategic Information Systems
Theory of Probability and its Applications
Zeitschrift für Betriebswirtschaft (Germany)